Biology and Synthetic Biotechnology of *Aureobasidium* spp.

Zhe Chi • Guang-Lei Liu • Zhen-Ming Chi

Biology and Synthetic Biotechnology of *Aureobasidium* spp.

Springer

Zhe Chi
College of Marine Life Sciences
Ocean University of China
Qingdao, Shandong, China

Guang-Lei Liu
College of Marine Life Sciences
Ocean University of China
Qingdao, Shandong, China

Zhen-Ming Chi
College of Marine Life Sciences
Ocean University of China
Qingdao, Shandong, China

ISBN 978-981-92-2606-1 ISBN 978-981-92-2607-8 (eBook)
https://doi.org/10.1007/978-981-92-2607-8

This Springer imprint is published by the registered company Springer Nature Singapore Pte Ltd.
The registered company address is: 152 Beach Road, #21-01/04 Gateway East, Singapore 189721, Singapore

Preface

Aureobasidium spp. are the black yeast-like polymorphic fungi with remarkable phenotypic and morphological plasticity, genome, gene and species diversity, and multi-bioproducts (Chap. 1). Because of the morphological and growth mode plasticity (Chap. 4), gene and species diversity (Chaps. 3 and 5), and multi-bioproducts (Chaps. 7, 8, 9, 10, 11 and 12), including various enzymes which can degrade a wide range of components from plants (Chap. 6), they are widely distributed in every corner of the world, especially harsh environments (Chaps. 2 and 3) and are regarded as polyextremotolerant generalists (Chap. 4). So, it is very easy for researchers to isolate and screen different strains of *Aureobasidium* spp. for production of various industrial bioproducts and for studies on evolutionary cell biology and phenotypic plasticity as the non-model organisms. It can be clearly read from the contents of Chaps. 7, 8, 9, 10, and 11 that different strains of *Aureobasidium* spp. have different kinds of signaling pathways. Therefore, it is very easy for them to sense, transfer, and respond to all the signals in the surroundings for them to adapt to different environments and to regulate metabolic pathways. Due to available annotated genomes, efficient transformation methods and genome editing tools (Chap. 5), all the synthetic pathways, their regulation mechanisms, molecular weights of polymers, and overproduction of industrial bioproducts can be easily elucidated, clarified, and edited by metabolic engineering, genome editing and synthetic biology (Chaps. 7, 8, 9, 10 and 11). Several strains of *A. pullulans* and *A. melanogenum* can be used as alternatives to synthetic pesticides to promote plant growth, protect wood, crops, and fruits from fungal pathogenic infection and keep food and feed security (Chap. 12). Because *Aureobasidium* spp. can be widely used in the sustainable economy, they have been received great attention in the academic circle and biotechnical companies. So, the authors make it sure that the book provides an extensive overview of *Aureobasiidum* spp., covering their biology and metabolite overproduction for potential alternative applications and is very useful for researchers studying the properties and unique biological life of *Aureobasidum* spp. and for biotechnical companies to produce new sustainable products or apply them in various sectors of biotechnology, agricultural, food, and pharmaceutical industries.

Qingdao, Shandong, China

Zhe Chi
Guang-Lei Liu
Zhen-Ming Chi

Acknowledgments

We thank the financial support by the National Key R&D Program of China, China (2021YFC2103200) and National Natural Science Foundation of China, China (Grant Nos. 31970058 and 31770061).

Competing Interests The authors have no competing interests to declare that are relevant to the content of this manuscript.

Contents

Abbreviations

1,8-DHN	1,8-dihydroxynaphthalene
AGC	Acetyl-glutamate cycle
Ags2	α-Glucan synthase 2
AMT	*Agrobacterium*-mediated gene transfer
ANI	Average nucleotide identity
BTU2	Beta-tubulin genes
CCR	Carbon catabolite repression
CEN DNA	Centromere DNA
CFU	Colony formation unit
CRISPR	Clustered regularly interspaced short palindromic repeat system
ELO2	Elongase-encoding gene
EMSA	Electrophoretic mobility shift assay
EPS	Exopolysaccharide
FA	Fumaric acid
FASs	Fatty acid synthetases
FOSs	Fructooligosaccharides
FsC	Fusigen
G418 gene	Geneticin resistance gene
HPT gene	Hygromycin resistance gene
HR-PKS	Highly reducing polyketide synthase
IOS	Inulooligosaccharides
ITS	Internal transcribed spacer
L-Orn	L-Ornithine
L-Piz	Piperazic acid
LSU	Large ribosomal subunit
M40Y	Harrold's malt yeast extract with 40%
MAT	Mating type
MEA	Malt extract agar
mtLSU	Mitochondrial large subunit
mtSSU	Mitochondrial small subunit
Mw	Molecular weight

NAT gene	Nourseothricin resistance gene
NCR	Nitrogen catabolite repression
NR-PKS	Non-reducing polyketide synthase
NRPS	None ribosome peptide synthase
NsdD	Never in sexual development D
OA	Oatmeal agar
ORF	Open reading frame
OUC	Ornithine urea cycle
PDA	Potato dextrose agar
PMA	Polymalic acid
PMAs	The PMA synthetase
POD	Peroxidase
PPO	Polyphenol oxidase
PYC	Pyruvate carboxylase
ROS	Reactive oxygen species
RPB2	RNA polymerase II second largest subunit
SOD	Superoxide dismutase
TAGs	Triacylglycerols
TCA	Tricarboxylic acid
T-DNA	Transfer DNA
TEF1α	Translation elongation factor 1-α
TEs	The transposable elements
UV	Ultraviolet rays
VHb	Vitreoscilla hemoglobin
VOCs	Volatile organic compounds
XOS	Xylooligosaccharides
YPD	Yeast-peptone-dextrose

Chapter 1
Introduction

Abstract An introduction to the whole book is given. It has been defined that *Aureobasidium* spp. are the black yeast-like polymorphic fungi with remarkable phenotypic and morphological plasticity to adapt to different environments and can produce multiple sustainable bioproducts which have many applications in various sectors of Biotechnology. Moreover, many questions on the basic biology of *Aureobasidium* spp., their detailed metabolisms and molecular regulations, remain to be explored. Therefore, different species of *Aureobasidium* have been receiving great attention in recent years because they can be used as non-model eucaryotes for studying evolution, fundamental biology, and synthetic biotechnology. This is the first time for this book to provide a comprehensive summary and review of ecology, taxonomy, basic biology, genomes and their editing techniques, enzymes, synthetic biotechnology, and applications of *Aureobasidium* spp. This book can be used as a new textbook for global researchers and biocompanies to plan and promote the studies on *Aureobasidium* spp.

Keywords *Aureobasidium* spp. · Yeast-like fungi · Versatility · Summary and review

Aureobasidium spp. are the yeast-like polymorphic fungi with remarkable phenotypic and morphological plasticity, and they have different cell morphologies, including yeast-like cells, blastoconidia, swollen cells, septate swollen cells, arthroconidia, chlamydospore, and hyphae. Hyphae of *Aureobasidium* spp. are hyaline, smooth- and thin-walled, and become dark-brown and thick-walled when they get old. Because most of them can produce melanin, which is accumulated in their cell walls, they are also called black yeasts. Now, they are regarded as polyextremotolerance because they can resist toward high salt concentrations, high sugar concentration, H_2O_2, UV radiation, heavy metals and draught, acidic and basic conditions, and a temperature spectrum from polar, desert to tropical condition (Dielentheis-Frenken et al. 2024) and are widely distributed in air, space, glacial ice, fuel tanks, salterns, water bodies, soil, wood, honey, and plant materials, as well as inorganic materials such as rocks, desert, and marble. They can be saprobes, pathogens, or

Z. Chi et al., *Biology and Synthetic Biotechnology of Aureobasidium spp.*,
https://doi.org/10.1007/978-981-92-2607-8_1

endophytes of plants and form a biofilm on plastics and any other solid surfaces. So, *Aureobasidium* spp. are the ubiquitous black yeast-like fungi with a wide variety of morphologies and growth modes, including "typical" single-budding yeast, larger multinucleate yeast that can make multiple buds in a single cell cycle, and multinucleate hyphae with variable morphology. The morphological flexibility may help to explain how *Aureobasidium* spp. can thrive in a very wide variety of environments, including harsh environments, in soil, in the ocean, and on plants as a saprophyte or endophyte. In very recent years, the versatile *Aureobasidium* spp. have shown an excellent promise as a model system for investigating mechanisms of cell adaptation to different environments, the regulation of cell shape and molecular metabolisms, cell division, remarkable phenotypic and metabolic plasticity, and elucidation of mechanisms of growth modes (Chi et al. 2022; Colarusso et al. 2025). In addition, the study of *Aureobasidium* spp. can greatly benefit from detecting new variants with unique genetic traits. Therefore, this ubiquitous polyextremotolerant fungi can be used as the non-model organisms for studies on evolutionary cell biology and phenotypic plasticity. Because *Aureobasidium* spp. can be widely used in the sustainable economy, they have received great attention in the academic circle and various sectors of biotechnology (Rensink et al. 2024; Xiao et al. 2024).

Taxonomically, the genus *Aureobasidium* spp. belongs to the Family Aureobasidiaceae, the order Dothideales inside the Ascomycetes and subdivision of Pezizomycotina. DNA-identified *Aureobasidium* species include *A. pullulans*, *A. melanogenum*, *A. subglaciale*, *A. namibiae*, *A. castaneae*, *A. pini*, *A. castaneae*, *A. insectorum*, *A. planticola*, *A. motuoense*, *A. intercalariosporum*, *A. iranianum*, *A. mangrovei*, *Aureobasidium* sp. TeO12, *A. mustum*, *A. vineae*, *A. uvarum*, *A. caatingaense*, *A. santateresinhaense*, *A. aubasidani, A. acericola*, *A. thailandense*, *A. khasianum*, *A. leucospermi*, *A. proteae*, *A. tremulum*, *A. welwitschiae*, *A. faidherbiae, A. microtermitis*, *A. aerium*, *A. bupleuri*, *A. zeae*, *A. hainanensis*, *A. albui*, *A. cavalettoi*, *A. diazvalderramae*, *A. ellingtonae*, *A. essambe*, *A. fermentans*, *A. mahoniae*, *A. musti*, *A. peruvianum*, *A. rubi*, *A. toomeae*, *A. vanuatuense*, *A. aleuritis, A. apocryptum*, *A. dalgeri*, *A. prunicola, A. ribis*, *A. sanguinariae*, *A. umbellulariae*, *A. vaccinii*, *A. epipinicola*, *A. aurantiacum*, and so on (Wu et al. 2023; Rensink et al. 2024; Xiao et al. 2024; da Silva et al. 2024), and the totally identified species are 55. Various loci (i.e., the internal transcribed spacer (ITS) rDNA, translation elongation factor-1α, β-tubulin, large ribosomal subunits (LSU, 18S rDNA and 26S rDNA), RNA polymerase II, mitochondrial DNA and even the whole genomic DNA) have been used for molecular identification of different strains of *Aureobasidium* spp. in addition to colony and cell morphology as well as assimilation profiles. They are phylogenetically and morphologically related to different strains of *Kabatiella* spp. and *Aspergillus* spp.

Among them the best known *Aureobasidium* species include *A. pullulans, A. aubasidani*, and *A. melanogenum.* They can synthesize and secrete amylase, cellulase, xylanase, pectinase, inulinase, mannanase, esterases, lipase, proteinase, laccases, and tannase so that they can utilize a wide range of natural products including various wastes from industries (Wang et al. 2022) and occur globally. It has been well known that different native strains of *Aureobasidium* spp. can synthesize pullulan, polymalate, β-1,3 and 1,6-glucan, fructooligosaccharide, liamocin, massoia lactone, intracellular lipids, fumarate, gluconic acid, siderophore, L-ornithine, melanin, and so on, some of which can replace non-sustainable chemicals and have many applications in food, environmental, agricultural, pharmaceutical, and cosmetic industries and nano-biotechnology. The synthetic pathways of pullulan, polymalate, liamocin, intracellular lipids, gluconic acid, fumarate, siderophore, melanin, and their molecular regulations have been basically elucidated and have been thoroughly reviewed (Wei et al. 2021; Kang et al. 2022; Qi et al. 2021; Chi et al. 2022; Tint et al. 2025).

Over 70 genomes of *Aureobasidium* spp. have been sequenced and annotated (Gostinčar et al. 2014, 2019; Jia et al. 2023). Although the genomic DNAs harbor the mating loci (*MAT1-1-1* and *MAT1-2-1*), the sexual reproduction of haploid and diploid cells has not been observed and reported for all the strains of *Aureobasidium* spp. Their genomic DNA can be edited using CRISPR-mediated genome editing, the efficient Cre/loxP site-specific recombination system, and *Agrobacterium*-mediated gene transfer (AMT). Plasmids or DNA fragments can be introduced into host cells via chemical transformation, protoplast transformation, and electroporation (Zhang et al. 2019a; Zhang et al. 2019b). Now, pullulan, polymalate, β-1,3 and 1, 6-glucan, fructooligosaccharide, liamocin, massoia lactone, intracellular lipids, fumarate, gluconic acid, siderophore, L-ornithine, melanin, L-piperazic acid, and putrescine have been overproduced by metabolic engineering and synthetic biology, and molecules of pullulan, polymalate, liamocin, and L-ornithine have been edited by synthetic biology (Wei et al. 2021; Kang et al. 2022; Qi et al. 2021; Chi et al. 2022; Tint et al. 2025).

Moreover, certain strains of *Aureobasidium* spp. can serve as alternatives to synthetic pesticides to promote plant growth and protect wood, crops, and fruits from fungal pathogenic infection.

Although bioproducts such as pullulan and polymalate from *Aureobasidium* spp. are widely used in industries, the biology of *Aureobasidium* spp. remains poorly understood. Therefore, this book provides the first comprehensive summary and review of ecology, taxonomy, basic biology, genomes and their editing techniques, enzymes, synthetic biotechnology, and applications of *Aureobasidium* spp. We believe that this book will help researchers from diverse backgrounds to understand the significance of *Aureobasidium* spp. and to utilize the information from the book for planning and promoting studies on *Aureobasidium* spp.

References

Chi Z, Kong CC, Wang ZZ, Wang Z, Liu GL, Hu Z, Chi ZM (2022) The signaling pathways involved in metabolic regulation and stress responses of the yeast-like fungi *Aureobasidium* spp. Biotechnol Adv 55:107–898

Colarusso A, Williams A, Gladfelter AS, Wirshing AC, Lew DJ (2025) Optimized vectors for genetic engineering of *Aureobasidium pullulans*. bioRxiv 27:634885

da Silva AC, Ferro LO, Alves VCS, Barbosa RN, Souza-Motta CM, Bezerra JDP (2024) Phylogenetic evidence revealed new *Aureobasidium* species (Saccotheciaceae, Dothideales) from Brazil. Mycol Prog 23:70

Dielentheis-Frenken MRE, Wibberg D, Blank LM, Tiso T (2024) Draft genome sequence and annotation of the polyextremotolerant polyol lipid producing fungus *Aureobasidium pullulans* NRRL 62042. BMC Genom Data 25:75

Gostinčar C, Ohm RA, Kogej T, Sonjak S, Turk M, Zajc J, Zalar P, Grube M, Sun H, Han J, Sharma A, Chiniquy J, Ngan CY, Lipzen A, Barry K, Grigoriev IV, Gunde-Cimerman N (2014) Genome sequencing of four *Aureobasidium pullulans* varieties: biotechnological potential, stress tolerance, and description of new species. BMC Genomics 15:1–29

Gostinčar C, Turk M, Zajc J, Gunde-Cimerman N (2019) Fifty *Aureobasidium pullulans* genomes reveal a recombining polyextremotolerant generalist. Environ Microbiol 21:3638–3652

Jia SL, Zhang M, Liu GL, Chi ZM, Chi Z (2023) Novel chromosomes and genomes provide new insights into evolution and adaptation of the whole genome duplicated yeast-like fungus TN3-1 isolated from natural honey. Funct Integr Genom 23:206

Kang XX, Jia SL, Wei X, Zhang M, Liu GL, Hu Z, Chi Z, Chi ZM (2022) Liamocins biosynthesis, its regulation in *Aureobasidium* spp., and their bioactivities. Crit Rev Biotechnol 42:93–105

Qi CY, Jia SL, Liu GL, Chen L, Wei X, Hu Z, Chi ZM, Chi Z (2021) Polymalate (PMA) biosynthesis and its molecular regulation in *Aureobasidium* spp. Int J Biol Macromol 174:512–518

Rensink S, van Nieuwenhuijzen EJ, Sailer MF, Struck C, Wösten HAB (2024) Use of *Aureobasidium* in a sustainable economy. Appl Microbiol Biotechnol 108(1):202

Tint KMM, Wei X, Wang P, Liu GL, Zhang M, Chi ZM, Chi Z (2025) Biotechnological application of *Aureobasidium* spp. as a promising chassis for biosynthesis of ornithine-urea cycle-derived bioproducts. Crit Rev Biotechnol 45(3):591–605

Wang P, Jia SL, Liu GL, Chi Z, Chi ZM (2022) *Aureobasidium* spp. and their applications in biotechnology. Process Biochem 116:72–83

Wei X, Liu GL, Jia SL, Chi Z, Hu Z, Chi ZM (2021) Pullulan biosynthesis and its regulation in *Aureobasidium* spp. Carbohyd Polym 251:117076

Wu F, Feng Z, Wang M, Wang Q (2023) Proposal of four new *Aureobasidium* species for exopolysaccharide production. J Fungi 9(4):447

Xiao D, Driller M, Dielentheis-Frenken M, Haala F, Kohl P, Stein K, Blank LM, Tiso T (2024) Advances in *Aureobasidium* research: paving the path to industrial utilization. Microb Biotechnol 17(8):e14535

Zhang Y, Feng J, Wang P, Xia J, Li X, Zou X (2019a) CRISPR/Cas9-mediated efficient genome editing via protoplast-based transformation in yeast-like fungus *Aureobasidium pullulans*. Gene 709:8–16

Zhang Z, Lu Y, Chi Z, Liu GL, Jiang H, Hu Z, Chi ZM (2019b) Genome editing of different strains of *Aureobasidium melanogenum* using an efficient Cre/loxp site-specific recombination system. Fung Biol 123:723–731

Chapter 2
Distribution of *Aureobasidium* spp. in Natural Environments

Abstract Different strains of *Aureobasidium* spp. are widely distributed in various environments, especially in harsh environments, such as natural honey, hypersaline waters in salterns, rocks, monuments, glacial/subglacial ice, mangrove ecosystems, phyllosphere of different plants, the carposphere of various fruits and vegetables, inner tissues of plants, tropical environments, seawater, deep-sea environments, deserts, and even air space. It has been confirmed that they are indeed polyextremotolerant generalists and have versatility in different environments and bioindustries by synthesizing melanin, exopolysaccharide, liamocin, massoia lactone, siderophore, trehalose, glycerol, antibiotics, and polymalate. The polyextremotolerance is also closely related to their remarkable phenotypic, morphological, and genomic plasticity. This makes it easy for researchers to gain different strains, even new novel species of *Aureobasidium* spp., in every corner of global natural environments to study their fundamental biology, evolution, and adaptation, and for biotechnological companies to develop and genetically modify more bioproducts from *Aureobasidium* spp. and their genomes by synthetic biotechnology.

Keywords Harsh environment · Distribution · Fungal ecology · Polyextremotolerance

2.1 Introduction

Aureobasidium spp. are widely distributed across diverse environments, including the phyllosphere, bathrooms, food, and feed. They also inhabit hypersaline waters in salterns, rocks, monuments, glacial/subglacial ice, mangrove ecosystems, seawater, deep-sea environments, deserts, and even space. Their high genetic recombination rates contribute to the effective dispersal, exceptional phenotypic plasticity, and polyextremotolerance, making them ecological generalists.

Z. Chi et al., *Biology and Synthetic Biotechnology of Aureobasidium spp.*,
https://doi.org/10.1007/978-981-92-2607-8_2

2.2 Desert, Rocks, Monuments, and Space

Many strains of *A. melanogenum* can be found in the Taklimakan Desert, the largest mobile desert in China. The desert called the "Dead Sea" is located in the hinterland of the Tarim Basin in China because there are few organisms which can exist in such a harsh environment characterized by very limited rainfall (10–50 mm), strong evaporative potential (2100–3400 mm), and extreme temperatures (the highest temperature in summer is 67.2 °C and the lowest temperature in winter is −20 °C). The main and most important stress factors in such a desert region are: the extremely limited availability of water, the extreme temperatures, the big change in temperature, the limited availability of organic carbon, high ultraviolet and infrared irradiation, and osmotic stress in the soil or on the rock surfaces (Jiang et al. 2016) (Fig. 2.1). For example, *A. melanogenum* XJ5-1 isolated from such a desert can produce a large amount of melanin which is accumulated in its cell wall (Jiang et al. 2016). Its cell size is larger than that of *A. melanogenum* P5 strain isolated from a mangrove ecosystem. The melanized cells of the yeast strain XJ5-1 have higher tolerance to UV radiation, oxidation (200.0 mM H_2O_2), heat treatment (40 °C), salt shock (200.0 g/L NaCl), desiccation and strong acid hydrolysis (6.0 M HCl) at high temperature (80 °C) than the non-melanized cells of the same yeast-like fungal strain XJ5-1 grown in the same medium with 2.0% of $(NH_4)_2SO_4$. At the same time, trehalose also plays an important role in the adaptation of the yeast-like fungal strain XJ5-1 to heat shock, high oxidation, and high desiccation. This means that the produced melanin and trehalose are closely related to its occurrence in such an extreme environment (Jiang et al. 2018). Another *A. melanogenum* 13-2 isolated from the same Taklimakan desert can produce 73.3 ± 2.3 g/L and 78.1 ± 3.5 g/L of pullulan within 5 days at a flask level and during a 10-Liter fermentation, respectively. The produced pullulan can play an important role in resistance to UV light, high NaCl concentration, heat treatment, strong oxidation of H_2O_2, and desiccation (Jiang et al. 2019).

A. melanogenum strains such as *A. melanogenum* XJ5–1 can grow at temperatures up to 37 °C (Fig. 2.2) because it can produce melanin in greater amounts than any other *Aureobasidium* species. Furthermore, growth at higher temperatures in flasks and fermenters can greatly reduce cooling water, contributing significantly to

A B

Fig. 2.1 The sand (**A**) and high salt soil (**B**) in the Taklimakan Desert

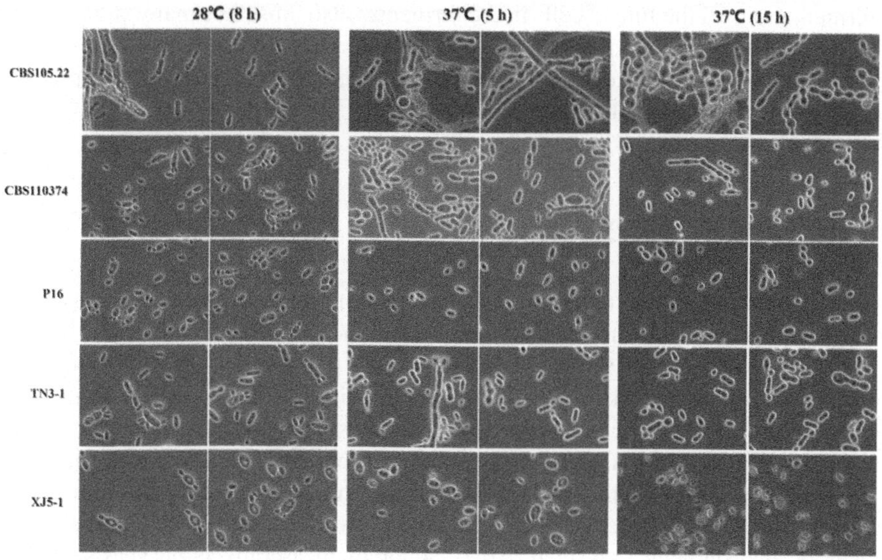

Fig. 2.2 Cell growth of different strains of *A. melanogenum* from different sources at 28 °C and 37 °C

lowering the energy demand for bio-industrial processes. In addition, the purified melanin from the black yeasts has a variety of pharmaceutical, cosmetic, and environmental applications because it has antimicrobial, antiviral, antitumor, antioxidant, anti-aging, and anti-inflammation activities and has a high capacity to absorb broad-spectrum electromagnetic radiation ranging from visible light and ultraviolet radiation to X-rays and bind heavy metals and radionuclides. The produced pullulan also has many applications in food, pharmaceutical, cosmetic, and agricultural industries.

Aureobasidium namibiae, which minimal growth temperature is 10 °C, optimal growth temperature is 25 °C, and maximal growth temperature is 30 °C, was isolated from dolomitic marble in the Namib Desert, Africa.

A. pullulans strains also can be obtained from rocks and monuments situated in climate areas with high humidity and lower exposure to sun irradiation, and they are associated with evident biodeterioration patterns (mainly dark rock surfaces). Their presence is quite variable, being dependent on the presence of a more suitable organic substratum in organic close vicinity of rocks and monuments (Urzı et al. 1999).

A. melanogenum strain W12, which has two nuclei in each of several cells, a genome size of 53,160,883 bp, and 90.8% of duplicated genes, was isolated from the samples collected inside an aircraft (Wagner et al. 2024), while *A. aerium* was obtained from air at Xiangshan Mountain, Haidian District, Beijing, China (Wang et al. 2022). Therefore, different strains of *Aureobasidium* spp. distributed in the desert, rocks, monuments, and space indeed can resist the stressed environments by producing melanin, trehalose, exopolysaccharides, and other substances. During fermentation, all the fungal cells in the fermenter also must face many stresses, such as pH and temperature fluctuations. So, the unique properties of *Aureobasidium* spp. obtained from the desert, rocks, monuments, and space have high potential applications in fermentation industries.

2.3 Natural Honey

It has been well known that natural honey, a common sweet food, contains a very high concentration of fructose and glucose, and the osmotic pressure in the honey is very high (Xue et al. 2019). It contains about 70% fructose and glucose of the sugars and 1.3% sucrose and other sugars (about 12%) with a water content of about 17.2% (Fig. 2.3). Therefore, the natural honey is rich in sugars and is a hostile substrate to most of microorganisms because of its low pH and high sugar content, leading to growth of only osmophilic and osmotolerant microorganisms. However, many osmophilic yeasts and yeast-like fungi could grow or survive in such natural hyper-osmotic environments (Xue et al. 2019). So far, many ascomycetous yeasts, obligate xerophiles, and xerotolerant species have been found in the honey. For example, the cell sizes, cell walls, and the number of small vacuoles of *A. melanogenum* TN3-1

Fig. 2.3 Honeycomb and honey sample

isolated from natural honey produced by the bee collected from Taian City (N36°11′, E117° 07′) of Shandong province, China, are much larger, thicker, and higher, respectively, than those of *A. melanogenum* P16 isolated from the mangrove ecosystems (Jia et al. 2020). At the same time, the glycerol, trehalose, and glycogen contents in *A. melanogenum* TN3-1 cells are also higher than those of *A. melanogenum* P16 from the mangrove ecosystems and *A. melanogenum* TN1-2 cells isolated from the same honey (Jiang et al. 2018). Most of the cells in the culture of the strain TN3-1 are arthroconidia, while all of the yeast-like fungal cells of *A. melanogenum* P16 cultivated under the same conditions are blastoconidia. *A. melanogenum* TN3-1 cells can actively transform 140.0 g/L of glucose into 110.29 ± 2.17 g/L of pullulan during the fermentation, whereas *A. melanogenum* P16 and TN1-2 convert 140.0 g/L of glucose into only 45.81 ± 1.7 g/L and 48.7 ± 2.6 g/L of pullulan. This means that *A. melanogenum* TN3-1 cells isolated from the natural honey can tolerate very high concentration of glucose (Xue et al. 2019). Intracellular glycerol and trehalose contents and the numbers of the small vacuoles in the cells also have close relations to high pullulan synthesis and high osmotic tolerance (Chen et al. 2020). Furthermore, it has been found that the yeast genera detected among flowers, bee gut mycobiota, and bee products are *Aureobasidium*, *Filobasidium*, *Meyerozyma*, and *Metschnikowia* (Agarbati et al. 2024). Furthermore, the natural honey was found to be a good bioresource for isolation of novel obligate xerophiles and xerotolerant fungi because the bees would collect pollen with different fungi from flowers of different species of plants in different native environments and such fungi have high potential applications in biotechnology. The osmophilic and osmotolerant fungi from honey are suitable for overproduction of bioproducts, including pullulan, from high concentration of glucose in the medium. Such a very high gravity fermentation technique can increase plant efficiency, reduce workforce and energy consumption for product recovery, and increase product yield per unit of working volume of the bioreactor.

2.4 Mangrove Ecosystem

Mangroves are a group of salt-tolerant plant species that occur in tropical and subtropical intertidal estuarine regions, provide the unique environments linking land and sea, and sustain the ecosystems containing a variety of plants, offering a habitat for mammals, birds, fish, and insects (Fig. 2.4) (Chi et al. 2012). Although all the mangrove trees live in such hostile environmental conditions, some of the conditions in these ecosystems are suitable for growth of the mangrove-associated microorganisms, including fungi, because the temperature (around 30 °C) is constant and the ecosystems are rich in nutrients and water (Jia et al. 2020). Different species of *Aureobasidium* spp. are also found to be widely distributed in such ecosystems. For example, the high pullulan-producing *A. melanogenum* P16, intracellular lipid-producing *A. melanogenum* P10, high liamocin-producing *A. melanogenum* 9-1, heavy oil-producing *A. melanogenum* P5, polymalate-producing *A. pullulans* var. *pullulans* MCW, and gluconic acid-producing *A. pullulans* P25 have been isolated

Fig. 2.4 Mangrove ecosystem in Hainan, China

from the leaves of the mangrove plant *Acanthus ilicifolius* from DongZaiGang, Haikou, Hainan Province, China (Jia et al. 2020). Three strains of *A. pullulans* were obtained from roots, fruits, and leaves of the mangrove trees in China (Jia et al. 2020), while *A. mangrovei* was isolated from healthy leaves collected from the mangrove tree *Avicennia marina* growing in the marsh zone in the Persian Gulf, Iran (Nasr et al. 2018).

A. hainanensis isolated from the mangrove system (the leaf of the mangrove plant, *Kandelia candel*) at DongZaiGang, Haikou, Hainan Province, China, was found to be able to produce high level of Ca^{2+}-gluconic acid and a β-fructofuranosidase with a high inulin-hydrolyzing activity (Jia et al. 2020). A thermotolerant strain, *A. melanogenum* DA22, which exhibits sustained active cell growth and high PMA production (24.6 g/L) at 35 °C, was isolated from the mangrove leaves in Hainan, China (Yang et al. 2026).

Therefore, different strains of *Aureobasidium* spp. obtained from the mangrove ecosystems can produce a wide range of bioactive products, which have many uses in biotechnology.

2.5 Marine Environments

The ocean covers over 70% of our planet, and marine environments also have rich bioresources of fungi (Fig. 2.5). For example, *Aureobasidium* sp., isolated from the deep-sea oil reserves in the Gulf of Mexico, can utilize alkane and alkene as the sole carbon sources, confirming that the deep-sea fungal taxa are valuable genetic

Fig. 2.5 Ocean environment

resources with potential use in bioremediation (Velez et al. 2020). *A. melanogenum* LUO5 collected from the deep-sea at a depth of 8225 m in the North Pacific can produce C_{10} and C_{12} aliphatic δ-lactones, fatty acid methyl esters, and eight known compounds (Dong et al. 2024). *A. pullulans* isolated from an Okinawan marine sponge can produce two new diketopiperazines and one new 1,3-dihydroxyphenol derivative (Shigemori et al. 1998). The 12 strains belonged to the yeast-like species *A. pullulans* and *Exophiala dermatitidis* were obtained from cold habitats (superficial and deep sediments, ice cores, and melt-waters) of the Calderone Glacier (Italy), which is the southernmost glacier in Europe (Branda et al. 2010).

2.6 The Saltern Ponds

The ponds in the salterns are characterized by evaporating water with salinity ranging from 3.0% (w/v) to saturated solutions of NaCl (35.0% w/v), low oxygen concentration, high light intensity, ample nutrient availability, and neutral pH (Fig. 2.6). Such hypersaline water is also the natural ecological niche for *A. pullulans* (Gunde-Cimerman et al. 2000). *A. pullulans* isolated from such an environment can tolerate the highest salt, with a strain isolated from salterns in Slovenia tolerating salt concentrations up to 17.0% (Gunde-Cimerman et al. 2000). It has been reported that *A. pullulans* HN6.2 isolated from sea saltern at Yellow Sea can produce 1.1 mg of siderophore/mL of the culture (Wang et al. 2009). The siderophore produced by the yeast strain HN6.2 is able to inhibit cell growth of *Vibrio anguillarum* and *Vibrio parahaemolyticus* isolated from the diseased marine animals.

Such high salt-tolerant *Aureobasidium* spp. can be economically used in biotechnological industries because they can grow in the medium prepared with seawater, which is much more available than freshwater in this world. Moreover, they also grow in wastewaters released from industries which often contain high salt concentrations. Thus, such halotolerant *Aureobasidium* spp. also can potentially find application in the wastewater treatment.

Fig. 2.6 The saltern ponds

2.7 The Phyllosphere of Plants and the Carposphere of Various Fruits and Vegetables

The phyllosphere includes leaves, flowers, stems, fruits, and pollens and is one of the largest habitats for harboring microbial life. The phyllosphere is considered to be a relatively harsh and unpredictable environment and poor nutrient availability, with wild changes in moisture levels (the wetting and drying cycle), high UV levels from sunlight, and temperatures that hinder the growth of most microorganisms. However, this unique habitat also offers a continual source of water, secondary metabolites, volatile compounds and other nutrients from plants, and a relatively spacious accommodation (Fig. 2.7). So far, the most thoroughly studied habitats for *Aureobasidium* spp. have been the phyllosphere and the carposphere of various fruits and vegetables. For example, *A. pullulans* Y68 isolated from the leave of one plant in south of China can produce 52.47 ± 0.09, 40.54 ± 1.36, 38.83 ± 0.25, 25.54 ± 1.79, 23.26 ± 0.78, and 15.72 ± 3.98 g/L of pullulan from glucose, sucrose, dextrin, fructose, maltose, and xylose, respectively. The fumarate-producing *A. pullulans* var. *aubasidani* DH177 was obtained from the leaves of *Weigela florida* at Ya-Shan national forest park in Shandong Province of China (Wang et al. 2018), while *Aureobasidium pini* occurs on needles of *Pinus tabuliformis* from China. *A. pullulans* var. *aubasidani* also can be isolated from the sap *Betula* sp., Leningrad region, Russia (Yurlova and de Hoog 2022), while *Aureobasidium acericola* was also found on leaves of the plant *Acer pseudosieboldianum*. It was found that *Aureobasidium khasianum* is distributed on fallen leaf litter of the plant *Wightia speciosissima*, India. *Aureobasidium mustum*, *Aureobasidium vineae*, and *Aureobasidium uvarum* can grow on the surface of grapes (Onetto et al. 2020). *Aureobasidium thailandense* sp. nov. isolated from leaves of the plant *Cerbera*

Fig. 2.7 The forests

odollum and wooden surfaces in Thailand can produce a non-pullulan extracellular polysaccharide and 68.56 to >1000 mg/L of liamocins (Peterson et al. 2013; Rattanahadsadin et al. 2024). Furthermore, *Aureobasidium* spp. are a significant cause of paint deterioration on wood by destroying both exterior and interior paint film. In fact, the genus *Aureobasidium* sp. was first isolated on grape leaves. Even the northwestern Patagonian phylloplane microbial community of the plant *N. pumilio* is dominated by *A. pullulans*, and the northwestern Patagonian region in Argentina is characterized by low temperatures, high desiccation, and high levels of ultraviolet (UV) radiation most of the year (Parra et al. 2023). *A. pullulans* also reside on carposphere of withered grapes (low water activity and exosmosis) for Italian passito wine production and has inhibitory capacity against phytopathogenic fungi (Lorenzini and Zapparoli 2019). It has been reported that *A. pullulans* and *Metschnikowia pulcherrima* occur regularly in the blossom samples of different fruit trees at three locations in southwest Slovakia (Vadkertiová et al. 2012).

The five strains of *A. pullulans* from 52 tree bark samples of the Medaram and Srisailam forest areas of Andhra Pradesh, India, can produce endo-1-4-β-xylanase similar to that of *A. pullulans* var. *melanigenum* on a xylan-agar medium (Bhadra et al. 2008). So, such yeast-like fungi could play an important role in hydrolysis of hemicelluloses into simple sugars.

One strain of *A. pullulans* is present in all the samples from different wine grape varieties in Denmark and is the dominating species, representing 40–80% of the isolates in these samples (Lederer et al. 2013). *A. pullulans* is the dominating yeast-like fungus on immature grapes, but it is never present on fully ripened grapes, whereas another strain of *A. pullulans* has high amounts on mature berries. This ascomycete is known to produce high amounts of exopolysaccharides, which form a biofilm on fruit surfaces, and it is able to decrease acidity in wine to a serious level, potentially allowing growth of spoilage microorganisms.

A. leucosperm, *A. castaneae*, and *A. pini* were obtained from the plant *Leucospermum* leaves, leaves of the plant *Castanea henryi*, and pine needle surface of the plant *Pinus tabuliformis* in China, respectively, while *A. iranianum* was isolated from bamboo in Iran, and *A. leucospermi* was also identified based on the isolates from leaves of the plant *Leucospermum conocarpodendron* in South Africa. In fact, *A. leucospermi* has been reported on different hosts and substrates (e.g., *Guapira gracilifora*, *Leucospermum* sp., *Mycocepurus goeldii*, among others) in different countries (Brazil, Portugal, Indonesia, and Korea). In addition, *A, santateresinhaense* was encountered in branches of the plant *Anadenanthera colubrina* (Fabaceae) at RPPN Fazenda Tamanduá in Brazil.

However, some strains of *Aureobasidium* spp. can cause leaf diseases. For example, *Aureobasidium proteae* and *Aureobasidium leucospermi* can cause leaf spots and are considered plant pathogens.

All the results mentioned above show that the phyllospheres are indeed suitable for living of different strains of *Aureobasidium* spp., and different strains of *Aureobasidium* spp. isolated from phyllospheres of different plants and mangrove trees have highly potential applications in biotechnology. In fact, the diversity of *Aureobasidium* spp. can be influenced by plant morphology, different levels of nutrients (carbohydrates, calcium, and phenolic compounds), and environmental factors (e.g., temperature, rainfall, wind, and solar radiation).

2.8 Endophytic Fungi of Plants

Endophytic fungi, including ascomycetes, basidiomycetes, and zygomycetes, located in the intercellular and intracellular space of host plant, can play a crucial role in both natural and agricultural ecosystems by producing antimicrobial agents, plant hormones, and hydrolytic enzymes. For example, the endophytic fungus *A. pullulans* KB3 was isolated from the bark of Koe-lreuteria paniculata in Quancheng Park, Jinan City, Shandong Province, China, and was identified as a new subspecies of *A. pullulans* and can produce a novel compound, Sphaerolone, a polyketone compound (Wang et al. 2025). An endophytic fungus, *A. pullulans* BSS6, secreting extracellular enzymes and auxin, could create a distinct niche for ecological adaptation during the symbiosis with the host Frankincense tree (Khan et al. 2016). *Phoma glomerata*, *A. proteae*, and *Sarocladium kiliense* isolated from inner tissues of wheat plants can inhibit the growth of *Fusarium graminearum* and *Fusarium culmorum*, which can cause head blight, a major disease on wheat crops worldwide (Comby et al. 2017). The endophytic *Aureobasidium* sp. strain isolated from roots and stems of the plant *Carya illinoinensis* can stimulate *Arabidopsis* growth by up-regulation of peroxidase and glutathione S-transferase activity, can enhance production of auxin/indole acetic acid, and is able to promote cytokinin signaling by direct contacting the seedlings. At the same time, it can produce many soluble metabolites, cell-free supernatant, and volatile organic compounds for plant growth promotion (Jiang et al. 2025). The fungal endophyte—*A. pullulans* CY.OS

13 isolated from the mango trees (*Mangifera indica* L.)—has a significant increase in seed germination and growth parameters of chili plants grown under pot-scale conditions (Sonsiam et al. 2024). The endophytic *Aureobasidium* sp. BGP-C02 isolated from healthy cardamom leaves in eastern Nepal has high antagonistic acvity against the pathogenic fungus *F. equiseti* in tomato plants (Bir Rai et al. 2026). So, the endophytic fungi of *Aureobasidium* spp. have positive relations with their plant hosts and have many benefits in agricultural industries.

2.9 Tropical Environments

A tropical environment refers to a region characterized by persistently high temperature and relative humidity (Fig. 2.8). It was found that 46 yeast-like strains (*A. pullulans*) isolated from water, soil, insect, and plant materials collected from the tropical environments of three Brazilian rain forests (Tijuca, Serra Capivari, and Pau de Fome) have the extracellular starch-degrading activity, esterase activity, lipase activity, protease activity, pectinase activity, and chitinase activity (Buzzini and Martini 2002). Three isolates of *A. pullulans* from leaves and fruit of native Brazilian plants *Byrsonima crassifolia* and *Eugenia dysenterica* at Cerrado conservation areas of Brazil display high potential for the control of citrus green mold (*P. digitatum*) producing the mycotoxin patulin, reduce the incidence of disease, and increase the storage time of fruit (Sperandio et al. 2015). Some isolates of *A. pullulans* obtained from fruits of diverse indigenous tropical plants from areas around

Fig. 2.8 Tropical forest

Bulawayo, Zimbabwe, can produce significant extracellular cellulolytic activity (Kudanga and Mwenje 2005). *Aureobasidium* sp. from two freshwater aquatic ecosystems, including the Cauca and Meléndez rivers, a wastewater treatment plant (PTAR), and a drinking water treatment plant in Cali, Colombia, can accumulate uncommon fatty acids such as petroselinic acid under the conditions induced by glycerol. This fatty acid, which has a double bond in position 6,7 and a melting point of 33 °C, has its potential as an alternative to margarine production, as well as a precursor to sophorolipids, estolide esters, soaps, and plastics (Ramirez-Castrillon et al. 2025). Furthermore, *A. caatingaense* was isolated from Brazilian tropical dry forest, Caatinga. This demonstrates that the tropical environments also have a good bioresource of *Aureobasidium* spp., which have high potential applications.

2.10 Cold Environments

In the cold environments, including the Antarctic (south pole), Arctic (north pole), high mountains, most of parts of the deep sea, and refrigerators (Fig. 2.9), there is also a wide distribution of *Aureobasidium* spp. For example, *A. subglaciale* can be found in Kongfjorden glacial and subglacial ice, seawater, in moss in the colder part of the year, or even in cool human-made environments, such as refrigerators (Zajc et al. 2022). In addition to its ability to grow at low ambient temperatures (4 °C), *A. subglaciale* also tolerates elevated salinity, high UV radiation, heavy metal contamination, and even gamma irradiation (Zajc et al. 2022). Furthermore, the cold-loving yeast-like fungus *A. subglaciale* can significantly reduce rotting of apples caused by *Botrytis cinerea*, *Penicillium expansum*, and *Colletotrichum acutatum* at low temperatures (Zajc et al. 2022). *A. pullulans* 116–1 strain (the accession number of D1/D2 26S rDNA is EU598220) was isolated from Antarctic sea sediments where salt is 2.8% and temperature is 1.5 °C (Zhang et al. 2012).

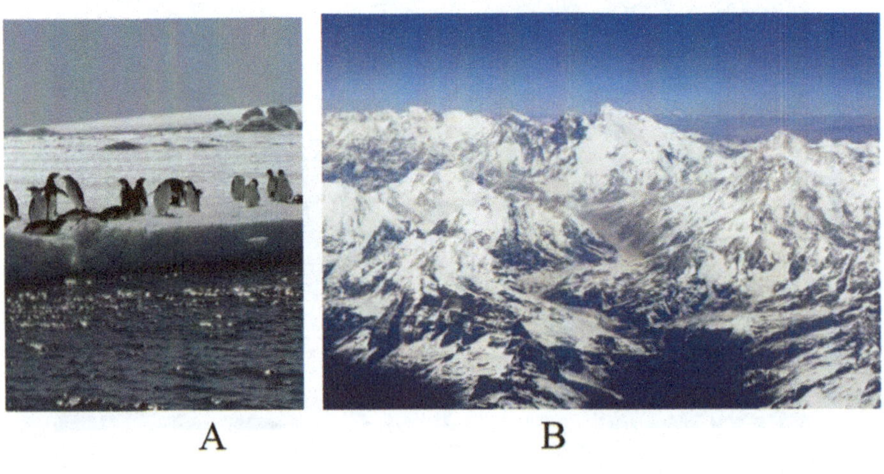

Fig. 2.9 The Antarctic environment (**A**) and glacial ice of the Himalayas (**B**)

Besides their low growth temperature, strains of the species also tolerate high salt concentrations, heavy-metal stress, and radiation, indicating that *A. subglaciale* can resist environmental influences.

2.11 The Clinical Samples

A. pullulans is also regularly isolated in clinical samples, even though it likely does not cause disease in humans. However, it is strange that some agricultural *A. pullulans* isolates are highly tolerant to the antifungal agents, and the clinical *Aureobasidium* isolates are more fungicide sensitive than the agricultural isolates (Magoye et al. 2023).

2.12 Conclusions and Perspectives

All the results mentioned above have strongly demonstrated that *Aureobasidium* spp. are widely distributed in different environments, especially in extreme environments, because they can produce various enzymes to decompose natural amylose, lipid, protein, cellulose, xylan, lignin, mannan, pectin, and tannin from plant materials and various synthetic chemicals from industries. This is also related to their ability to outcompete any other species by fast growth, to greatly adapt to and colonize different novel environments through high recombination rate, high dispersal ability, many cell morphologies and multi-growth modes; to tolerate a plethora of different stresses by producing melanin, trehalose, polysaccharides, glycerol, cryoprotectant, and anti-oxidant; to have nutritional versatility because of the wide enzymatic profiles and secretome composition; to produce antimicrobial compounds such as siderophores inhibiting bacterial pathogen *Vibrio* spp. in marine animals, liamocin killing bacterial pathogen *Streptococcus* spp., Massoia lactone killing all the fungal pathogens and killer toxin with proteolytic activity, β-1,3-glucanase and chitinase against plant pathogens *Penicillum digitatum* and *Geotrichum citriaurantii* (Moura et al. 2021); and to form biofilm on solid surface. So, *Aureobasidium* spp. are considered to be a polyextremotolerant generalist (Gostinčar et al. 2019).

References

Agarbati A, Gattucci S, Canonico L, Ciani M, Comitini F (2024) Yeast communities related to honeybees: occurrence and distribution in flowers, gut mycobiota, and bee products. Appl Microbiol Biotechnol 108:175

Bhadra B, Sreenivas Rao R, Singh PK, Sarkar PK, Shivaji S (2008) Yeasts and yeast-like fungi associated with tree bark: diversity and identification of yeasts producing extracellular endoxylanases. Curr Microbiol 56(5):489–494

Bir Rai G , Kayastha S, Dahal S, Subba B, Malla S (2026) *Aureobasidium* sp. from cardamom plant as a potential biocontrol for *Fusarium equiseti* infecting tomato plants. Biol Cont 219:106077
Branda E, Turchetti B, Diolaiuti G, Pecci M, Smiraglia C, Buzzini P (2010) Yeast and yeast-like diversity in the southernmost glacier of Europe (Calderone Glacier, Apennines, Italy). FEMS Microbiol Ecol 72:354–369
Buzzini P, Martini A (2002) Extracellular enzymatic activity profiles in yeast and yeast-like strains isolated from tropical environments. J Appl Microbiol 93:1020–1025
Chen L, Wei X, Liu GL, Hu Z, Chi ZM, Chi Z (2020) Glycerol, trehalose and vacuoles had relations to pullulan synthesis and osmotic tolerance by the whole genome duplicated strain *Aureobasidium melanogenum* TN3–1 isolated from natural honey. Int J Biol Macromol 165:131–140
Chi ZM, Liu TT, Chi Z, Liu GL, Wang ZP (2012) Occurrence and diversity of yeasts in the mangrove ecosystems in Fujian, Guangdong and Hainan provinces of China. Ind J Microbiol 52(3):346–353
Comby M, Gacoin M, Robineau M, Rabenoelina F, Ptas S, Dupont J, Profizi C, Baillieul F (2017) Screening of wheat endophytes as biological control agents against Fusarium head blight using two different in vitro tests. Microbiol Res 202:11–20
Dong HY, Wang Y, Zhang XY, Zhang M, Yang LH, Zou ZB, Li Y, Xie MM, Yang XW, Wang B (2024) Chemical constituents from the deep-sea-derived fungus *Aureobasidium melanogenum* LUO5. Chem Biodivers 21:e202400507
Gostinčar C, Turk M, Zajc J, Gunde-Cimerman N (2019) Fifty *Aureobasidium pullulans* genomes reveal a recombining polyextremotolerant generalist. Environ Microbiol 21:3638–3652
Gunde-Cimerman N, Zalar P, de Hoog S, Plemenitaš A (2000) Hypersaline waters in salterns-natural ecological niches for halophilic black yeasts. FEMS Microbiol Ecol 32:235–240
Jia SL, Chi Z, Liu GL, Hu Z, Chi ZM (2020) Fungi in mangrove ecosystems and their potential applications. Crit Rev Biotechnol 40:852–864
Jiang H, Ma Y, Chi Z, Liu GL, Chi ZM (2016) Production, Purification, and gene cloning of a β-fructofuranosidase with a high inulinhydrolyzing activity produced by a novel yeast *Aureobasidium* sp. P6 isolated from a mangrove ecosystem. Mar Biotechnol 18(4):500–510
Jiang H, Xue SJ, Li YF, Liu GL, Chi ZM, Hu Z, Chi Z (2018) Efficient transformation of sucrose into high pullulan concentrations by *Aureobasidium melanogenum* TN1-2 isolated from a natural honey. Food Chem 257:29–35
Jiang H, Chen TJ, Chi Z, Hu Z, Liu GL, Sun Y, Zhang SH, Chi ZM (2019) Macromolecular pullulan produced by *Aureobasidium melanogenum* 13-2 isolated from the Taklimakan desert and its crucial roles in resistance to the stress treatments. Int J Biol Macromol 15(135):429–436
Jiang Z, Peng F, Jinping Yu J, Li Q (2025) Plant growth-promoting effects and possible mechanisms of a plant endophytic fungus *Aureobasidium* sp. JRF1. Plant Physiol Biochem 222:109724
Khan AL, Al-Harrasi A, Al-Rawahi A, Al-Farsi Z, Al-Mamari Z, Waqas M, Asaf S, Elyassi A, Mabood F, Shin JH, Lee IJ (2016) Endophytic fungi from frankincense tree improves host growth and produces extracellular enzymes and indole acetic acid. PLoS ONE 11:e0158207
Kudanga T, Mwenje E (2005) Extracellular cellulase production by tropical isolates of *Aureobasidium pullulans*. Can J Microbiol 51(9):773–776
Lederer MA, Nielsen DS, Toldam-Andersen TB, Herrmann JV, Arneborg N (2013) Yeast species associated with different wine grape varieties in Denmark. Acta Agri Scandinavica Sect B Soil Plant Sci 63(1):89–96
Lorenzini M, Zapparoli G (2019) Yeast-like fungi and yeasts in withered grape carposphere: Characterization of *Aureobasidium pullulans* population and species diversity. Int J Food Microbiol 289:223–230
Magoye E, Nägeli L, Bühlmann A, Hilber-Bodmer M, Keller P, Mühlethaler K, Riat A, Schrenzel J, Freimose FM (2023) Clinical *Aureobasidium* isolates are more fungicide sensitive than many agricultural isolates. Microbiol Spect 11(2):1–5
Moura VS, Pollettini FL, Ferraz LP, Mazzi MV, Kupper KC (2021) Purification of a killer toxin from *Aureobasidium pullulans* for the biocontrol of phytopathogens. J Basic Microbiol 61:77–87

Nasr S, Mohammadimehr M, Geranpayeh Vaghei M, Ali Amoozegar M, Fazeli SAS (2018) *Aureobasidium mangrovei* sp. nov., an ascomycetous species recovered from Hara protected forests in the persian gulf, Iran. Antonie Van Leeuwenhoek 111(9):1697–1705

Onetto CA, Schmid SA, Roach MJ, Borneman AR (2020) Comparative genome analysis proposes three new *Aureobasidium* specie s isolated from grape juice. FEMS Yeast Res 20:foaa052

Parra M, Diego Libkind D, Todd Hittinger C, Álvarez L, Bellora N (2023) Assembly and comparative genome analysis of a Patagonian *Aureobasidium pullulans* isolate reveals unexpected intraspecific variation. Yeast 40:197–213

Peterson SW, Manitchotpisit P, Leathers TD (2013) *Aureobasidium thailandense* sp. nov. isolated from leaves and wooden surfaces. Int J Syst Evol Microbiol 63:790–795

Ramirez-Castrillon M, Benavides-León TA, Arcos-Velasco LV, Pantoja-Pulido KD, Lopez-Parra LL, Bolaños-Rojas AC, Osorio-Cadavid E (2025) Tropical lakes as a novel source of oleaginous yeasts with lipid profiles for biodiesel, oleochemical, and nutraceutical applications. World J Microbiol Biotechnol 41:105

Rattanahadsadin N, Lotrakul P, Bankeeree W, Punnapayak H, Prasongsuk S (2024) Production and characterization of extracellular liamocins produced from fungal strains of *Aureobasidium* spp. World J Microbiol Biotechnol 40:253

Shigemori H, Tenma M, Shimazaki K, Kobayashi J (1998) Three new metabolites from the marine yeast *Aureobasidium pullulans*. J Nat Prod 61:696–698

Sonsiam K, Sonesouphap O, Sangkaew A, Chulaka P, Sihanonth P, Yompakdee C (2024) Roles of endophytic fungi isolated from *Mangifera indica* L. in promoting plant growth. J Microbiol Biotechnol 34(9):1857–1866

Sperandio EM, Martins Do Vale HM, Monteiro Moreira GA (2015) Yeasts from native Brazilian Cerrado plants: Occurrence, diversity and use in the biocontrol of citrus green mould. Fung Biol:984–993

Urzı C, De Leo F, Lo Passo C, Criseo G (1999) Intra-specific diversity of *Aureobasidium pullulans* strains isolated from rocks and other habitats assessed by physiological methods and by random amplified polymorphic DNA (RAPD). J Microbiol Meth 36:95–105

Vadkertiová R, Molnárová J, Vránová D, Sláviková E (2012) Yeasts and yeast-like organisms associated with fruits and blossoms of different fruit trees. Can J Microbiol 58(12):1344–1352

Velez P, Gasca-Pineda J, Riquelme M (2020) Cultivable fungi from deep-sea oil reserves in the Gulf of Mexico: Genetic signatures in response to hydrocarbons. Mar Environ Res 153:104816

Wagner DN, Varaljay VA, Lyon WJ, Crouch AL, Allex-Buckner C, Bifffnger JC, Crookes-Goodson WJ, Kelley-Loughnane NK, Stamps BW (2024) Draft genome sequence of potentially dikaryotic black fungus *Aureobasidium melanogenum* isolated from aircraft. Environ Microbiol 13(3):1–3

Wang W, Chi ZM, Chi Z, Li J, Wang XH (2009) Siderophore production by the marine-derived *Aureobasidium pullulans* and its antimicrobial activity. Biores Technol 100:2639–2641

Wang GY, Bai T, Miao Z, Ning W, Liang W (2018) Simultaneous production of single cell oil and fumaric acid by a newly isolated yeast *Aureobasidium pullulans* var. *aubasidani* DH177. Bioproc Biosyst Eng 41:1707–1716

Wang CB, Jiang N, Tu Y, Zhu YQ, Xue H, Li Y (2022) *Aureobasidium aerium* (Saccotheciaceae, Dothideales), a new yeast-like fungus from the air in Beijing, China. Phytotax 544(2):185–119

Wang S, He JN, Wang YJ, Zhao WY, Yang QX, Wang YN, Zhang Y, Zhang LP, Liu HW (2025) Metabolome and genome analysis of a novel endophytic fungus *Aureobasidium pullulans* KB3: discovery of polyketones and polyketone biosynthesis pathway. Biochem Genet 63(4):3047–3064

Xue SJ, Chen L, Jiang H, Liu GL, Chi ZM, Hu Z, Chi Z (2019) High pullulan biosynthesis from high concentration of glucose by a hyperosmotic resistant, yeast-like fungal strain isolated from a natural comb-honey. Food Chem 286:123–128

Yang Y, Dai L, Li F, Xu X, Zou X (2026) Identification of a thermotolerant strain of *Aureobasidium melanogenum* DA22 and functional analysis of Spt23 as a key transcriptional regulator in hightemperature polymalic acid fermentation. Biores Technol 457:134956

Yurlova NA, de Hoog GS (2022) A new variety of *Aureobasidium pullulans*, Chracterized by exopolysaccharide structure, nutritional features. Antonie Leeuwenhoek 1997 72:141–147

Zajc J, Černoša A, Sun X, Fang C, Gunde-Cimerman N, Song Z, Gostinčar C (2022) From glaciers to refrigerators: the population genomics and biocontrol potential of the black yeast *Aureobasidium subglaciale*. Microbiol Spectr 10(4):e0145522

Zhang X, Hua M, Song C, Chi ZM (2012) Occurrence and diversity of marine yeasts in Antarctica environments. J Ocean Univ China 11(1):70–74

Chapter 3
Taxonomy of *Aureobasidium* spp.

Abstract So far, more than 55 species of *Aureobasidium* have been clearly identified based on their cultural characteristics, cell morphology, assimilation profiles, and genomic DNA sequences. Among them, *A. pullulans* and *A. melanogenum* are mostly found, are able to produce various useful products in biotechnology, and have been thoroughly studied. In addition, we found that most of the reported *A. pullulans* so far are *A. melanogenum*. Moreover, only the genomes of *A. pullulans*, *A. melanogenum, A. aubasidani, A. hainanensis*, and *A. thailandense* have been characterized, mined, and edited. Most of them are non-fermenters and grow under aerobic conditions, and only a few species, such as *A. fermentans*, *A. essambe*, *A. vanuatuense*, *A. peruvianum,* and *A. albui*, can ferment some sugars; especially, *A. peruvianum* can ferment melibiose, D-glucose, maltose, sucrose, trehalose, cellobiose, melezitose, raffinose, and weakly ferment inulin. So far, the genomic genes responsible for the fermentation, fermentation products, and their synthetic pathways and molecular regulations have not been clarified.

Keywords Taxonomy · Phylogenetics · Cell morphology · *Aureobasidium* spp

3.1 Introduction

The genus *Aureobasidium* was established by Viala and Boyer in 1891, with *Aureobasidium vitis* as the type species. *A. vitis* is now considered a synonym of *A. pullulans*. As stated in the last chapter, *Aureobasidium* species commonly exhibit plant-associated lifestyles, including saprobic, epiphytic, endophytic, or pathogenic interactions. Also as mentioned in the last chapter, they have been isolated from diverse environments, such as air, deep sea, glacier ice, insects, juice, rocks, soil (including desert and polar regions), water (fresh, marine, and hypersaline), preserved food (frozen and salt-cured), fuel tanks, artificial surfaces (e.g., man-made materials and indoor environments), honey, romaine lettuce, dogs, and hospital laboratories. Consequently, they are globally distributed across temperate, tropical, and polar regions.

Z. Chi et al., *Biology and Synthetic Biotechnology of Aureobasidium spp.*,
https://doi.org/10.1007/978-981-92-2607-8_3

3.2 The Methods for Identification of *Aureobasidium* spp.

All the yeast-like fungal strains exhibit growth at incubation temperatures between 4 °C and 30 °C, with the optimal growth observed at either 25 °C or 28 °C, depending on the strain. Notably, a few strains—*A. melanogenum*, *A. ellingtonae*, *A. peruvianum*, *A. toomeae*, and *A. musti*—are also capable of growth at 37 °C, primarily in filamentous form, albeit at a slower rate compared to lower temperatures. The best overall average growth across all the species is recorded on Malt Extract Agar (MEA) at 25 °C. The most commonly used media for cultivating *Aureobasidium* species are Yeast-Peptone-Dextrose (YPD), Potato Dextrose Agar (PDA), Oatmeal Agar (OA), Malt Extract Agar (MEA), and Harrold's malt yeast extract with 40% (M40Y).

Aureobasidium species exhibit distinct colony morphologies on different growth media. These fungi demonstrate high morphological variability, producing exclusively asexual forms with diverse structures. Their salinity tolerance suggests potential as effective colonizers of osmotic environments, including salt marshes, halophytic habitats, intertidal zones, and certain indoor ecosystems.

The cell morphology of *Aureobasidium* spp. is characterized by: (1) Primary and secondary conidia that may transform into budding cells; (2) Conidiogenous hyphae that produce conidia or synchronously generate conidia, chlamydospores, and transversely septate vegetative hyphae.

Although most *Aureobasidium* species are generally poor or non-fermenters, some exceptions exist. *A. fermentans*, *A. essambe*, *A. vanuatuense*, *A. peruvianum*, and *A. albui* can ferment sucrose, with only *A. albui* capable of fermenting melibiose. The three species (*A. fermentans*, *A. essambei*, and *A. vanuatuense*) also weakly ferment D-glucose and raffinose. Notably, *A. peruvianum* is an exception, as it can ferment various sugars, including melibiose, D-glucose, maltose, sucrose, trehalose, cellobiose, melezitose, raffinose, and weakly ferment inulin.

Eight genetic loci—the internal transcribed spacer (*ITS*), large subunit (*28S rDNA*) of the nrDNA repeat, translation elongation factor 1-α gene (*TEF1α*), elongase-encoding gene (*ELO2*), RNA polymerase II second largest subunit gene (*RPB2*), beta-tubulin gene (*BTU2*), mitochondrial large subunit (*mtLSU*), and mitochondrial small subunit (*mtSSU*)—are used for molecular identification of *Aureobasidium* species. Previous studies have confirmed that the *ITS* region serves as a reliable DNA barcoding marker for fungal species-level identification. In contrast, the *BTU2* region exhibits lower conservation than *ITS* and thus provides subspecies-level discrimination.

For example, *A. melanogenum* strain W12, isolated from the polymer-coated surface of an aircraft, shows >99% sequence identity to *A. melanogenum* CBS 110374 based on *ITS* and *BTU2* gene sequences.

As increasing numbers of *Aureobasidium* spp. genomes are sequenced and annotated, the whole genome data are now being employed for molecular species identification. Since the whole genomes encompass the complete genetic information of a species, they offer significantly higher accuracy and reliability for phylogenetic classification than any other methods. This approach utilizes phylogenetic trees constructed from the whole-genome data to delineate species relationships.

Average nucleotide identity (ANI) analysis is a method for assessing nucleotide-level genomic similarity between two genomes. Based on pairwise alignment, ANI quantifies the evolutionary distance between the two genomes. The calculation involves dividing a genome into fragments, which are then aligned against a reference genome. ANI is derived as the mean identity of all the BLASTN matches meeting the following criteria: >30% overall sequence identity and alignment coverage of ≥70% of the fragment length. Among *Aureobasidium* species, ANI values typically range from 77% to 98%. Within *A. pullulans* strains, ANI exceeds 97%, while *A. melanogenum* strains show ANI values between 86% and 96%. For instance, the ANI between *A. melanogenum* W12 (isolated from a polymer-coated aircraft surface) and *A. melanogenum* CBS is 89.1% (Wagner et al. 2024).

The primers used for molecular identification of *Aureobasidium* spp. are listed in Table 3.1.

The genus *Aureobasidium* belongs to the phylum Ascomycota, order Dothideales, subdivision Pezizomycotina, and family Aureobasidiaceae. These fungi are commonly referred to as "black yeasts" due to their ability to accumulate melanin in their cell walls. As of recent studies (Wu et al. 2023; da Silva et al. 2024), the genus comprises 55 species as shown in the following.

Table 3.1 The primers for molecular identification of *Aureobasidium* spp.

Gene name	Primer names	Sequence 5′-3′
ITS	ITS-F	CTTGGTCATTTAGAGGAAGTAA
	ITS-R	TCCTCCGCTTATTG ATATGC
28S rDNA	LSU-F	GTACCCGCTGAACTTAAGC
	LSU-R	ATCCTGAGGGAAACTTC
ELO2	ELO2-F	CACTCTTGACCGTCCCTTCGG
	ELO2-R	GCGGTGATGTACTTCTTCCACCAG
BTU2	Btu-F	GGTAACCAAATCGGTGCTGCTTTC
	Btu-R	ACCCTCAGTGTAGTGACCCTTGGC
mtLSU	mLSU-F	GTACTTTTG C ATAATG G GTCAGC
	mLSU-R	TATGTTTCGTAGAAAACCAGC
mtSSU	mSSU-F	CAGCAGTCAAGAATATTAGTCAATG
	mSSU-R	GCGGATTATCGAATTAAATAAC
TEF1-α	EF1-F	CYCCYGGHCAYCGTGAYTTYAT
	EF1-R	ATGACACCRA-CRGCRACRGTYTG
RPB2	RPB2-F	GAYGAYMGWGATCAYTTYGG
	RPB2-R	CCCATRGCTTG YTTRCCCAT

3.3 Taxonomy of *Aureobasidium* spp.

3.3.1 Aureobasidium pullulans

Cultural Characteristics The fungal colonies of *A. pullulans* cultivated on MEA/PDA exhibit glassy, slimy, or sticky textures with pinkish to yellowish color. After prolonged incubation, dark melanized filaments or conidia produce black sectors within the fungal colonies.

Cell Morphology The fungal cells include primary and secondary conidia, as well as conidia transforming into budding cells. The conidia-generating filaments produce conidia either simultaneously or consecutively. Additional features include chlamydospores and transversely septate, vegetative filaments. The primitive conidia-generating cells are either intercalated or terminal on the hyaloid filament. Its conidia range from hyaloid to black brown, with hyaloid conidia being one-celled, two-celled, or multicelled, glassy, and spheroid. The secondary hyaloid conidia are smaller than the primary ones. The maximum tolerated salt concentration is 15.0% (w/v) NaCl. Growth temperatures: Minimum 4 °C, optimum 25 °C, maximum 30 °C. However, weak cell growth occurs at 37 °C.

Assimilation Profiles Glucose, fructose, sucrose, glucuronate, inositol, soluble starch, lactose, and citrate are utilized by *A. pullulans,* and gelatin is liquefied. Absence of fermentation is observed.

Molecular Identification The accession numbers for the genetic sequences of *A. pullulans* are as follows: *ITS* (OR334642), *LSU* (OR334843), *RPB2* (OR452387), *TEF1* (PV067690), and *TUB2* (PV067681), respectively.

Compared to any other species in the genus, some strains of *A. pullulans* can tolerate salt concentrations up to 17.0% (w/v) (Zalar et al. 2008). High salt tolerance of *A. pullulans* is economically profitable for industrial bioprocesses because it is able to use seawater, which is much more plentiful than freshwater. This significantly reduces industrial fermentation costs and preserves freshwater, given the substantial water needs in industrial bioprocesses. Additionally, wastewater from industry often contains high salt concentrations, making the halotolerant *Aureobasidium* spp. potential candidates for wastewater treatment. It is well documented that various *A. pullulans* species produce pullulan and β-glucan, and some strains are also effective in controlling pathogenic fungi on a wide range of fruits, vegetables, and plants as described in Chap. 12 in this book (Zalar et al. 2008).

A. pullulans is mainly derived in the phyllosphere and carposphere of various fruits and vegetables, where it also reproduces as an endophyte in numerous plant species. Its ecology is remarkably diverse and global, inhabiting environments as varied as hypersaline waters (e.g., salterns), glacial ice, polluted water, preserved food (e.g., frozen or salt-treated), household surfaces, synthetic polymers, aviation fuel tanks, anthropogenic environments, croplands, atmospheric samples, and soils.

3.3.2 *Aureobasidium melanogenum*

Cultural Characteristics The fungal colonies of *A. melanogenum* cultivated on MEA/PDA media exhibit glassy, slimy, or sticky textures due to a large amount of sporulation and exopolysaccharides formation. Their coloration ranges from olive-like-brownness to dark at the center, transforming into mustard-like yellow toward the edge, which appears yellowish-white. Aerial mycelium is partially developed in some colony regions.

Cell Morphology The filaments of *A. melanogenum* are located in the center of colonies, appearing dark brown, smooth to slightly coarse, with heavy walls. They are transversely septate, limited at the septa, incorporated in extracellular polymeric substances (EPS), and are separated into one- to two-celled, dark brown chlamydospores. The conidia-generating cells are primitive, intercalated, or terminally localized on transparent filaments. Its conidia are produced simultaneously in crowded groups and are either transparent or dark brown. The dark brown conidia are one- to two-celled, glassy, and spheroid. The unilateral, bilateral, and multilateral budding of transparent conidia are commonly seen. The maximum tolerated salt concentration is 10.0% (w/v) NaCl.

Assimilation Profiles Glucose, fructose, xylose and sucrose are utilized by *A. melanogenum*. Absence of fermentation is observed.

Molecular Identification The sequences of *A. melanogenum* are deposited in GenBank with the following accession numbers: *ITS* (NR_159589), *LSU* (NG_056960), *RPB2* (KT693972), *TEF1* (FJ657887), and *TUB2* (FJ157858).

As its name indicates, most strains of *A. melanogenum* are well known to produce larger quantities of melanin than any other *Aureobasidium* species. It can also metabolize hydrocarbons and utilize them as the sole carbon and energy source. Unlike any other species, *A. melanogenum* strains can reproduce at temperatures up to 37 °C, though their growth is less efficient at lower temperatures. During fermentation, this thermotolerant characteristic can reduce cooling requirements, significantly lowering energy consumption in industrial bioprocesses when *A. melanogenum* is applied to fermentation industry.

Primarily found in oligotrophic and water environments, *A. melanogenum* is identified as a novel species based on *A. pullulans* var. *melanogenum*. It also keeps biotechnological importance due to its ability to produce pullulan, polymalate, liamocin, siderophores, intracellular lipids, gluconic acid, and melanin (da Silva et al. 2024).

A. pullulans can be different from *A. melanogenum* based on melanin production and differences in growth with galactitol, glucono-δ-lactone, creatine, creatinine, and gelatin liquefaction. Their separation is further supported by SNPs in the rDNA internal transcribed spacer region, partial translation elongation factor-1α, β-tubulin, RNA polymerase II sequences, and the whole genomic DNA sequences. In fact, we found that most of *A. pullulans* strains reported in the literature belong to *A. melanogenum*, and most native *Aureobasidium* spp. strains isolated from various environments are *A. melanogenum,* not *A. pullulans* (Jia et al. 2023).

In addition, *A. melanogenum* exhibits hemolytic activity, oligotrophy, polymorphism, siderophore production, and pigmentation. These characteristics facilitate its differentiation from the biotechnologically important relative *A. pullulans*, as mentioned by Černošal et al. (2025).

3.3.3 Aureobasidium subglaciale

Cultural Characteristics The fungal colonies of *A. subglaciale* cultivated on MEA/PDA media appear smooth and flat due to a large amount of sporulation, exhibiting a pinkish coloration. The central areas of the fungal colonies keep this pinkish coloration, while the edges are transformed to dark brown when it is old. The edge consists of thick, undulating, superficial, and immersed branched filaments.

Cell Morphology Its vegetative filaments are transparent, glassy, and thin-walled, with transverse septa. In older cultures, they may locally transform into dark brown, heavy-walled filaments. The conidia-generating cells are predominantly primitive, occurring either intercalated or terminally on filaments. Its conidia are produced simultaneously in dense clusters from small denticles, as well as continuously on short lateral branches. The conidia vary in coloration, ranging from transparent to dark brown. Its transparent conidia are one-celled, glassy, and spheroid, exhibiting significant morphological plasticity. The black conidia are one- to two-celled, and their budding is frequently noticed. The maximum tolerated NaCl concentration is 10.0% (w/v). The growth temperature ranges from a minimum of 4 °C to an optimum and a maximum of 25 °C.

Assimilation Profiles Glucose, fructose, and sucrose are utilized by *A. subglaciale*. Absence of fermentation is observed.

Molecular Identification The accession number for molecular identification is MB897700.

A. subglaciale is a psychrotolerant species that is able to multiply at temperatures between 0 and 30 °C, indicating a preference for low temperature environments. In addition to low growth temperatures, strains of this species exhibit tolerance to high salt concentrations, heavy-metal stress, and radiation (Zalar et al. 2008). *A. subglaciale* primarily inhabits the glacial ecosystems in the world.

3.3.4 *Aureobasidium namibiae*

Cultural Characteristics On the MEA plate, the fungal colonies of *A. namibiae* exhibit a glassy, glossy, leatherlike texture. The coloration is predominantly pinkish with a brownish in the center and a white edge formed by superficial aerial filaments. On the PDA plate, its colonies are glassy and shiny due to a large amount of sporulation, exhibiting an orange-white hue with an olive-like-brown center.

Cell Morphology The vegetative filaments of *A. namibiae* are similar to those of *A. subglaciale*. Some filaments locally transform into dark-brown, heavy-walled structures. Its conidia-generating cells are primitive and arise either intercalated (within filaments) or terminally (on transparent filaments or larger transformed conidia). Its conidia develop simultaneously in dense clusters from small denticles and later produce continuously on short lateral branches. Its conidia display two types: transparent conidia: one-celled, glassy, spheroid, highly variable in shape and size, often with an indistinct hilus and dark-brown conidia: one- to two-celled, encircled by a grainy extracellular polysaccharide. If two-celled, they are limited at the septum. Cell budding happens frequently. The maximum tolerated salt concentration is 10.0% (w/v) NaCl. Its growth temperatures: Minimum at 10 °C, optimum at 25 °C, and maximum at 30 °C (Zalar et al. 2008).

Assimilation Profiles Glucose, fructose, and sucrose are utilized by *A. namibiae*. Absence of fermentation is observed.

Molecular Identification The molecular identification results are as follows: the accession numbers are FJ150875 (*ITS*), FJ150937 (LSU), KT693973 (*RBP2*), and FJ157863 (*TUB2*), respectively.

The species *A. namibiae* was obtained from marble in the Namib Desert. This strain can resist NaCl concentrations up to 10.0% (w/v) and multiplies best at temperatures between 10 °C and 30 °C (Zalar et al. 2008).

3.3.5 *Aureobasidium castaneae*

A. castaneae (MycoBank: MB 838314) was derived from its plant host genus, *Castanea* sp. This species exists as an endophyte or pathogen on the plant *Castanea henryi* leaves in China.

Cultural Characteristics The non-sexual morphology can be seen on PDA plate. Its fungal colonies cultivated on PDA display rapid growth, forming glassy, even spreads that transition to an olive-like black coloration. The fungal colony edges appear unordered and dark green, with slippery masses of conidia covering the surface.

Cell Morphology The filament of *A. castaneae* is submerged. The filaments are glassy, thin-walled, septate, and transparent to brown in coloration. Its conidia-generating cells are primitive, appearing either transparent or brown, and are mostly intercalated with rare terminal occurrences. Conidia display color difference (transparent or brown) matching their parent conidia-generating cells. They are spheroid, upright, and vary in size, featuring a rounded to subtruncate base with an even basal hilus. Its conidia are thin-walled and contain two or more spots.

Assimilation Profiles Glucose, fructose, and sucrose are utilized by *A. castaneae*. Absence of fermentation is observed.

Molecular Identification The accession numbers for the *ITS* and *LSU* sequences of *A. castaneae* are MW364284 and MW364275, respectively.

A. castaneae obtained from the plant *Castanea henryi* is phylogenetically related to *A. thailandense* based on *ITS* and *LSU* sequence data. However, *A. castaneae* can be easily distinguished from *A. thailandense* based on their conidial size (*A. castaneae*: 7.1–10.6 × 2.7–4.7 μm vs. *A. thailandense*: 3–10 × 5–12 μm).

3.3.6 Aureobasidium pini

A. pini (Mycobank no: MB828664) was obtained from needles of the plant *Pinus tabuliformis* in China (Jiang et al. 2019).

Cultural Characteristics The fungal colonies of *A. pini* cultivated on MEA medium display mildly fast and are spreading, glassy, flatness, quickly turning to olive-like black, with dark green, unordered edges, covered with conidia.

Cell Morphology The filaments of *A. pini* are semi-hyaline, glassy, thin-walled, and septate, finally becoming brown and heavy-walled. The conidia-generating cells are primitive, transparent, and interposed. Its conidia are primitive and are produced simultaneously in dense clusters from small denticles, transparent, variable in shape and size, ellipsoid to spherical to oval-like. The secondary conidia are yielded by terminal or sub-terminal, mono-or bipolar budding of the primary conidia. Its chlamydospores are frequently yielded, dark, one- to two-celled.

Assimilation Profiles Glucose, fructose, and sucrose are utilized by *A. pini*. Absence of fermentation is observed.

Molecular Identification The accession numbers of *ITS* and *LSU* of *A. pini* are MK184533.

A. iranianum is different from *A. pini* based on the dimensions of chlamydospore (11.8–15 × 8.5–11.5 μm in *A. pini* vs. 10–12 × 6–7 μm in *A. iranianum*). *A. pini* is also different from *A. namibiae* in conidia sizes (5.1–10.5 × 2.5–4.8 μm in *A. pini* vs. 7–17 × 3.5–7 μm in *A. namibiae* (Jiang et al. 2019).

3.3.7 *Aureobasidium castaneae*

A. castaneae (MycoBank: MB 838314) is isolated from the plant host genus, *Castanea henryi*, and is endophytic or pathogenic on *C. henryi* leaves (Jiang et al. 2021).

Cultural Characteristics Its non-sexual morphology is seen on PDA medium. The filament of *A. castaneae* is submerged, and there is no aerial mycelium. Its fungal colonies cultivated on PDA medium are spreading, glassy, flat, quickly turning to olive-like black, with dark-green, unordered edges, covered with conidia.

Cell Morphology The filaments of *A. castaneae* are glassy, slim-walled, septate, transparent, or brown. The conidia-generating cells are primitive, transparent or brown, intercalated. Its conidia are transparent or brown based on the color of the conidia-generating cells, variable in size, spheroid, upright, slim-walled, with two or more spots.

Assimilation Profiles Glucose, fructose, and sucrose are utilized by *A. castaneae*. Absence of fermentation is observed.

Molecular Identification The accession numbers of *ITS* and *LSU* of *A. castaneae* are MW364284 and MW364275, respectively.

A. castaneae is phylogenetically related to *A. thailandense* based on phylogenetic inference of *ITS* and *LSU* sequences. However, *A. castaneae* can be easily differentiated from *A. thailandense* based on conidial size (7.1–10.6 × 2.7–4.7 μm in *A. castaneae* vs. 3–10 × 5–12 μm in *A. thailandense*) (Jiang et al. 2021).

3.3.8 *Aureobasidium insectorum*

A, insectorum (FN 571251) was obtained from the insect cicada (Wu et al. 2023).

Cultural Characteristics The fungal colonies of *A. insectorum* are cultivated mildly on PDA/MEA/OA media as mentioned above, respectively. Its colonies cultivated on PDA medium are flat, glassy, bituminous black with white felty edges; the colonies cultivated on MEA medium are flat, felty, and greenish-black with white fimbriate edges; and its colonies cultivated on OA medium are flatness, whitish, and olive-like black in the center, with sparse aerial mycelium. Its optimal growth temperature is 28 °C. *A. insectorum* multiplies better on concentrations of 15.0% (w/v) NaCl than its closely related species.

Cell Morphology The filament of *A. insectorum* is transparent, glassy, and slim-walled. The conidia-generating cells are completely cleaved, glassy, and tubular. Conidia at polar or bipolar buds are transparent, non-septate, glassy-walled, spherical to elongate-spherical. Its chlamydospores are black, glassy, globular to spherical, septate or non-septate, and limited at the septa.

Assimilation Profiles Glucose, fructose, and sucrose are utilized by *A. insectorum*. Absence of fermentation is observed.

Molecular Identification The accession numbers of *ITS*, *LSU*, *RPB2*, *TEF1*, and *TUB2* of *A. insectorum* are OR334644 3, OR334845, OR452385, PV067688, and PV067680, respectively. The phylogenetic positions of this species are shown on the deep and major branches.

This species is phylogenetically close to *A. aerium*, *A. intercalariosporum*, *A. leucospermi*, and *A. khasianum*. These five species of *Aureobasidium* spp. can be differentiated based on their conidia size differences (4.2–7.4 × 1.7–3.5 μm in *A. insectorum*, vs. 12.8–19.5 × 7.9–11.9 μm in *A. aerium*, vs. 8–13 × 5–9/8–24 × 2–10 μm in *A. leucosperm*, vs. 3–4 × 2–40 μm in *A. khasianum*, vs. 10.5–17.1 × 10.5–12.9 μm in *A. intercalariosporum*) (Wu et al. 2023).

3.3.9 *Aureobasidium planticola*

A. planticola was derived from the plant leaf of China (Wu et al. 2023).

Cultural Characteristics The fungal colonies of *A. planticola* develop mildly on PDA, MEA, and OA media, respectively. Its fungal colonies on PDA medium are flat, bituminous black, and pale gray in the center, while those cultivated on MEA medium are flat, felty, and brown, with white fimbriate at the edge and those cultivated on OA medium have a glassy edge and are flat, olive-like black, and solid. Its optimal growth temperature is 28 °C. *A. planticola* propagates well on MEA medium supplemented with 10.0% (w/v) NaCl, while *A. iranianum* reproduces more weakly under the same conditions.

Cell Morphology The filament of *A. planticola* is pigmented, glassy, heavy-walled, and branched. The conidia-generating cells are completely cleaved, located laterally or terminally, single or in clusters, gray-dark to dark. Its conidia are hyaloid, non-septate, even-walled, and ellipsoid to egg-shaped.

Assimilation Profiles Glucose, fructose, and sucrose are utilized by *A. planticola*. Absence of fermentation is observed.

Molecular Identification The accession numbers of *ITS* and *LSU* of *A. planticola* are OP856711 and OP857212, respectively.

This species is phylogenetically related to *A. thailandense* and *A. castanea,* but they have filament color differences (transparent or brown in *A. castaneae*, vs. transparent in *A. thailandense*, vs. dark black in *A. planticola*) (Wu et al. 2023).

3.3.10 Aureobasidium motuoense

Its fungal name number is FN 571263.

Cultural Characteristics Like *A. planticola*, the fungal colonies of *A. motuoense* develop mildly on PDA, MEA, and OA media, respectively. Its colonies on PDA medium are dark brown, with unordered black zones, and sparse aerial mycelium, while those cultivated on MEA medium are flat, brownish olive-like, and white near the edge, and the fungal colonies cultivated on OA medium are flat, solid, and bitumen black. Its optimal growth temperature is 30 °C. *A. motuoense* can resist concentrations of up to 15.0% (w/v) NaCl.

Cell Morphology The filament of *A. motuoense* is transparent to dark brown, glassy-walled, and branched. The conidia-generating cells are 8.7 × 3.9 μm. Its conidia are transparent, glassy-walled, terminal, and mono- or bipolar budding. Its chlamydospores are black, glassy, and globular to egg-shaped.

Assimilation Profiles Glucose, fructose, and sucrose are utilized by *A. motuoense*. Absence of fermentation is observed.

Molecular Identification The accession numbers of *ITS* and *LSU* of *A. motuoense* are OP856710 and OP857211, respectively.

This species is phylogenetically related to *A. melanogenum* and *A. acericola* (Wu et al. 2023).

3.3.11 Aureobasidium intercalariosporum

Its fungal name number is FN 571252.

Cultural Characteristics Like *A. motuoense* and *A. planticola*, the fungal colonies of *A. intercalariosporum* develop mildly on PDA, MEA, and OA media, respectively. Its colonies cultivated on PDA medium have a complete edge, and are floccose, greenish black, and white at the edge while those cultivated on MEA medium have a wavy edge, and are flatness, brownish olive-like, gray at the center and white near the edge, with sparse aerial filament and those cultivated on OA medium are flatness, pale, and olive-like brown to white from the middle to the edge. The optimal growth temperature of *A. intercalariosporum* is 28 °C. It can be multiplied on MEA medium supplemented with 15.0% (w/v) NaCl, stronger than *A. mangrovei*, *A. pullulans*, and *A. planticola*.

Cell Morphology The filament of *A. intercalariosporum* is composed of branched, septate filament that occurs singly, and is verrucous to glassy and slim-walled. The conidia-generating cells are primitive, glassy, and tubular. Its conidia are glassy-walled transparent, non-septate, egg-shaped, and spheroid or elongated spheroid. The chlamydospores of *A. intercalariosporum* are deeply melanized, glassy, heavy walled, and globular to spheroid.

Assimilation Profiles Glucose, fructose, and sucrose are utilized by *A. intercalariosporum*. Absence of fermentation is observed.

Molecular Identification The accession numbers of *ITS* and *LSU* of *A. intercalariosporum* are OP856704 and OP857205, respectively.

This species is phylogenetically related to *A. pini*, *A. planticola*, *A. khasianum*, *A. leucospermi*, *A. namibiae*, *A. pullulans*, and *A. proteae*. The major differences between the eight species are conidia size (10.5–17.1 × 10.5–12.9 μm in *A. intercalariosporum*, vs. 4.2–7.4 × 1.7–3.5 μm in *A. planticola*, vs. 6.2–8.5 × 3.6–4.2 μm in *A. pini*, vs. 7–17 × 3.5–7 μm in *A. namibiae*, vs. 8–11 × 4–5 μm in *A. leucospermi*, vs. 7.5–16 × 3.5–7 μm in *A. pullulans* (Wu et al. 2023).

3.3.12 Aureobasidium iranianum

Mycology Bank number of *A. iranianum* is MB800705, and it was derived from stems of *Bambusa* sp. (Poaceae), Iran (Arzanlou and Khodaei 2012).

Cultural Characteristics The fungal colonies of *A. iranianum* on MEA medium develop moderately fast and are spreading, glassy, flat, quickly turning to olive-like black, with dark green, unordered edge, covered with conidia.

Cell Morphology The filaments of *A. iranianum* are sub-transparent, glassy, slim-walled, septate, soon being brown, heavy-walled. The conidia-generating cells are primitive, transparent, intercalated, or sometimes terminal. Its conidia are produced simultaneously in dense clusters from small denticles and are blastic, transparent to dark brown, glassy, one-celled, thin-walled, with several spots, variable in shape and size, ellipsoid to spherical to oval-like, with a rounded to subtruncate base and flat basal hilus. Its secondary conidia are yielded by terminal or sub-terminal, mono- or bipolar budding of the primary conidia. Arthroconidia commonly produced are dark and one- to two- or rarely four-celled. Its conidia and secondary conidia have a scar at the separation site.

Assimilation Profiles Glucose, fructose, and sucrose are utilized by *A. iranianum*. Absence of fermentation is observed.

Molecular Identification The accession numbers of *ITS* and *LSU* of *A. iranianum* are KM093738 and KM093739, respectively.

A. iranianum is different from any other known species of the genus *Aureobasidium* by a large amount of darkly melanized arthroconidia and conidial dimensions (Arzanlou and Khodaei 2012).

3.3.13 *Aureobasidium mangrovei*

A. mangrovei sp. nov. was obtained from healthy leaves collected from hara mangrove trees (*Avicennia marina*) growing in the marsh zone in the Persian Gulf, Iran (Nasr et al. 2018).

Cultural Characteristics The fungal colonies of *A. mangrovei* early are pinkish with shiny wet surface on YPD medium and finally the fungal colonies are darkly melanized when it gets old.

Cell Morphology *A. mangrovei* cells are multiplied by polar budding and simultaneous budding. Its chlamydospores have separate or are intercalated, one- or two-celled, sub-globular to oval-shaped, of variable size, transparent, and dark brown.

Assimilation Profiles D-glucose, sucrose, D-xylose, maltose, cellobiose, trehalose, raffinose, melibiose, melezitose, L-arabinose, L-rhamnose, ribitol, D-mannitol, myo-inositol imidazole, and glucosamine are utilized by *A. mangrovei*, and L-galactose, lactose, L-arabinose, D-ribose, succinic acid, and glycerol are infirmly utilized. Growth of *A. mangrovei* also exists at 34 and 37 °C in the medium without vitamin, containing 50.0 and 60.0% (w/v) glucose and in the presence of 15.0% (w/v) NaCl. Absence of fermentation is observed.

Molecular Identification The accession numbers of *ITS* and *LSU* of *A. mangrovei* are KT089085 and KT089084, respectively.

3.3.14 *Aureobasidium sp. TeO12*

Aureobasidium sp. TeO12 derived from the soils planted with tea in the northern regions of Iran can convert caffeine into theophylline (Ashengroph 2017).

Cultural Characteristics Its fungal colonies cultivated on PDA medium develop fast, are glassy, and covered with conidia. The colonial surface is early white and gradually turns dark due to its chlamydospore production. The conidia, arthroconidia, and a large amount of hyphal elements producing conidia of various shapes and sizes can be observed.

Assimilation Profiles Gelatin and urea are degraded by *Aureobasidium* sp. TeO12. It can be cultivated in the presence of 10.0% (w/v) NaCl and 5.0% (w/v) glucose and at 20–35 °C and pH 4–10 (the optimal growth temperature is 30 °C and the optimal pH is 7.0). Glucose, fructose, and sucrose are utilized by *Aureobasidium* sp. TeO12. Absence of fermentation is observed.

Molecular Identification The accession number of *ITS1–5.8S–ITS2* sequence region (620 bp) of *Aureobasidium* sp. TeO12 is KT439072.

3.3.15 *Aureobasidium mustum*

A. mustum (MycoBank: MB836845) was obtained from grape juice. The *ITS2* regions of *A. pullulans* and *A. mustum* (accession number: PRJEB38309) vary by only a single base (Onetto et al. 2020).

Assimilation Profiles Glucose, fructose, and sucrose are utilized by *A. mustum*. Absence of fermentation is observed.

3.3.16 *Aureobasidium vineae*

Like *A. mustum, A. vineae* (MycoBank: 836849) was derived from grape juice (Onetto et al. 2020).

Assimilation Profiles Glucose, fructose, and sucrose are utilized by *A. vineae*. Absence of fermentation is observed.

Molecular Identification Accession numbers of *ITS*, *LSU*, *TUB2*, and *TEF1* of *A. vineae* are CAIJEN010000012, CAIJEN000000000, CAIJEN010000007, and CAIJEN010000007, respectively.

3.3.17 *Aureobasidium uvarum*

Like *A. vineae* and *A. mustum*, *A. uvarum* (MycoBank: MB836846) was obtained from grape juice (Onetto et al. 2020).

Cultural Characteristics The fungal colonies of *A. uvarum* on YPD medium develop tardily and are pale-yellow, not pigmented, coarse, desiccated, with unordered edge.

Cell Morphology The conidia of *A. uvarum* are oval-shaped to ellipsoid, transparent, slim-walled, and bud terminally.

Assimilation Profiles Glucose, fructose, and sucrose are utilized by *A. uvarum*. Absence of fermentation is observed.

Molecular Identification Accession numbers of *ITS*, *LSU*, *TUB2*, and *TEF1* of *A. uvarum* are CAINUL010000013, CAIJEN000000000, CAINUL010000003, and CAINUL010000003, respectively.

A. uvarum is similar to *A. insectorum,* but they vary in their conidia sizes (8–10.0 × 3–5 μm vs 4.2–7.4 × 1.7–3.5 μm in *A. insectorum*) and lack of pigment (vs *A. insectorum* which is almost completely pigmented on the four tested standard media as mentioned above) (Onetto et al. 2020; Jumbam et al. 2025).

3.3.18 *Aureobasidium caatingaense*

A. caatingaense was derived from branches of the plant *Anadenanthera* in Brazilian tropical dry forest, Caatinga (da Silva et al. 2024).

Culture Characteristics Its fungal colonies developed on PDA medium are wine buffer while the fungal colonies developed on MEA medium are yellowish-green, edges slightly flocculent, white, with sparse aerial mycelia.

Cell Morphology The filaments of *A. caatingaense* have septate, are glassy, slim-walled and transparent, and its filaments are pigmented and heavy-walled when it gets old. The conidia-generating cells are globular to sub-globular, solitary and grouped, intercalated on the transparent to light brown filaments. Its conidia are transparent, light brown when it is old, do not have septa, glassy-walled and are upright, spheroid, oval, and guttate. The chlamydospores of *A. caatingaense* are black-brown, glassy-walled, heavy-walled, non-septate, globular to sub-globular, intercalated, and terminal.

Assimilation Profiles Glucose, fructose, and sucrose are utilized by *A. caatingaense*. Absence of fermentation is observed.

Molecular Identification The accession numbers of *ITS*, *LSU*, *TEF1*, and *TUB2* of *A. caatingaense* are OR340179, OR340237, PQ140158, and PP840908, respectively.

A. caatingaense is phylogenetically related to *A. santateresinhaense*. *A. caatingaense* can be different from *A. santateresinhaense* by having slightly less wide filaments, larger conidia, smaller chlamydospores, lack of budding cells, elongated conidia-generating cells, and colorful colonies. *A. caatingaense* is also closely related to *A. thailadense*, *A. castaneae*, and *A. microtermitis*. But it varies from these species by having thinner filaments, slightly larger conidia, larger chlamydospores, and untypical, sparse aerial mycelia that exist sometimes in any other *Aureobasidium* species (da Silva et al. 2024).

3.3.19 Aureobasidium santateresinhaense

A. santateresinhaense was derived from branches of the plant *Anadenanthera colubrina* (da Silva et al. 2024).

Culture Characteristics The fungal colonies of *A. santateresinhaense* developed on PDA medium are olive-like black and velvet-like in the center, while those grown on MEA medium are olive-like buff and leathery in the center, with sectors of aerial mycelia whitish along the edge.

Cell Morphology The filaments of *A. santateresinhaense* have septate, are glassy, thin-walled, transparent when it is young and melanized and heavy-walled and toruliform when it is old. The conidia-generating cells are primitive and differentiated from filaments, globular to subglobular, alone or clustered, intercalated on filament, and brown. The conidia of *A. santateresinhaense* are transparent and light brown when it is old, do not septate, are glassy-walled, variable in shape and size, upright to slightly curved, ellipsoid, globular to subglobular, droplike, and cell budding is sometimes noticed. The chlamydospores of *A. santateresinhaense* are dark brown, glassy-walled, heavy-walled, do not have septate, are globular to subglobular, intercalated, and terminal.

Assimilation Profiles Glucose, fructose, and sucrose are utilized by *A. santateresinhaense*. Absence of fermentation is observed.

Molecular Identification The accession numbers of *ITS*, *LSU*, *TUB2*, and *TEF1* of *A. santateresinhaense* are OR340177, OR340235, PP840910, and PQ140157, respectively.

A. santateresinhaense and *A. caatingaense* and the related isolates/species have been described above. Additionally, *A. santateresinhaense* sometimes produce elongated conidia-generating cells which are not present in any other related species (da Silva et al. 2024).

3.3.20 *Aureobasidium aubasidani*

Assimilation Profiles *A. aubasidani* cannot assimilate methyl-D-glucoside and lactose and can produce aubasidan-like EPS with sodium nitrate, rather than the pullulan produced by *A. pullulans* and *A. melanogenum*, and can yield a large amount of fumarate and gluconate in the presence of $CaCO_3$ (Wei et al. 2023). Glucose, fructose, xylose and sucrose are utilized. Absence of fermentation is observed.

Molecular Identification The accession numbers of *ITS*, *28SrDNA*, *Tef1*, and *Tub2* of *A. aubasidani* are FJ150909, FJ150952.1, FJ157906, and FJ157864.1, respectively.

A. aubasidani is a sister of *A. ellingtonae* and *A. proteae*. But *ITS* sequences of *A. aubasidani* vary in those of *A. ellingtonae* by 1/486 bp, but are not different from *A. proteae* and *A. pullulans*. The *28S rDNA* sequence of *A. aubasidani* varies from that of *A. proteae* and *A. ellingtonae* by 1/566 and 1/489 bp, respectively. Sequences of the *ELO2* gene are different from those of *A. pullulans* at 12/699 base pairs and different from those of *A. ellingtonae* at 10/692 bp (Jumbam et al. 2025).

3.3.21 *Aureobasidium acericola*

A. acericola was obtained from leaves of the plant *Acer pseudosieboldianum* in Korea (Lee et al. 2021).

Cultural Characteristics The fungal colonies of *A. acericola* on MEA medium develop moderately fast, are stretched, glassy, flat, quickly turning to olive-like black, with dark green, unordered edges, covered with conidia. Its mycelium is submerged.

Cell Morphology The filament of *A. acericola* is subhyaline, glassy, slim-walled, and septate, and finally becomes brown and heavy-walled. The conidia-generating cells are primitive and transparent. Its conidia are blast-like, transparent, variable in shape and size, ellipsoid to spherical to oval-like, upright, with a rounded to subtruncate base and slim-walled, with several spots. Its secondary conidia are yielded by terminal or sub-terminal, bipolar budding of the primary conidia. Black chlamydospores are frequently produced by *A. acericola*.

Assimilation Profiles Glucose, fructose, and sucrose are utilized by *A. acericola*. Absence of fermentation is observed.

Molecular Identification The accession numbers of *ITS* and *LSU* of *A. acericola* are MT863788 and MT863787, respectively.

A. acericola is obtained from the plant host genus *Acer* and is phylogenetically closely related to *A. melanogenum*. However, *A. acericola* can be different from *A. melanogenum* based on the conidial sizes (3.2–3.9 × 5.9–9.2 mm in *A. acericola* vs. 3.5–5 × 8–30 mm in *A. melanogenum*) and dimensions of chlamydospore (5.4–7.7 6.9–8.4 mm in *A. acericola* vs. 8–12 × 13–16 mm in *A. melanogenum*) (Lee et al. 2021).

3.3.22 *Aureobasidium thailandense*

A. thailandense (Mycobank number: MB 801148) was isolated from the leaves of the plant *Cerbera odollum* Gaertn in Thailand (Peterson et al. 2013).

Cultural Characteristics The fungal colonies of *A. thailandense* developed on PDA medium are reddish brown in the center, shiny from conidia and exopolysaccharide production, edges unordered.

Cell Morphology The short, broad filament of *A. thailandense* yields conidia terminally or laterally. Its conidia are subglobular, oval-like, and glassy-walled. The individual cells of *A. thailandense* proliferate by polar budding.

Assimilation Profiles Glucose, fructose, and sucrose are utilized by *A. thailandense*. Absence of fermentation is observed.

Molecular Identification The accession numbers of *ITS* and *TUB2* of *A. thailandense* are JX462674 and EU719407, respectively.

A. thailandense is different from *A. pullulans* by having reddish brown pigmented filament after growth on PDA medium, while *A. pullulans* keeps white to pinkish and its varieties darken in different colors (Peterson et al. 2013).

3.3.23 *Aureobasidium khasianum*

A. khasianum was isolated from the decomposing fallen leaves of the plant *Wightia speciosissima* in India (Ashish and Pratibha 2018).

Cultural Characteristics The fungal colonies of *A. khasianum* produced on the leaf surface are ring-shaped, hairy, and dark brown to black. Its filament is partly immersed in the substrate. The fungal colonies on MEA medium develop fast, are spreading, and are covered with conidia, white when young and dark green when old.

Cell Morphology The filaments of *A. khasianum* are septate, branchy, pale brown to brown, and glassy-walled. The setae produced by *A. khasianum* are unbranched, erect, upright, dark brown, glassy, and slim-walled. The conidiophores are microworm-like to semi-macroworm-like, mono worm-like, adhering to the sides of the setae and forming conidia-generating cells which are mono blast-like. Its conidia arising in acropetal, branched chains are fragmenting, non-septate, glassy, transparent to pale olive-like. Its filament grown on MEA medium is thick, glassy, septate, becoming brown and heavy-walled. The arthroconidia yielded are integrated, intercalated, and ellipsoid. The conidia-generating cells on filaments are phialophora-like, lateral, terminal, or intercalated. Its conidia are yielded simultaneously in dense clusters and are transparent, glassy, one-celled, variable in shape and size. Its secondary conidia are yielded from the primary conidia by budding.

Assimilation Profiles Glucose, fructose, and sucrose are utilized by *A. khasianum*. Absence of fermentation is observed.

Molecular Identification The accession numbers of *ITS* and *LSU* of *A. khasianum* are MH188305 and MH188306, respectively.

3.3.24 Aureobasidium leucospermi

A. leucospermi (MycoBank: MB560556) is derived from the plant host genus *Leucospermum* (Crous et al. 2011) and was obtained from leaves of *Leucospermum conocarpodendron* in South Africa and branches of *Anacardium occidentale* (da Silva et al. 2024).

Culture Characteristics The fungal colonies of *A. leucospermi* are flat, stretching with featherlike edges. The fungal colonies of *A. leucospermi* developed on MEA medium are rose-colored while those cultivated on OA medium are polished. The fungal colonies of *A. leucospermi* cultivated on PDA medium are gray-dark, with patches of beige, while those grown on the leaves are sub-round, ununiform, occurring on both surfaces of a leaf, necrotizing, submerged, pale to medium brown with an elevated, black-brown edge.

Cell Morphology Its filament produced is submerged while its conidia-carrying structures are heap-like to sporodochium, amphigenous, substomatous, subepidermic, pulvinated, desiccated or crystallized, pale brown, and separated. The matrix of *A. leucospermi* can be observed in substomatous hole, are dark brown, and mainly composed of elongated pseudo-parenchymatous cells with a large blade, becoming transparent, thinner-walled at the mycelial peak. The conidia-generating cells are cylindric, claviform or globular, merged, terminal, conidial ontogenesis wholly blastic, with a large amount of simultaneously produced conidia. Its conidia in vitro are alone, non-septate, ellipsoid to round.

Assimilation Profiles Glucose, fructose, and sucrose are utilized by *A. leucospermi*. Absence of fermentation is observed.

Molecular Identification The accession numbers of *ITS*, *LSU*, and *RPB2* of *A. leucospermi* are NR_156246, MH877257, and KT693970, respectively.

A. leucospermi is different from *A. khasianum* by 3 bp in *ITS* and 1 bp in *LSU* (da Silva et al. 2024).

3.3.25 Aureobasidium proteae

A. proteae (MycoBank: MB560557) was derived from a leaf of a plant *Protea cynaroides* seedling in South Africa (Crous et al. 2011).

Cultural Characteristics The fungal colonies of *A. proteae* produced on the leaves are unordered, occurring on the leafstalk and base of blade, extending up the leaves, necrotizing, submerged, pale to medium brown with an eleviform, dark brown edge. The colonies of *A. proteae* cultivated on MEA medium have medium to sparse aerial mycelium, are superficially olivelike-gray, radial stria appearing in the agar, and can be observed from the bottom of the fungal colonies..

Cell Morphology The filament produced by *A. proteae* is submerged. The conidia-carrying structures yielded are concretional sporodochium, amphigenous, substomatous, subepidermic, pulvinated, desiccated or crystallized, pale brown, and separated. Its matrix can be observed in substomatous hole, is dark brown, mainly composed of elongated pseudo-parenchymatous cells with a large blade, and becomes transparent and thinner-walled at the mycelial peak. The conidia-generating cells of *A. proteae* are cylindric, claviform or globular, merged, terminal, conidial ontogenesis wholly blastic, with a large amount of simultaneously produced conidia. The conidia in vivo are solitary, non-septate, ellipsoid to spherical. The conidia in vitro are similar.

Assimilation Profiles Glucose, fructose, and sucrose are utilized by *A. proteae*. Absence of fermentation is observed.

Molecular Identification The accession numbers of *ITS*, *LSU*, and *RPB2* of *A. proteae* are JN712491, JN712557, and KJ693974, respectively.

3.3.26 Aureobasidium tremulum

Culture Characteristics Its early growth on PDA medium is milky and white, later turning brown to dark brown. Its fungal colonies appear to be coarse and desiccated. Each colony is oval with a bulging rise from a cross-section, and the edges appear to be wavy.

Cell Morphology The cells of *A. tremulum* are commonly oblong-shaped, with very few cells assuming an unordered appearance. Its cell budding happens uniformly.

Assimilation Profiles The optimal growth temperature of *A. tremulum* is 20–25 °C. D-glucose, D-xylose, maltose, saccharose, trehalose, melezitose, L-arabinose, and raffinose are utilized by *A. tremulum*, while adonitol, xylitol, D-galactose, methyl-alpha-D-glucopyranoside, and cellobiose are infirmly utilized. Absence of fermentation is observed.

Molecular Identification The accession numbers of *ITS* and *LSU* sequences of *A. tremulum* are MK503657 and MK503660, respectively, and its mycoBank is MB829941.

A. tremulum is unique and is not phylogenetically related to any known species of *Aureobasidium* spp. The phylogenetic tree based on the *ITS* sequences shows that *A. tremulum* forms a sister branch to *A. thailandense* and *A. mangrovei*. In the phylogenetic tree based on the *LSU* sequences, it is not closely related to any known species but is placed at equal evolutionary match with *A. caulivorum* (GenBank: FJ150944) (Crous et al. 2019).

3.3.27 *Aureobasidium welwitschiae*

The name of *A. welwitschiae* comes from the plant host *Welwitschia mirabilis*, and *A. welwitschiae* was obtained from the dead leaf tips of the plant *W. mirabilis* (Crous et al. 2023).

Culture Characteristics The fungal colonies of *A. welwitschiae* are flat, extending, with sparse aerial mycelium and glassy, lobated edge. Its colonies cultivated on MEA and PDA media are outer layer and bottom layer dark mouse-like gray while its colonies cultivated on Oatmeal Agar (OA) medium are superficially deep brown in the center, pale golden-yellow at outer location.

Cell Morphology The filament produced by *A. welwitschiae* consists of transparent, glassy, ramous, septate hyphae that become dark brown, heavy-walled, verrucose, covered in adhesive, yielding ungregarious or a group of globular chlamydospores when it is old. The conidia-generating cells are reduced to conidiogenous loci, which are unnoticeable phialic openings on filament, yielding conidia that concentrate in a mucoidal mass. Its conidia are ungregarious, non-septate, transparent, glassy, guttate, spindle-shaped-ellipsoid, upright to slightly curved, the mycelial peak not sharp, base narrowed, hilus sub-monotonous.

Assimilation Profiles Glucose, fructose, and sucrose are utilized by *A. welwitschiae*. Absence of fermentation is observed.

Molecular Identification The accession numbers of *ITS*, *LSU*, *ACTA*, *TEF1* (second part), and *TUB2* of *A. welwitschiae* are OQ628484.1, OQ629066.1, OQ627933.1, OQ627959.1, and OQ627965.1, respectively, and its MycoBank is MB 848078.

A. welwitschiae is closely related to *A. iranianum* and *A. thailandense*, but it is phylogenetically different (Crous et al. 2023).

3.3.28 *Aureobasidium faidherbiae*

A. faidherbiae name comes from the plant host *Faidherbia albida*, and *A. faidherbiae* was derived from the dead leaf of the plant (Crous et al. 2023).

Culture Characteristics The fungal colonies of *A. faidherbiae* are flat, extending, with sparse aerial mycelium and a glassy, lobed edge. Its colonies cultivated on the MEA, PDA, and OA media are outer layer and bottom layer gray dark.

Cell Morphology The filament produced by *A. faidherbiae* consists of transparent, glassy, branchy, septate hyphae, commonly with nodulous swellings. The filaments produced turn dark brown, becoming heavy-walled, verrucose, covered in adhesive, and limited at septa when it is old. The conidia-generating cells are reduced to unnoticeable phialophoric loci on the filament. Its conidia are transparent, glassy, non-septate, guttate, spindle-shaped, ellipsoid, curved to hooked, widest in the center, the mycelial peak subobtuse, tapered to truncated hilus. The primary conidia produce several flat-tipped dentate apex loci that yield a group of secondary conidia.

Assimilation Profiles Glucose, fructose, and sucrose are utilized by *A. faidherbiae*. Absence of fermentation is observed.

Molecular Identification The accession numbers of *ITS*, *LSU*, *RPB2* (first part), and *TEF1* (first part) of *A. faidherbiae* are OQ628485.1, OQ629067.1, OQ627945.1, and OQ627952.1, respectively, and its MycoBank is MB 848079.

A. faidherbiae is similar to *A. thailandense*, *A. bupleuri*, and *A. welwitschiae*, but is phylogenetically different (Crous et al. 2023).

3.3.29 *Aureobasidium microtermitis*

A. microtermitis name comes from the insect host termite genus *Microtermes* sp., and it was obtained from the gut of the insect termite *Microtermes* sp. feeding on wooden logs in a forest area in India (Crous et al. 2021).

Culture Characteristics The fungal colonies of *A. microtermitis* cultivated on PDA medium are stretching, with unordered edge, milky white marginally and turning olive-like black in center. Its colonies cultivated on YPD medium are ring-shaped, slightly bulging, glassy, shiny, with a whole edge and a pigmented loop in the center.

Cell Morphology Its conidia cultivated on PDA medium are subglobular to oval. The simultaneous polar budding is noticed when it is cultivated on PDA and YPD media. A few dark brown, short filament with intercalated chlamydospores can be observed after growth on PDA medium. Its pigmented chlamydospores have separate or intercalated, subglobular to globular.

Assimilation Profiles Glucose, fructose, and sucrose are utilized by *A. microtermitis*. The optimal growth temperature is 25–28 °C. Some growth can be observed at 15 °C and 30 °C, whereas no growth is observed at 4 °C and 37 °C; however, significant growth on high sugar media can be observed. It is also able to resist high salt concentrations (5.0–15.0% w/v NaCl). Absence of fermentation is observed.

Molecular Identification The accession numbers of *ITS* and *LSU* of *A. microtermitis* are MW276135 and MW276136, respectively, and its MycoBank is MB 839078.

Based on *LSU* and *ITS* sequences, *A. microtermitis* is phylogenetically related to *A. thailandense,* but they have differences based on colony morphology and conidia size. The conidia of *A. thailandense* are slightly bigger (3–10 × 5–12 μm) than those of *A. microtermitis* (Crous et al. 2021).

3.3.30 *Aureobasidium aerium*

A. aerium (MycoBank no: MB843527) was obtained from the air (Wang et al. 2022).

Cultural Characteristics The fungal colonies of *A. aerium* cultivated on PDA medium exhibit mild characteristics. Its fungal colonies are darkly olive-like, flat, glassy, green-yellow, and fringed edges, often covered with conidia.

Cell Morphology The filaments of *A. aerium* are septate, glassy, slim-walled, soon becoming pigmented and thick-walled. The conidia are transparent, non-septate, glassy-walled, ellipsoid to oval, different in size, with oval to sub-shorten base and a flat, unclear hilus. Its chlamydospores are dark brown, glassy to lightly coarse-walled, septate to non-septate, globular to ellipsoid, one- to two-celled.

Assimilation Profiles Glucose, fructose, and sucrose are utilized by *A. aerium*. Absence of fermentation is observed.

Molecular Identification The accession numbers of *ITS* and *LSU* of *A. aerium* are ON007058 and ON007708, respectively.

A. aerium is phylogenetically similar to *A. subglaciale* obtained from subglacial ice in seawater, *A. leucosperm* obtained from the plant *Leucospermum* leaves, *A. namibiae* obtained from dolomitic marble in the Namib Desert, and *A. pini* obtained from the plant pine needle surface. However, these five species can be easily different from their conidia sizes (12.8–19.5 × 7.9–11.9 μm in *A. aerium* vs. 5.5–28 × 2–6.5 μm in *A. subglaciale* vs. 8–13 × 5–9/8–24 × 2–10 μm in *A. leucosperm* vs. 7–17 × 3.5–7 μm in *A. namibiae* vs. 5.1–10.5 × 2.5–4.8 μm in *A. pini*) (Wang et al. 2022).

3.3.31 Aureobasidium bupleuri

A. bupleuri name comes from the plant host *Bupleurum gibraltarium* and was obtained from the stems and leaves of the plant *B. gibraltarium* in Spain (Bills et al. 2012).

Cultural Characteristics The fungal colonies of *A. bupleuri* cultivated on MEA medium are feather-like, ropy to coarsely fringed, frequently producing well clarified, appressed to immersed, curved or crooked radial filaments, slightly elevated and grainy toward the center. The filaments of *A. bupleuri* in the colonies are immersed, radially spreading, sometimes collecting moist to grainy masses of the pigmented yeast-like cells with the black and transparent falciform conidia toward the colony center, shiny to wet at start of the cultivation, but a large amount of immersed to appressed hyphae without conidia when it gets old.

Cell Morphology The filament of *A. bupleuri* is transparent to pale pink. Early the filament is pigmented and the immersed filaments have a large amount of pigmented conidia and yeast-like cells. Finally, the filaments are produced at extremes of older fungal colonies.

Assimilation Profiles Glucose, fructose, and sucrose are utilized by *A. bupleuri*. Absence of fermentation is observed.

Molecular Identification The accession number of *ITS* of *A. bupleuri* is KT693738.

3.3.32 Aureobasidium zeae

Cultural Characteristics The fungal colonies of *A. zeae* cultivated on MEA medium have two kinds of colonies: light (18%) and dark (82%) pigmented. All the colonies have a very coarse and leather-like growth, with tightly formed edges, and yield a grayish aerial mycelium and no conidia.

Assimilation Profiles Glucose, fructose, and sucrose are utilized by *A. zeae*. The optimal growth temperature for most of the strains is 24 °C (Reifschneider 1980). Absence of fermentation is observed.

The accession number of its Mycology Bank is MB283372.

3.3.33 Aureobasidium hainanensis

Genomic Characteristics The whole genome size of *A. hainanensis* derived from the mangrove system (the leaf of the mangrove plant, *Kandelia candel*) at DongZaiGang, Haikou, Hainan Province, China, is 25.41 Mb, and the G + C content of its genome is 50.69%. Totally 6098 protein-coding genes and 264 non-coding RNA genes were predicted. The annotation results show that the yeast-like fungal strain has complete metabolic pathways of TCA cycle, EMP pathway, pentose phosphate pathway, glyoxylic acid cycle, and other central metabolic pathways. Furthermore, the inulinase activity associated with β-fructofuranosidase and high glucose oxidase activity in this strain have been demonstrated. It was found that *A. hainanensis* is located at the root of most species of *Aureobasidium* spp. and at a separate cluster of all the phylogenetic trees. *A. hainanensis* was predicted to contain three NRPS gene clusters, five type-I PKS gene clusters, and one type-I NRPS/PKS gene cluster via analysis at the antiSMASH Website. It may synthesize epichloenin A, fusaric acid, elsinochromes, and fusaridione A. Based on its unique DNA sequence, taxonomic position in the phylogenetic tree, and evolutional position, *A. hainanensis* is identified as a novel species of *Aureobasidium* spp. and has highly potential applications.

Assimilation Profiles Glucose, fructose, and sucrose are utilized by *A. hainanensis*. Absence of fermentation is observed.

Molecular Identification The accession numbers of *ITS* and *28S rDNA* of *A. hainanensis* are KF260961.1 and KF260961, respectively, and the accession number of the draft genome data of it is RZIQ01000000.

A. hainanensis was found to be able to produce high levels of Ca^{2+}-gluconic acid and a β-fructofuranosidase with a high inulin-hydrolyzing activity (Jia et al. 2019).

3.3.34 *Aureobasidium albui*

A. albui occurs in various plant phylloplanes in East Baton Rouge, Louisiana.

Culture Characteristics The cells of *A. albui* multiply well on MEA, PDA, and M40Y media, respectively. The fungal colonies cultivated on MEA medium are pigmented and greenish brown at the beginning with a heavy yellow edge, while those grown on PDA medium are pale yellow and yeast-like in the center, becoming brownish at the edges. Whereas the colonies cultivated on M40Y medium are pale yellow in the center, then pink to dark to brick red toward edges and starch-like. The fungal colonies cultivated on PDA medium are pale pink and monotonous. Its filaments are produced in all the culture media.

Cell Morphology The vegetative filaments of *A. albui* are glassy, slim-walled, and septate. The conidia-generating filaments are transparent, simultaneously producing a large amount of conidia both terminally and laterally. Its conidia are transparent, glassy, slim-walled, oval-like to ellipsoid or sometimes lengthened. The conidial cells propagate by polar budding. Its chlamydospores are oval-like (one-celled) to subglobular, pale brown. *A. albui* produces intercalated arthrospores, one- to two-celled, limited at the septum.

Assimilation Profiles D-glucose, L-arabinose, sodium succinate, and quinic acid are firmly utilized by *A. albui*. Galactitol and sodium gluconate are mildly utilized while L-sorbose, gluconic acid lactone, and mannitol are infirmly utilized. A pinkish pigmentation is produced while L-sorbose is utilized by *A. albui*. A light pink pigmentation is yielded when *A. albui* is cultivated in mannitol, light orange pigmentation appears when it is cultivated in gluconic acid lactone, and purple pigmentation exists when it is cultivated in erythritol. Infirm fermentation of sucrose and melibiose can be observed. So, presence of fermentation is observed. *A. albui* firmly utilizes potassium nitrate, potassium nitrite, L-lysine, ethylamine, and D-glucosamine, but infirmly utilizes cadaverine as the sole source of nitrogen.

Growth Temperature and Salt Tolerance The cell growth of *A. albui* between 4 and 30 °C and on 10.0% (w/v) NaCl is optimistic; the optimal growth temperature is 28 °C. Its growth cultivated on the osmotic media is very slow, yellow with pale brown regions.

The cell appearances of *A. albui* are similar to those of *A. vanuatuense*, but they vary in the cell growth at 30 °C (vs weak yeasty growth in *A. vanuatuense*). Additionally, the cell growth of *A. albui* is almost fully pigmented on PDA medium, whereas the limited pigmentation of *A. vanuatuense* is observed when it is cultivated on the same medium.

Its Mycology Bank number is MB848255.

3.3.35 *Aureobasidium cavalettoi*

A. cavalettoi was obtained from weeping willow leaves of plant on the campus of Purdue University of USA.

Culture Characteristics The fungal colonies of *A. cavalettoi* cultivated on MEA medium are pink with darker areas in the center and shiny and their edges become greenish while those grown on PDA medium are white developing dark portions at the center, monotonous and no change. The colonies of *A. cavalettoi* cultivated on M40Y medium are greenish center, outer 1/2 milky, shiny. When it is cultivated on MEA, PDA, and M40Y media, the filaments yield at edges, and its colony surfaces are ribbed or coarse with lighter radial mycelial groups, making the colony a star-like and radiating look.

Cell Morphology The vegetative filaments of *A. cavalettoi* are glassy, slim-walled, and septate. The conidia-generating filaments are primitive and transparent, simultaneously yielding conidia both terminally and laterally. The conidia of *A. cavalettoi* are transparent, glassy, and epllipsoid. The conidial cells propagate by polar and bipolar budding. Its chlamydospores are common, are produced as inter-calated arthrospores, ellipsoid to subglobular, and transparent and become black brown, one- to three-celled when they get old.

Assimilation Profiles Sodium gluconate and D-glucose are firmly utilized by *A. cavalettoi*, while L-sorbose, L-arabinose, salicin, ribitol, mannitol, and sodium succinate are unfirmly utilized. Potassium nitrate and L-lysine are firmly utilized as sole nitrogen sources. Cadaverine, D-glucosamine, and tryptophan are infirmly utilized by *A. cavalettoi*. Absence of fermentation is observed.

Growth Temperature and Salt Tolerance Its good cell growth occurs between 4 and 30 °C and on 15.0% (w/v) NaCl, and the optimal growth temperature is 28 °C.

A. cavalettoi is phylogenetically related to *A. insectorum*, but varies in its slower growth rate on PDA medium and bigger conidia.

3.3.36 *Aureobasidium diazvalderramae*

A. diazvalderramae was obtained from air in Cajamarca Department, Peru.

Culture Characteristics The fungal colonies of *A. diazvalderramae* cultivated on MEA medium are superficially monotonous, pink while those cultivated on M40Y medium are ream. Its colonies cultivated on PDA medium are pigmented, white in center, then producing a dark brown circle encircled by thin groups of milky and brown, outer 1/2 to edge white with dispersed brownish dots, shiny. Its filaments are produced at edges on all the culture media.

Cell Morphology The vegetative filaments produced by *A. diazvalderramae* are glassy, slim-walled, and septate. Its conidia-generating filaments are primitive and transparent, simultaneously yielding conidia mostly terminally. The conidia produced by *A. diazvalderramae* are transparent, glassy, ellipsoid, and conidial cells propagate by polar and bipolar budding. Its chlamydospores are commonly produced, producing intercalated arthrospores, one- to two-celled, ellipsoid to subglobular, light brown when they get old.

Assimilation Profiles D-glucose, L-arabinose, and sodium succinate are mildly utilized as sole carbon sources by *A. diazvalderramae*, while α-rhamnose, D-salicin, erythritol, D-mannitol, myo-inositol, and quinic acid are infirmly utilized. Potassium nitrate, ethylamine, L-lysine, and cadaverine are utilized as sole nitrogen sources by *A. diazvalderramae*. Absence of fermentation is observed.

Growth Temperature and Salt Tolerance Its cell growth is well at 4–30 °C and on 10.0% (w/v) NaCl; its optimal growth temperature is 28 °C.

A. diazvalderramae is similar to *A. namibiae* but varies in having lengthier and slimmer conidia and larger one-celled chlamydospores.

3.3.37 Aureobasidium ellingtonae

A. ellingtonae was derived from the interior of the plant phylloplanes in Indiana, USA.

Culture Characteristics The fungal colonies of *A. ellingtonae* cultivated on MEA, PDA, and M40Y media are shiny, edge glassy, filaments are yielded at edges.

Cell Morphology The vegetative filaments formed by *A. ellingtonae* are glassy, thin-walled, and septate. The conidia-generating filaments are primitive and transparent, simultaneously yielding conidia both terminally and laterally. The conidia of *A. ellingtonae* are transparent, glassy, oval-like to ellipsoid or sometimes sub-spindle-shaped, reproducing by polar and bipolar budding. Its chlamydospores are ellipsoid to subglobular, transparent, mostly one-celled.

Assimilation Profiles D-glucose is mildly metabolized, and L-arabinose and sodium gluconate are infirmly utilized by *A. ellingtonae*. L-lysine, potassium nitrite, and ethylamine are firmly utilized. Potassium nitrate and cadaverine are mildly utilized, while glucosamine and tryptophan are infirmly utilized by *A. ellingtonae*. Absence of fermentation is observed.

Growth Temperature and Salt Tolerance Its cell growth is well at 4–37 °C and on 15.0% (w/v) NaCl; the optimal growth temperature is 25 °C.

A. ellingtonae is close to *A. pullulans* but varies in smaller and thinner conidia and the lack of two-celled chlamydospores.

3.3.38 Aureobasidium essambe

Culture Characteristics The fungal colonies of *A. essambe* cultivated on MEA medium become completely black, with edges whitish, and yielding filaments, while those cultivated on PDA medium are milky, glassy, shiny, with no filaments. The colonies cultivated on M40Y medium are greenish brown, monotonous, superficially reticular with whitish edges, yielding filaments at edges.

Cell Morphology The vegetative filaments of *A. essambe* are glassy, thin-walled, and septate. The conidia-generating filaments are transparent. The conidia of *A. essambe* are developed terminally and laterally, are occasionally sub-mycelial, peak, transparent, glassy, mostly ellipsoid, sometimes oval-like or spindle-shaped. Its conidia reproduce by polar budding while its chlamydospores are yielded as intercalated arthrospores, deep brown one- to two-celled, sometimes three-celled, limited at the septum.

Assimilation Profiles D-glucose, sodium gluconate, sorbose, arabinose, and quinic acid can be utilized by *A. essambe*. D-glucose, sucrose, melezitose, and raffinose are infirmly fermented. So, presence of fermentation is observed. Potassium nitrate, ethylamine, L-lysine, cadaverine, and D-glucosamine can be utilized as the sole nitrogen source while D-tryptophan is infirmly utilized by *A. essambe*.

Growth Temperature and Salt Tolerance Its cell growth is well between 4 and 30 °C, and the optimal growth temperature is 28 °C. But its cell growth on 15.0% (w/v) NaCl is not common, slow, becoming grayish black and pigmented.

A. essambe is close to *A. microtermitis,* but varies in more quickly pigmented and grows more slowly.

3.3.39 Aureobasidium fermentans

A. fermentans is different from any other *Aureobasidium* species by its ability to ferment glucose, sucrose, fructose, mannose, and raffinose. So, it is a fermenter. In contrast, *A. pullulans* has no ability to ferment any sugar. *A. fermentans* is phylogenetical to *A. xishuangbannaense* and *A. pini*, but *A. fermentans* and *A. xishuangbannaense* vary at 5/891 bp (*28S rDNA*), and 1/506 bp (*ITS*); *A. fermentans* and *A. pini* vary at 1/588 bp (*28S rDNA*) but have 100% match at the *ITS* sequences. *A. fermentans* is different from *A. xishuangbannaense* by its lengthier conidia, chlamydospores yielded by *A. fermentans* vs lack in *A. xishuangbannaense*, and the lack of aerial filament while aerial filament is yielded in *A. xishuangbannaense*. *A. fermentans* is different from *A. pini* by the bigger conidia (Jumbam et al. 2025).

3.3.40 *Aureobasidium mahoniae*

A. mahoniae was obtained from a variety of host plant leaf surfaces in tropical and subtropical regions of North America and Australia.

Culture Characteristics The fungal colonies of *A. mahoniae* cultivated on MEA medium are orange in the center, followed by a yellow zone, black brown and shiny at the edge while its colonies cultivated on PDA medium are purple in the center, followed by a light brown zone, and monotonous while the colonies cultivated on M40Y medium are orange in the center, followed by a wet shiny zone and whitish edge that becomes pigmented. The colony edges grown in all the media are glassy and uniform with filaments.

Cell Morphology The vegetative filaments of *A. mahoniae* are glassy, slim-walled, and septate. The conidia-generating filaments are primitive and transparent, simultaneously yielding conidia mostly terminally. Its conidia are transparent, glassy, oval-like to ellipsoid, with hilus. The conidial cells reproduce mostly by simultaneous polar and bipolar budding. Its chlamydospore is yielded as intercalated, one- to two-celled, ellipsoid to subglobular to unordered in shape, and transparent.

Assimilation Profiles Sodium gluconate is firmly utilized by *A. mahoniae*; D-glucose and arabinose are mildly utilized; sodium succinate and quinic acid are infirmly utilized. L-lysine as a sole nitrogen source is mildly utilized, while potassium nitrite, potassium nitrate, ethylamine, cadaverine, D-glucosamine, and D-tryptophan are mildly utilized by *A. mahoniae*. Absence of fermentation is observed.

Growth Temperature and Salt Tolerance Its cell growth is well at 4–37 °C and on 10.0% (w/v) NaCl, and its optimal growth temperature is 25 °C.

A. mahoniae is close to *A. melanogenum* and *A. motuoense*, but it varies in being orange to purple to deep red in its culture cultivated on MEA and PDA media (Jumbam et al. 2025).

3.3.41 *Aureobasidium musti*

A. musti was derived from a variety of host plant leaf surfaces in tropical and subtropical regions of North America and South Pacific Australia.

Culture Characteristics The fungal colonies of *A. musti* cultivated on MEA medium are orange in the center, followed by a yellow zone, dark brown, and shiny at the edge. The colonies cultivated on PDA medium are purple in the center, followed by a light brown zone, and dull, while those cultivated on M40Y medium are orange in the center, followed by a wet shiny zone and whitish edge that is pigmented. The colony edges grown in all the culture media are glassy and uniform with filaments.

Cell Morphology The vegetative filaments of *A. musti* are glassy, slim-walled, and septate. The conidia-generating filaments are primitive and transparent, simultaneously yielding conidia mostly terminally. Its conidia are transparent, glassy, oval-like to ellipsoid with hilus. The conidial cells propagate mostly by simultaneous polar and bipolar budding. Its chlamydospore is yielded as intercalated, one- to two-celled, ellipsoid to subglobular to unordered in shape, and transparent.

Assimilation Profiles Sodium gluconate is firmly utilized by *A. musti*; D-glucose and rabinose are mildly utilized; sodium succinate and quinic acid are mildly utilized. L-lysine as a sole nitrogen source is mildly utilized, while potassium nitrite, potassium nitrate, ethylamine, cadaverine, glucosamine, and tryptophan are infirmly utilized by *A. musti*. Absence of fermentation is observed.

Growth Temperature and Salt Tolerance The cell growth is well at 4–37 °C, optimally at 25 °C and on 10.0% (w/v) NaCl.

A. musti is close to *A. melanogenum* and *A. motuoense*, but *A. melanogenum* and *A. motuoense* are different from *A. musti* cultivated on MEA and PDA media in being orange to purple to deeply red in their cultures (Jumbam et al. 2025).

3.3.42 Aureobasidium peruvianum

Culture Characteristics The fungal colonies of *A. peruvianum* cultivated on MEA, PDA, and M40Y media are white to milky, shiny to monotonous, edges glassy with filaments yielded.

Cell Morphology The vegetative filaments of *A. peruvianum* are glassy, thin-walled, and septate. Its conidia-generating filaments are transparent, simultaneously yielding conidia both terminally and laterally. Its conidia are transparent, glassy, bowing to bean-shaped. Its conidia reproduce by polar budding. Its chlamydospores are oval-like to ellipsoid to subglobular and are yielded as intercalated. The chlamydospores are pale brown, one- to two-celled, sometimes three-celled.

Assimilation Profiles D-glucose is infirmly utilized by *A. peruvianum*. Melibiose is firmly fermented, while D-glucose, sucrose, raffinose, maltose, trehalose, cellobiose, and melezitose are mildly fermented, and inulin is infirmly fermented by *A. peruvianum*. So, presence of fermentation is observed. Potassium nitrate, L-lysine, glucosamine, and tryptophan as the sole source of nitrogen are firmly utilized by *A. mahoniae*. Sodium nitrite is mildly utilized, while cadaverine is infirmly utilized.

Growth Temperature and Salt Tolerance Its colonies are developed well at between 4 and 37 °C with the optimal growth at 28 °C and on 10.0% (w/v) NaCl.

A. peruvianum is close to *A. vanuatuense,* but it varies in the thinner and bowed to bean-shaped conidia, whereas the conidia yielded by *A. vanuatuense* are upright and ellipsoid (Jumbam et al. 2025).

3.3.43 Aureobasidium rubi

A. rubi was obtained from plant phylloplanes and air in the USA and Europe.

Culture Characteristics Its fungal colonies cultivated on MEA, PDA, and M40Y media are pink, shiny, edges glassy, filament production at the edges. The culture grown on PDA medium has white yeast-like cells encircled by a completely dark pigmented zone. The colony edges are whitish with brown flecking, edge unordered.

Cell Morphology The vegetative filaments yielded by *A. rubi* are glassy, slim-walled, and septate. The conidia-generating filaments are primitive and transparent. Its conidia are developed simultaneously, mostly terminal. The conidia of *A. rubi* are transparent, glassy, mostly oval-like to ellipsoid, sometimes subspindle-shaped. The conidia reproduce mostly by simultaneous polar budding. Its chlamydospores are yielded as intercalated, are one- to two (−three)-celled, ellipsoid to subglubular, and brown.

Assimilation Profiles D-glucose, arabinose, mannitol, and sodium gluconate as the sole source of carbon are mildly utilized by *A. rubi*, while alicin is only infirmly utilized. L-lysine and potassium nitrate as the sole source of nitrogen are firmly utilized by *A. rubi*, while ethylamine and cadaverine are infirmly utilized. Absence of fermentation is observed.

Growth Temperature and Salt Tolerance Its cell growth is well at 4–30 °C and on 10.0% (w/v) and 15.0% (w/v) NaCl. Its optimal growth temperature is 28 °C. Its Mycology Bank number is MB848262.

A. rubi is close to *A. intercalariosporum* but varies in the tardier growth rate, shorter conidia, and smaller one-celled chlamydospores of *A. rubi* (Jumbam et al. 2025).

3.3.44 *Aureobasidium toomeae*

Culture Characteristics The fungal colonies of *A. toomeae* cultivated on MEA medium are pigmented and are greenish brown in the center, white colony edges with filament yielded, wholly dark with crumple radiating from the center, edges greenish brown, and glassy, while those cultivated on PDA medium are not pigmented, desiccated, milky and yeast-like in the center, whitish outer zone, edge with filaments, nervure-like radiations originating from the center. The colonies grown on M40Y medium are whitish.

Cell Morphology The vegetative filaments yielded by *A. toomeae* are glassy, thin-walled, and septate. The conidia-generating filaments are transparent, simultaneously yielding conidia both terminally and laterally. The produced conidia are transparent, glassy, ellipsoid, spindle-shaped, or vessel-shaped. Its conidia propagate by polar and bipolar budding, while the chlamydospores are subglobular to ellipsoid and yielded as intercalated arthrospores, brownish, one–three-celled.

Assimilation Profiles Sodium gluconate and D-quinic acid are mildly utilized; D-glucose, erythritol, arabinose, and mannitol are infirmly utilized by *A. toomeae*. Cadaverine as the sole nitrogen source is mildly utilized, ethylamine, and L-lysine are infirmly utilized by *A. toomeae*. Absence of fermentation is observed.

Growth Temperature and Salt Tolerance Its colonies develop well at 4–37 °C and on 10% (w/v) NaCl; the optimal growth temperature is 25 °C.

Its Mycology Bank number is MB848262.

A. toomeae is close to *A. lini*, but it varies in the smaller and thinner conidia. *A. toomeae* cultivated on MEA medium yields greenish brown colonies and vessel-shaped conidia, whereas *A. lini* cultivated on the same medium have whitish gray colonies with curved conidia (Jumbam et al. 2025).

3.3.45 *Aureobasidium vanuatuense*

Culture Characteristics The fungal colonies of *A. vanuatuense* cultivated on MEA medium are pigmented, turn greenish brown at the center, pale yellow outer zone, greenish brown flecking at the edge, and filaments at edge pinkish, while those cultivated on PDA medium are milky orange yeast-like cells in center and edge, middle zone greenish and pigmented. Its filaments are yielded at edge. The colonies of *A. vanuatuense* cultivated on M40Y medium are purplish without pigmented color and to have filaments at edge reddish.

Cell Morphology The vegetative filaments developed by *A. vanuatuense* are glassy, thin-walled, and septate. The conidia-generating filaments are transparent or slightly purple in coloration, yielding conidia mostly laterally pear-shaped. The conidia are transparent, glassy, and ellipsoid. Its conidia reproduce by polar budding. The chlamydospores are ellipsoid to subglobular, intercalated, deep brown, one- to two-celled.

Assimilation Profiles D-glucose, sodium gluconate, sodium succinate, L-arabinose, and quinic acid are firmly utilized by *A. vanuatuense*, while erythritol and mannitol are tardily utilized. An orange coloration is yielded when quinic acid is utilized by *A. vanuatuense*. The colonies yield a light orange coloration when *A. vanuatuense* is cultivated on L-arabinose and sodium succinate. Those colonies are pinkish coloration when it is cultivated on erythritol. D-glucose, sucrose, melezitose, and raffinose are infirmly fermented by *A. vanuatuense*. So, presence of fermentation is observed. Potassium nitrate, ethylamine, and L-lysine are utilized; sodium nitrite, cadaverine, D-glucosamine and D-tryptophan are infirmly utilized.

Growth Temperature and Salt Tolerance Its cell growth is well at 4–30 °C and on 15% (w/v) NaCl and the optimal growth temperature is 28 °C.

Its Mycology Bank number is MB848260.

A. vanuatuense is close to *A. castaneae* and *A. albui*, but *A. vanuatuense* cultivated on PDA medium is different from them in lower growth rate, the production of orange coloration, the lack of pigmentation, and very low yeast-like cell growth at 30 °C (Jumbam et al. 2025).

3.3.46 Aureobasidium aleuritis

Culture Characteristics *A. aleuritis* is a plant parasite on the leaves of the plant *Aleurites* spp. The fungal colonies on leaf are big, unordered, dark reddish brown on ventral superficies, pale reddish brown on ventral side.

Cell Morphology The conidia-generating cells are upright, claviform to subcylindric, tightly crowded. Its conidia are one-celled, glassy, transparent, spindle-shaped to subcylindric with upright or slightly curved sides. The conidia are yielded simultaneously only at the mycelial peak of the conidia-generating cells on a thin denticulate (Jumbam et al. 2025).

Assimilation Profiles Glucose, fructose, and sucrose are utilized by *A. aleuritis*. Absence of fermentation is observed.

3.3.47 *Aureobasidium apocryptum*

Culture Characteristics *A. apocryptum* is parasitic on the leaves of plant *Acer* species. The colonies on leaf are unordered, usually coalescent.

Cell Morphology The conidia-generating cells are tightly crowded, upright, claviform to sub-cylindric. Its conidia are glassy, transparent, ellipsoid, one-celled, uncommonly slightly curved, and are produced at the same time at mycelial peak of the conidia-generating cells (Jumbam et al. 2025).

Assimilation Profiles Glucose, fructose, and sucrose are utilized by *A. apocryptum*. Absence of fermentation is observed.

3.3.48 *Aureobasidium dalgeri*

Culture Characteristics *A. dalgeri* is close to *A. aleuritis,* but it varies in yielding bigger, conidiomata, sometimes intercalated conidia-generating cells and in the shape of the mycelial peak of the conidia-generating cells. The colonies on leaf are big, unordered, pale brown with a purplish edge.

Cell Morphology The conidia-generating cells are upright, terminal or sometimes intercalated, sub-cylindric to claviform, tightly crowded. The conidia-generating cells with swollen mycelia are conidia-like-celled, terminally, yielding glassy, transparent, upright and ellipsoid conidia (Jumbam et al. 2025).

Assimilation Profiles Glucose, fructose, and sucrose are utilized by *A. dalgeri*. Absence of fermentation is observed.

3.3.49 *Aureobasidium prunicola*

Culture Characteristics *A. prunicola* is a parasite on leaves of the plant *Prunus virginiana*.

Cell Morphology It has filaments which are thin-walled, growing side by side. The conidia-generating cells are claviform to widely ellipsoid, upright, tightly crowded. The conidia yielded by *A. prunicola* are transparent, one-celled, glassy, ellipsoid, upright or sometimes slightly curved, and are yielded at the same time at the peak of the conidia-generating cells (Jumbam et al. 2025).

Assimilation Profiles Glucose, fructose, and sucrose are utilized by *A. prunicola*. Absence of fermentation is observed.

Its Mycology Bank number is MB309382.

3.3.50 Aureobasidium ribis

Culture Characteristics *A. ribis* is a parasite on the plant *Ribes nigrum*. Its fungal colonies on the leaf are unordered, grayish brown with dark brown edges, usually occurring on the edge of the plant leaf.

Cell Morphology The conidia-generating cells are crowdedly populated, upward or upright, claviform to sub-cylindric, circular mycelial peak, often branchy. Its conidia are one-celled, transparent, ellipsoid, upright, glassy, and are yielded at the same time at the mycelial peak portion of the conidia-generating cells (Jumbam et al. 2025).

Assimilation Profiles Glucose, fructose, and sucrose are utilized by *A. ribis*. Absence of fermentation is observed.

Its Mycology Bank number is MB309384.

3.3.51 Aureobasidium sanguinariae

Culture Characteristics *A. sanguinariae* is parasitic on the plant *Sanguinaria canadensis*. The fungal colonies appeared on the leaf are unordered, yellowish brown with darker edges.

Cell Morphology The conidia-generating cells are upright, broadly ellipsoid, densely crowded, claviform or sub-cylindric, often branched. Its conidia are transparent, one-celled, glassy, kidney-shaped to grasshook-shaped, sometimes cylindric with dull terminals, and are yielded at the same time at the mycelial peak of the conidia-generating cells (Jumbam et al. 2025).

Assimilation Profiles Glucose, fructose, and sucrose are utilized by *A. sanguinariae*. Absence of fermentation is observed.

The Mycology Bank number is MB309386.

3.3.52 *Aureobasidium umbellulariae*

Culture Characteristics *A. umbellulariae* is parasitic on leaves of the plant *Umbellularia californica*. The fungal colonies developed on the leaf are unordered, primarily on dorsal superficies, occurring as brown, mortified spots with dark edges, finally extending over the whole leaf.

Cell Morphology The conidia-generating cells are densely crowded, upright, claviform to cylindric, and are yielded at the same time at the mycelial peak (Jumbam et al. 2025).

Assimilation Profiles Glucose, fructose, and sucrose are utilized by *A. umbellulariae*. Absence of fermentation is observed.

Its Mycology Bank number is MB309388.

3.3.53 *Aureobasidium vaccinii*

A. vaccinii was obtained from leaves of the plant *Vaccinium* spp.

Cell Morphology The species yields conidiomata and conidia in a simultaneous manner (Jumbam et al. 2025).

Assimilation Profiles Glucose, fructose, and sucrose are utilized by *A. vaccinii*. Absence of fermentation is observed.

Its Mycology Bank number is MB126507,

3.3.54 *Aureobasidium epipinicola*

A. epipinicola (MycoBank Number: MB 858293) name comes from *Pinus nigra* needles and was derived from sooty mold community on the plant *Pinus nigra* needles, Poland.

Culture Characteristics Its fungal colonies cultivated on MEA medium are flatness, pinkish white, edge fringed while the colonies cultivated on PDA medium are flatness, pinkish white, edge fringed.

Cell Morphology Its filament consists of branchy septate, and is transparent or rarely pale brown to brown, glassy filaments, which are transformed into arthroconidia and conidia-generating cells. Its arthroconidia are globular, sub-globular or widely ellipsoid, limited at septa and transparent when it is young, brown to black brown, glassy, thick-walled, non-septate or one-septate, with oil guttules, intercalated, in un-branched or branched chains when it gets old. The conidia-generating cells are cylindric, club-shaped or widely ellipsoid, transparent, glassy, non-septate, with unremarkable conidia-generating loci, intercalated or ends on filament. Its conidia are ellipsoid to widely ellipsoidal, transparent, glassy, non-septate, and budding at polar areas.

Assimilation Profiles Glucose, fructose, and sucrose are utilized by *A. epipinicola*. Absence of fermentation is observed.

Molecular Identification The accession numbers of *ITS*, *LSU*, *TEF1*, and *TUB2* of *A. epipinicola* are OR334647, OR334848, PV067689, and PV067682, respectively.

A. epipinicola produces a free-standing phylogenetic cluster similar to *A. insectorum*. The sequence match between *A. epipinicola* and *A. insectorum* is low, especially in protein-coding genes: 98% in *ITS*, 99% in *LSU*, 90% in *TEF1*, and 90% in *TUB2*. *A. epipinicola* is distinguished from *A. insectorum* by having bigger filament, smaller but wider arthroconidia, and in general slightly wider conidia that are ellipsoid or widely ellipsoid in appearance (Czachura and Piątek 2025).

3.3.55 Aureobasidium aurantiacum

A. aurantiacum was derived from leaf tissue of the plant *S. natalensis*.

Molecular Identification The phylogenetic analysis of four gene regions based on *ITS* (OM417256), *LSU* (OM333570), *RPB2* (ON624213), and *TUB2* (ON995119) of *A. aurantiacum* shows it is a unique species sister to *A. mangrovei* mentioned above (Steinrucken et al. 2022).

Assimilation Profiles Glucose, fructose, and sucrose are utilized by *A. aurantiacum*. Absence of fermentation is observed.

3.4 Conclusions and Perspectives

Based on the available *ITS* sequences of different species of *Aureobasidium* spp., the phylogenetic tree constructed is shown in Fig. 3.1.

As stated above, most of species of *Aureobasidium* are obtained from the phyllospheres of different plants. This means that the phyllospheres of different plants are indeed the best niches for growth of species of *Aureobasidium*. This is closely associated with their ability to synthesize and secrete the enzymes such as amylase, xylanase, proteinase, lipase, mannase cellulase, pectinase, inulinase, esterases,

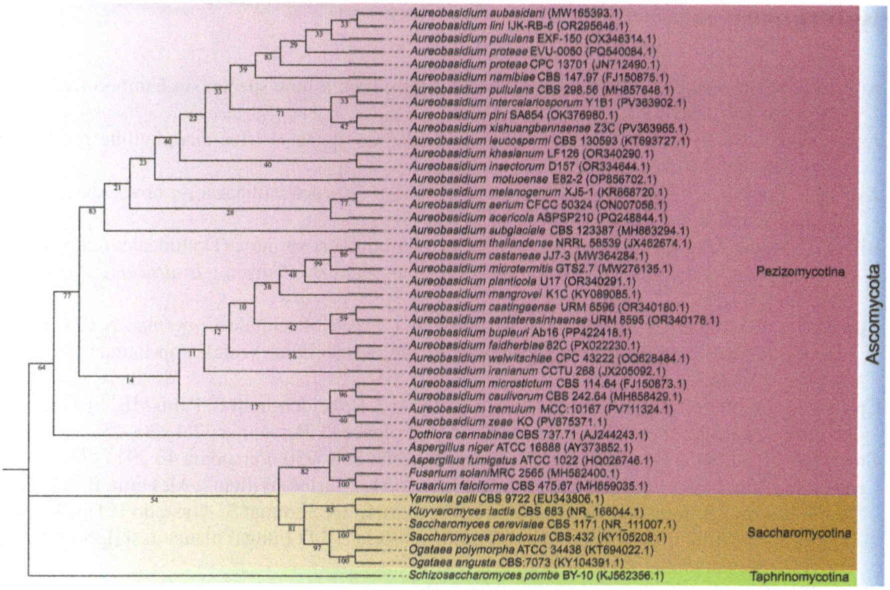

Fig. 3.1 The phylogenetic tree of most of species of *Aureobasidium* and other fungi based on available ITS sequences

lipase, laccases, and tannase that can hydrolyze starch, xylan, cellulose, lipid, protein, mannan, pectin, inulin, tannin, and phenolic compounds from plant tissues as shown in Chap. 6.

Although 55 species of *Aureobasidium* have been clarified, to date, only the genomes of *A. pullulans*, *A. melanogenum*, *A. aubasidani, A. hainanensis*, and *A. thailandense* have been characterized, mined, and edited. Furthermore, only biosynthetic pathways and molecular regulation modes of pullulan, polymalate, liamocin, melanin, and siderophore in *A. melanogenum*, fumarate biosynthetic pathway and molecular regulation modes in *A. aubasidani* and only biosynthetic pathway and molecular regulation mode of gluconic acid in *A. pullulans* have been elucidated. In addition, pullulan, polymalate, liamocin, and siderophore and its derives have been commercially produced in large scale, and only native *A. pullulans* and *A. melanogenum* isolated from natural environments can be used as effective antifungal agents in agricultures. Therefore, the genomic DNAs, fermentation products, and enzymes from any other species of *Aureobasdium* mentioned above are still being awaited to be explored and mined. So, *Aureobasidium* spp. have very high potential in biology and biotechnology.

Although most *Aureobasidium* species are generally non-fermenters, *A. fermentans*, *A. essambe*, *A. vanuatuense*, *A. peruvianum*, and *A. albui* can ferment some sugar; especially, *A. peruvianum* can ferment melibiose, D-glucose, maltose, sucrose, trehalose, cellobiose, melezitose, raffinose, and weakly ferment inulin. However, it is still unknown which genes are involved in the fermentation ability and which fermentation products can be yielded from these sugars.

References

Arzanlou M, Khodaei S (2012) *Aureobasidium iranianum,* a new species on bamboo from Iran. Mycosph 3(4):404–408

Ashengroph M (2017) A novel strain of *Aureobasidium* sp TeO12 for theophylline production from caffeine. 3 Biotech 7:176

Ashish P, Pratibha J (2018) *Aureobasidium khasianum* (Aureobasidiaceae) a novel species with distinct morphology. Phytota 374(3):257–262

Bills GF, Menéndez G, Platas VG (2012) Kabatiella bupleuri sp. nov. (Dothideales), a pleomorphic epiphyte and endophyte of the Mediterranean plant *Bupleurum gibraltarium* (Apiaceae). Mycol 104(4):962–973

Černošal A, Gostinčar C, Holcar M, Kostanjšek R, Lenassi M, Gunde-Cimerman N (2025) The impact of *Aureobasidium melanogenum* cells and extracellular vesicles on human cell lines. Sci Rep 15:1413

Crous PW, Summerell BA, Swart L, Denman S, Taylor J, Bezuiden-hout C, Palm ME, Marincowitz S, Groenewald JZ (2011) Fungal pathogens of Proteaceae. Persoonia 27(1):20–45

Crous PW et al (2019) Fungal planet description sheets: 868–950. Persoonia 42:291–473

Crous PW, Osieck ER, Jurjevi Ž, Boers J, Van Iperen AL, Starink-Willemse M, Dima B, Balashov S, Bulgakov TS, Johnston PR, Morozova OV, Pinruan U, Sommai S, Alvarado P, Decock CA, Lebel T, McMullan-Fisher S, Moreno G, Shivas RG (2021) Fungal planet description sheets: 1284–1382. Persoonia 47:178–374

Crous PW, Osieck ER, Shivas RG, Tan YP, Bishop-Hurley SL, Esteve-Raventós F, Larsson E, Luangsa-ard JJ, Pancorbo F, Balashov S, Baseia IG, Boekhout T, Chandranayaka S, Cowan DA, Cruz RHSF, Czachura P, De la Peña-Lastra S, Dovana F, Drury B (2023) Fungal planet description sheets: 1478–1549. Persoonia 50:158–310

Czachura P, Piątek M (2025) The genus *Aureobasidium* from sooty mould communities in Poland, including *A. epipinicola* sp. nov. and *A. insectorum*, new to Europe. Plant Pathol 74:1593–1611

da Silva AC, Ferro LO, Alves VCS, Barbosa RN, Souza-Motta CM, Bezerra JDP (2024) Phylogenetic evidence revealed new *Aureobasidium* species (Saccotheciaceae, Dothideales) from Brazil. Mycol Prog 23:70

Jia SL, Ma Y, Chi Z, Liu GL, Hu Z, Chi ZM (2019) Genome sequencing of a yeast-like fungal strain P6, a novel species of *Aureobasidium* spp.: insights into its taxonomy, evolution, and biotechnological potentials. Ann Microbiol 69:1475–1488

Jia SL, Zhang M, Liu GL, Chi ZM, Chi Z (2023) Novel chromosomes and genomes provide new insights into evolution and adaptation of the whole genome duplicated yeast-like fungus TN3-1 isolated from natural honey. Funct Integr Genom 23:206

Jiang N, Liang YM, Tian CM (2019) *Aureobasidium pini* sp. nov. from pine needle in China. Phytotaxa 402:199–206

Jiang N, Fan X, Tian C (2021) Identification and characterization of leaf-inhabiting fungi from *Castanea plantations* in China. J Fung 7(1):64

Jumbam B, Zasada IA, Bensch K, Crous PW, Aime MC (2025) A taxonomic revision of *Aureobasidium* (Saccotheciaceae, Dothideales) with new species, new names, and typification. Persoonia 54:147–195

Lee DH, Cho SE, Oh JY, Cho EJ, Kwon S (2021) A novel species of *Aureobasidium* (Dothioraceae) recovered from *acer pseudosieboldianum* in Korea. J Asia-Pac Biodivers 14:657–661

Nasr S, Mohammadimehr M, Geranpayeh Vaghei M, Ali Amoozegar M, Abolhassan S, Fazeli S (2018) *Aureobasidium mangrovei* sp. nov., an ascomycetous species recovered from Hara protected forests in the Persian Gulf. Iran Antonie van Leeuwenhoek 111:1697–1705

Onetto CA, Schmidt SA, Roach MJ, Borneman AR (2020) Comparative genome analysis proposes three new *Aureobasidium* species isolated from grape juice. FEMS Yeast Res 20:foaa052

Peterson SW, Manitchotpisit P, Leathers TD (2013) *Aureobasidium thailandense* sp. nov. isolated from leaves and wooden surfaces. Int J Syst Evol Microbiol 63(Pt_2):790–795

Reifschneider FJB (1980) Cultural and morphological variability of *Kabatiella zeae*. Trans Br Mycol Soc 75(2):239–241

Steinrucken TV, Vitelli JS, Holdom DG, Tan YP (2022) The diversity of microfungi associated with grasses in the *Sporobolus indicus* complex in Queensland, Australia. Front Fung Bio 3:956837

Wagner DN, Varaljay VA, Lyon WJ, Crouch AL, Allex-Buckner C, Biffinger JC, Crookes-Goodson WJ, Kelley-Loughnane N, Stamps BW (2024) Draft genome sequence of potentially dikaryotic black fungus *Aureobasidium melanogenum* isolated from aircraft. Microbio Res Announc 13(3):e0075623

Wang CB, Jiang N, Tu Y, Zhu YQ, Xue H, Li Y (2022) *Aureobasidium aerium* (Saccotheciaceae, Dothideales), a new yeast-like fungus from the air in Beijing. China Phytotaxa 544(2):185–192

Wei X, Zhang M, Wang GY, Liu GL, Chi ZM Chi Z (2023) The ornithine-urea cycle involves fumaric acid biosynthesis in Aureobasidium pullulans var. aubasidani, a green and eco-friendly process for fumaric acid production. Synth Syst Biotechnol 8:33–45

Wu F, Feng Z, Wang M, Wang Q (2023) Proposal of four new *Aureobasidium* species for exopolysaccharide production. J Fung 9:447

Zalar P, Gostinčar C, de Hoog GS, Sudhadham U, Gunde-Cimerma N (2008) Redeffnition of *Aureobasidium pullulans* and its varieties. Stud Mycol 61:21–38

Chapter 4
Fundamental Biology of *Aureobasidium* spp.

Abstract All the colony and cell morphology characteristics of *Aureobasidium* spp. are very similar. The colonies of *Aureobasidium* spp. are initially yellow, creamy, light pink, or light brown. After a day to a few weeks, their colonies become dark brown/black due to production of melanin-like pigments. *Aureobasidium* spp. display yeast-like cells, swollen cells, septate swollen cells, arthroconidia, chlamydospores, septate chlamydospores, and hyphae, so that they are polyextremotolerant generalists because of the cell morphological flexibility. Their life cycle consists of six sub-cycles. Their cell division and growth include single-cell budding and unique multiple budding in a single cell cycle. Although *Aureobasidium* spp. have haploid cells or diploid cells, only asexual propagation has been observed. The reasons why they lack sexual reproduction are still completely unknown. So far, only extracellular vesicles and vacuoles, along with their functions in *Aureobasidium* spp., have been investigated, whereas other cellular organelles and their functions remain to be explored.

Keywords Single budding · Multiple budding · Life cycle · Basic biology · Polyextremotolerance

4.1 Introduction

Although biosynthesis pathways of pullulan, polymarate, siderophore, melanin, fumarate, intracellular lipids, and their regulation at molecular level have been clearly elucidated and pullulan, polymarate, siderophore, melanin, fumarate, intracellular lipids, and *A. pullulans* cells have been widely applied in various sector of biotechnology, it has been little known about fundamental biology of *Aureobasidium* spp. so far.

Z. Chi et al., *Biology and Synthetic Biotechnology of Aureobasidium spp.*,
https://doi.org/10.1007/978-981-92-2607-8_4

4.2 Colony and Cell Morphology

As discussed in Chap. 3, the colonies of *Aureobasidium* spp. on agar surface are spread, opaque, and are often covered with the mucous materials of conidia and usually have a thin aerial mycelium or no mycelium. For example, colonies of *A. pullulans* initially appear as cream, yellow, pink, or light brown, but in the later growth stages, the colonies change their color to dark olivaceous or blackish due to the production of chlamydospores with melanin and have dark center, a typical feature of *A. pullulans*. Moreover, the colonies are initially smooth and later become covered in slime. The margin of colonies is yellowish white and dark-brown when they are old. However, these characteristics can be different based on the used culture growth media and among the different strains. The colony morphology can be obviously affected by the temperature, light, pH and sugar content, cultivation time, and different kinds of carbon and nitrogen sources. Even, different strains grown under the same conditions and on the same medium may display different colony morphology such as the color of the colony. Some colony morphologies of different strains of *Aureobasidium* spp. isolated from different environments in this laboratory are shown in Fig. 4.1. These results show that colony morphologies of different strains grown on different agar are diverse. The old cultures often produce melanin. This is why they are called black yeasts. Different colony morphologies of each species of *Aureobasidium* spp. have been described very clearly in Chap. 3.

As stated in Chap. 3, the mycelia of *Aureobasidium* spp. are usually hyaline, brown, yellow, pink, ochre, or black. The hyphae have cells that are longer than they are wide. Their cells are hyaline and septate, often turning brown and thick-walled after growing for a long time (Fig. 4.2). Uni- or bi-, or multicellular conidia can be produced from the hyphae (Fig. 4.2). The conidiogenic cells are localized laterally or terminally or intercalated on hyaline hyphae or thick-walled hyphae.

The yeast-like cells (blastoconidia or conidia) are produced simultaneously in dense groups. They are hyaline, smooth, and unicellular of various shapes and sizes. Uni-, bi-, or multicellular conidia also can be produced from primary conidia. The secondary conidia are smaller than the primary conidia and occur frequently on polar or bipolar.

Swollen cells can be developed from blastoconidia or hyphae by enlarging and by producing a thick cell wall. They can be globular and ellipsoid in shape and can be either septated or not.

The cylindrical-shaped arthroconidia are yielded through fragmentation of hyphae and are generally bigger than blastoconidia (Rensink et al. 2024).

Chlamydospores are thick- and melanized walled, one-celled or two-celled or cell chain asexually produced spores arising from the modification of a hyphal segment which contains intercalary chlamydospores and yeast-like cells. We also found that one chlamydospore can bud to form two-celled chlamydospores or a chain of chlamydospores (Fig. 4.3) (Hansali et al. 2026). Because they contain a large amount of melanin in their cell wall and have thick walls, chlamydospores can resist heat, acid, and desiccation, high levels of UV radiation, high H_2O_2 oxidation, high

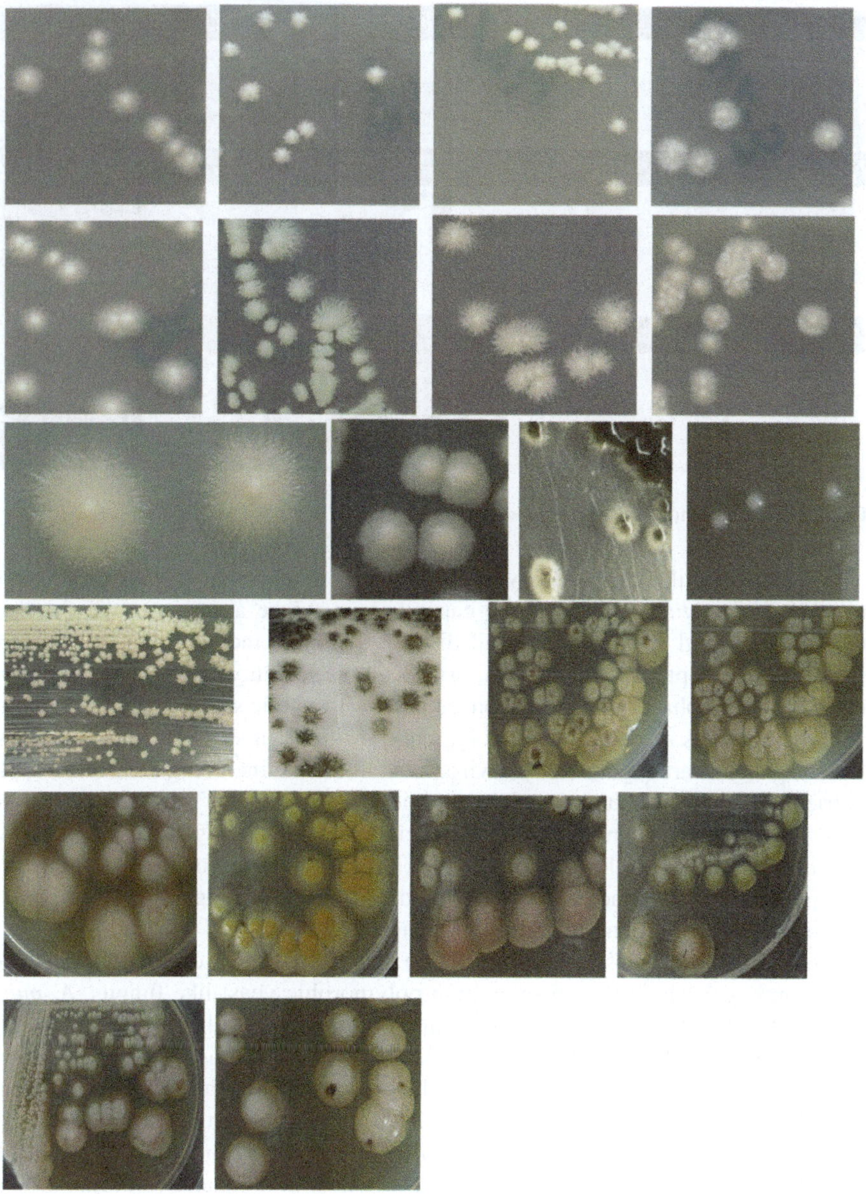

Fig. 4.1 Morphologies of colonies of different strains of *Aureobasidium* spp.

salts, and high sugars so that they can easily adapt and survive in harsh environments (Jiang et al. 2016, Chen et al. 2020).

Different cell morphologies of each species of *Aureobasidium* spp. have been described very clearly in Chap. 3, Figs. 4.2 and 4.3.

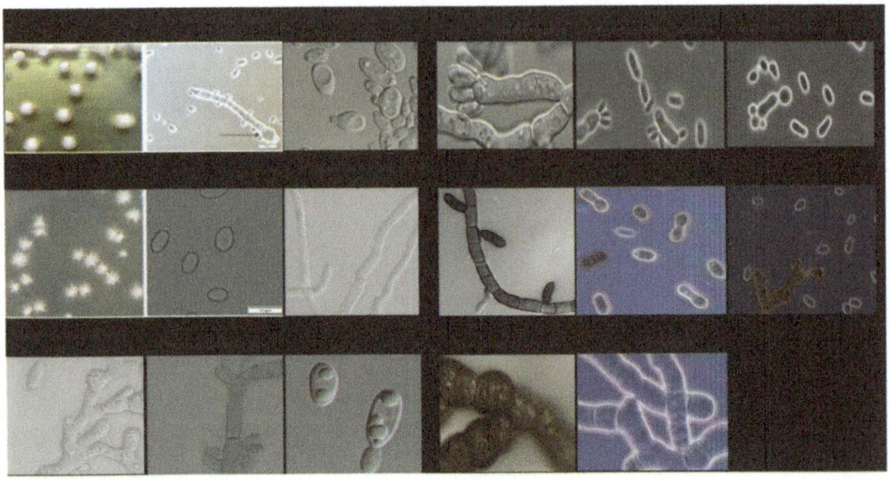

Fig. 4.2 Cell morphologies of *Aureobasidium* spp.

Therefore, unlike many other yeasts that maintain relative morphological uniformity, *Aureobasidium* spp. have the capacity to produce a variety of distinct cell types with varied shapes, sizes, and division patterns, including yeast-like cells, swollen cells, septate swollen cells, arthroconidia, chlamydospores, septate chlamydospores, hyphae, monilioid hyphae, and meristematic structures, all containing variable numbers of nuclei. Morphological transitions in such polymorphic yeast-like fungal cells are regulated by environmental factors, including temperature, pH, nutrient availability, cultivation time, and population density. Their cell morphologies are also determined by each specific species and genetic background. *Aureobasidium* spp. cell morphogenesis is also resource responsive, with the relative proportion of filamentous versus budding yeast-like fungal cell development related to growth medium and nutrient composition.

Furthermore, the substrate concentration in the liquid batch culture also regulates cell morphogenesis in the cosmopolitan polymorphic yeast-like fungus, *A. pullulans*, isolated from the pelagic open ocean. Its filamentous cell development is triggered only under high initial substrate conditions, suggesting that hyphal growth can be more advantageous under eutrophic conditions and may influence pelagic fungal interactions with particulate organic matter. Filamentous growth is proportionally declined before the exhaustion of substrate and before budding yeast-like fungal cell division enters stationary phase, possibly modulated by quorum sensing as previously evidenced in other polymorphic yeasts, suggesting the budding yeast-like fungal cell growth is more relevant to oligotrophic environments. The budding yeast-like fungal unicells decrease in size and become more elongated in shape in response to substrate depletion, resulting in higher cell superficial area to volume ratios, which can affect yeast dispersal and/or provide a nutrient uptake advantage under oligotrophic conditions, suggesting that unicellular fungal morphology itself is also modulated by resource availability (Diver et al. 2025).

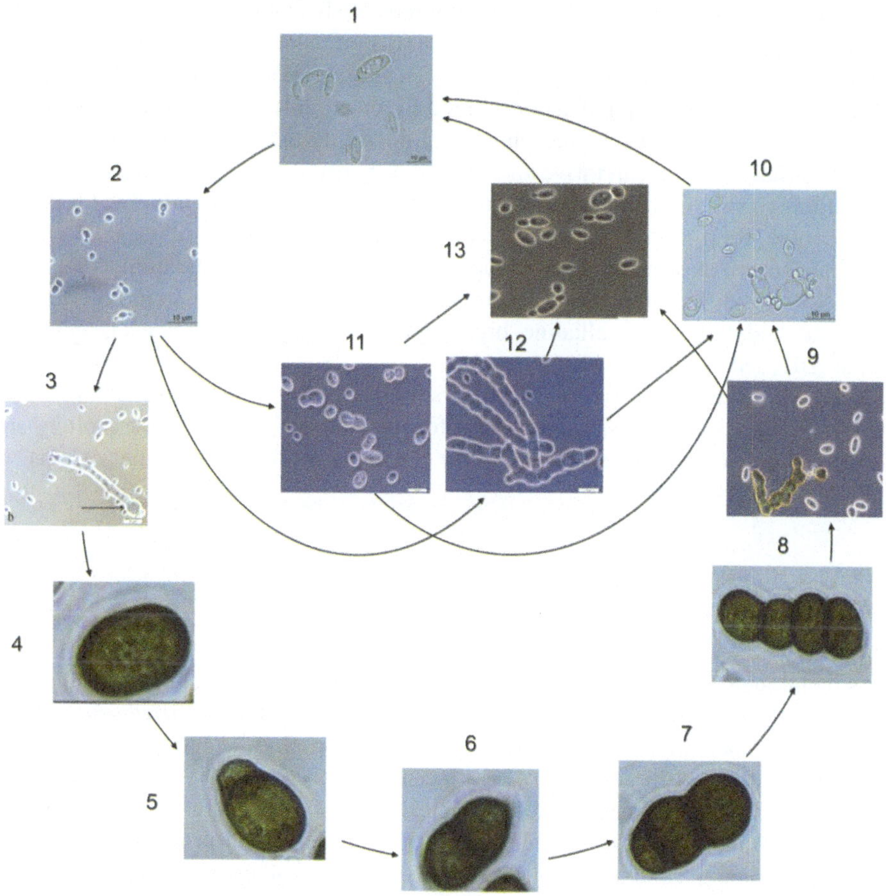

Fig. 4.3 Life cycle of *Aureobasidium* spp. 1. Blastoconidia are oval and yeast-like cells; a basal scar can be observed. Budding yields another secondary conidia under the favorable conditions (2); 2. Blastoconidia germinate, producing another secondary conidia or promycelium which are transformed into hyphae (3); 2. Blastoconidia also grow and are transformed into colorless chlamydospores with septa by germination (11), or blastoconidia also grow and are transformed into colorless arthroconidia chain with septa by germination (12). 3. Hypha with blastoconidia outgrowing laterally, but the blastoconidia can also be produced on termini of hyphae, terminally, with terminal swollen cells. 4. Chlamydospore with melanin in cell walls. 5. Germination of chlamydospore (a polarized chlamydospore with a pear-shaped form); 6. Two chlamydospores with a septum; 7. Germination of two chlamydospores, 8. A chain of chlamydospores; 9. Intercalary chlamydospores with rosary-like arrangement and melanin in cell walls. They serve as permanent spores, surviving unfavorable conditions, and melanin protects them against harmful effects of stress conditions; 10. Black chlamydospores budding to hyaline blastoconidia with multi-budding; 13. Black chlamydospores budding to hyaline blastoconidia with single-budding. 11. Blastoconidia (2) grow and are transformed into colorless chlamydospores (septate swollen cells); their contents are granular. 13. Colorless chlamydospores budding to hyaline blastoconidia with a single bud (13) or multi-buds (10). 12. Colorless arthroconidia chain budding to hyaline blastoconidia with single bud (13) or multi-buds (10) (Kockova-Kratochvilova et al. 1980)

The cultures of *A. pullulans* can simultaneously display multiple stages of a hypothetical pathway for eukaryotic evolution: uninucleate single cells, multinucleate single cells, and multicellular hyphae. The diverse cell morphospecies exhibit a high degree of phenotypic plasticity, and different strains of them have significantly different pigmentation. The morphological flexibility may help to explain how different strains of them can thrive and survive in a very wide variety of environments, including in soil, in the ocean, and on plants as a saprophyte or endophyte, salterns, bathrooms, food, honey, desert, feeds, water-damaged wood, polythermal glacier, and air space, as mentioned in Chap. 2. The morphological flexibility may also explain why different strains of *Aureobasidium* spp. are polyextremotolerant and tolerate cold, hot, acidic, alkaline, high saline, and sugar environments (Humphries et al. 2017; Samson et al. 2019).

Therefore, currently, many researchers think that *A. pullulans* and *A. melanogenum* can be utilized as a model system for the study of aquatic yeast ecology and evolutionary cell biology at a more fundamental level and the origins of multicellularity.

4.3 Life Cycle

As mentioned above, *Aureobasidium* spp. have polymorphism in having a filamentous, yeast-like cells and others. Different species of the genus *Aureobasidium* spp. exhibit a wide range of life styles, including saprophytes, endophytes, plant or human pathogens. In addition, many species of the genus of *Aureobasidium* are known to have significant biotechnological values such as pullulan, polymalate, liamocin, fumarate, β-glucan, Massoia lactone and melanin production and commercial potentials, and are currently used as a biocontrol agent against phytopathogenic fungi. They have various morphological states such as yeast-like budding form, scytalidium-like conidiogenesis with part of mycelia fragmenting and forming conidia, arthroconidia, and dark and thick-walled chlamydospores, as shown in Fig. 4.2. The genus typically forms aseptate, dark, and also sometimes hyaline conidia. They also have typical mycelial characters observed in many anamorphic ascomycetes of micronematous conidiophores and phialidic conidiogenesis.

The life cycle of *Aureobasidium* spp. usually consists of six subcycles: blastoconidia (conidia or yeast-like cells) produce other blastoconidia or form a promycelium. This promycelium is formed because the daughter cells of the blastoconidia do not separate from the mother cell and continue budding, creating a septate chain of this type of conidia.

The blastoconidia bud from the short lateral branches of hyphae, the conidiophore-like structures and the primary conidia. They are ellipsoidal, spindle-like, cylindrical, or lemon-like in shape. Blastoconidia (the unicellular yeast-like cells) are colorless, unicellular, and divide typically by bipolar budding or unipolar budding. Sometimes, two buds simultaneously emerge at two opposite sites of the single mother cell. Then, two buds sequentially emerge at one site of the single mother cell

(Goshima 2022). However, *A. pullulans* obtained by Mitchison-Field et al. (2019) has a round shape and has multiple buds at one site of the single mother cell. The unicellular yeast-like cells are beneficial for product formation during industrial fermentation processes because the yeast-like cells are more robust and less sensitive to shear forces compared to hyphal cell forms during the fermentation (Klement and Büchs 2013). Furthermore, the yeast-like cells enable higher oxygen and nutrient transfer due to lower viscosity, resulting in higher growth rates and more efficient product formation (Xiao et al. 2024). Blastoconidia are also produced by a budding process from the parent hypha. In early budding stages, the blastoconidia remain free of a mucilaginous coating. The outer wall layer is ruptured while the inner layer remains intact during the budding process. Blastoconidia, when presented with the favorable environment and nutrient, germinate and bud new blastoconidia. A mucus sheath envelops some of the hyphae. The hyphae continue to elongate and produce large numbers of blastoconidia, becoming septate and covered with melanin.

Blastoconidia also can differentiate into non-septated or septated and swollen cells. These swollen cells can bud off blastoconidia, differentiate into chlamydospores, which are black, larger, often monocellular, bicellular, and occasionally tricellular or multicellular, or produce germ tubes, giving rise to a septate mycelium and by producing a thick cell wall. This cell type is globular and ellipsoid in shape and can be either septated or not. The swollen cells not only produce blastoconidia, but also produce other (septated) swollen cells and chlamydospores. Swollen cells are produced from blastoconidia and are capable of budding new blastoconidia as well as becoming septate swollen cells or chlamydospores. The swollen cells without melanin are a transitional form between yeast-like cells and chlamydospores (Rensink et al. 2024). The swollen cells may be a stable cell type in the cell life cycle of *A. pullulans*, and transformation of swollen cells to chlamydospores may be regulated by nutritional or environmental factors (pH and temperature) or cultivation time (Rensink et al. 2024). Campbell et al. (2004) found that only swollen cells and chlamydospores, and neither hyphae nor unicellular blastospores, are responsible for pullulan biosynthesis. On the contrary, we found that the unicellular blastospores (yeast-like cells or conidia) of *A. melanogenum* are majorly responsible for pullulan, polymarate, liamocin, siderophore, and fumarate biosynthesis (Chen et al. 2019, 2020; Chi et al. 2009, 2012; Deshpande et al. 1992; Jiang et al. 2020; Lu et al., 2019; Ma et al. 2014; Wang et al. 2020; Wei et al. 2023; Zhang et al. 2021b, 2023). McNeil et al. (1989) also thought that yeast-like forms (batch culture) or filamentous form (continuous culture) of *Aureobasidium* spp. appeared to be the primary agents of pullulan production.

The cylindrical-shaped arthroconidia are generally bigger than blastoconidia, with a length and width of 6.5–22 and 4.5–13 μm, respectively. They form through fragmentation of hyphae (Rensink et al. 2024).

Filamentous growth starts from a single yeast-like cell. Then, septation and branching appear in the filaments, followed by many buddings from various sites on the filament (Goshima 2022). At low cell density, filamentous growth occurs, but at high cell density, the filament immediately starts multibuddings under the control of a quorum-sensing mechanism (Goshima 2022). It has been believed that filamentous growth with branching is helpful for them to reach to the unoccupied areas, whereas budding at high cell density allows their cells to be released to the free water and translocated to other areas (Goshima 2022). The filamentous mycelium is the cells from where individual hyphae often produce these unicells by budding.

The unicells described so far include the budding one-celled, two-celled, or multi-celled blastospores and single- and multi-celled swollen cells, some of which then convert to melanin-producing chlamydospores.

The chlamydospores with single or numerous septa can give rise to a mycelium, which in turn can produce blastoconidia and chlamydospores. Chlamydospores develop from swollen cells or from septate swollen cells or by hyphae and can be found in chains, as free cells, or firmly attached to hyphae. This cell type is characterized by having a very thick wall which can withstand air drying without collapse, and the presence of large amounts of melanin deposits over its cell walls. Some hyphae strands which produce chlamydospores become septate, thick-walled, and covered with melanin. The transition from yeast-like cells to chlamydospores of *A. pullulans* can be induced by higher than 3.0% (w/v) glucose in the medium with a limiting nitrogen source and a low buffer capacity. On the other hand, higher NH_4SO_4 concentrations actually stimulate the appearance of chlamydospore (Bermejo et al. 1981). Different chlamydospores are capable of withstanding desiccation and, resistant to ultraviolet irradiation. It is believed that the resistant nature of chlamydospores is due to the melanin in the cell wall, thick walls, as well as other molecules. The thick-walled chlamydospores covered with melanin are transformed from swollen cells (Rensink et al. 2024). We also found that chlamydospores can bud to form two-celled chlamydospores and a chain of chlamydospores (Hansali et al. 2026). Therefore, chlamydospore is not a resting cell. It is still unknown whether melanin accumulation in its cell wall or the thick wall causes chlamydospore formation with the resistant nature. Therefore, it is of significant to investigate the genetic and metabolic mechanisms for chlamydospore formation. This will be discussed in more detailed in Chap. 10.

The yeast-like cell walls contain only 13% glucose, but more mannose, galactose, and bound lipid than chlamydospore walls while glucose is the main component of chlamydospore walls and accounts for 36% of the dry weight and most of the glucan portion of chlamydospore walls is insoluble in dilute alkali and contains linear chains of $(1 \rightarrow 3)$ and $(1 \rightarrow 6)$ linked glucose (Brown et al. 1973).

Many factors affect this complex life cycle, with the nitrogen source (both organic and inorganic) and medium pH particularly influential.

The whole life cycle of *Aureobasidium* spp. is shown in Fig. 4.3. In summary, the blastoconidia produce other blastoconidia or form a mycelium. This mycelium is formed because the daughter cells of the blastoconidia do not separate from the mother cell and continue budding, creating an aseptate chain of this type of spore. Blastoconidia can also differentiate into (non-)septated swollen cells. These swollen cells can bud off blastoconidia, differentiate into chlamydospores, or produce germ tubes, giving rise to a septate mycelium. The chlamydospores can give rise to a mycelium, which in turn can produce blastoconidia and chlamydospores. The chlamydospore also can bud to form another chlamydospore.

Yeast chlamydospores and melanized filaments have an electron-dense, melanin-rich, granular, outer cell-wall layer, which yeast-like cells lack. This granular material is also found in cross walls of filaments and chlamydospores.

For example, the life cycle of *A. pullulans* NG isolated from strawberry fruit can be divided into six stages: blastoconidia, swollen cells, septate swollen cells, hyphae, meristematic structure (MS), and chlamydospores. In the early stage, yeast-like cells (blastoconidia) at pH $\geq$ 6.0 are the main cell morphology in its life cycle. Then, they gradually differentiate to form swollen cells (pH = 4.5) to adapt to the living environment with changes in environmental pH, and blastoconidia, as the main cell morphology of their lifecycle can produce a variety of important metabolites such as pullulans. At the end of their life cycle, swollen cells can also form septate swollen cells, chlamydospores, and mycelium. The swollen cells of *A. pullulans* NG are responsible for pullulan synthesis. The citric acid accumulated in the cells can regulate the cellular differentiation of *A. pullulans* NG to maintain growth in the swollen cell form and increase pullulan production (Zeng et al. 2023). The swollen cells in *A. pullulans* NG have much higher antioxidant capacity and stress defense capacity than yeast-like cells. The swollen cells have high amino acid metabolism, carbohydrate metabolism, and lipid metabolism based on the analysis of transcriptomics and metabolomics (Zeng et al. 2023). The glyoxylate metabolism in the swollen cells of *A. pullulans* NG plays a key role in the stress defense. This demonstrates that *A. pullulans* NG can use cell differentiation to help it adapt to its environment against adverse conditions (Zeng et al. 2023).

4.4 Cell Division and Growth

Aureobasidium spp. are the ubiquitous black yeast-like fungi with a wide variety of morphologies, and their growth modes include "typical" single-budding yeast, larger multinucleate yeast that can make multiple buds in a single cell cycle, and multinucleate hyphae that also can make multiple buds (Fig. 4.2). The morphological flexibility may help to explain how *Aureobasidium* spp. can thrive and survive in a very wide variety of environments, including in soil, in desert, in hypersaline waters, in cold environments, honey, in the ocean, and on plants as a saprophyte or endophyte and so on as stated in Chap. 2.

4.4.1 Single Cell Budding of Aureobasidium spp.

Saccharomyces cerevisiae, Kluyveromyces lactis, Candida albicans, Ustilago maydis, Cryptococcus neoformans, and most strains of *Aureobasidium* spp. only form a single bud in each cell cycle. However, during the pulmonary infections in human

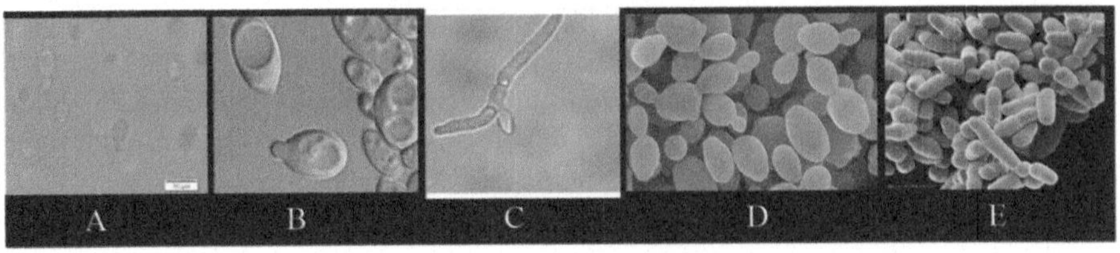

Fig. 4.4 Single budding of *Aureobasidium* spp. (**A**, **B** and **C**) and *S. cerevisiae* (**D**) and fission cells of *S. pombe* (**E**). Even the scar (**B**) on the primary conidia can be seen

cells, the pathogenic yeast *C. neoformans* first produces the titan cells that undergo DNA replication without budding to yield large polyploid cells. After that, the titan cells can bud and divide asymmetrically to make small haploid yeast cells. In contrast, the yeast-like fungal strains *A. pullulans, Paracoccidioides brasiliensis*, and *Mucor circinelloides* do not follow this critical rule happened in *C. neoformans*.

So, most budding yeast cells divide by making a single bud, and this single budding is thought to be critical to ensure faithful segregation of the genome and cellular contents when the cells are divided. It has been well confirmed that the yeast-like fungal cells of *Aureobasidium* spp. have two growth models: single budding and multiple budding. In the single budding yeast, a single bud is produced each cell cycle. During singularity in budding, after each mitosis, a daughter and mother each receive a single nucleus (Fig. 4.4A–D). So, the nuclear division and morphogenesis must be coordinated to make sure that each daughter cell receives only one nucleus. The multiple budding in *Aureobasidium* spp. is discussed in the next subsection.

It has been known that in *S. cerevisiae,* the cell division cycle begins with the emergence of a single, precisely positioned bud, and asymmetric growth enlarges this bud until cell division occurs. These growth events are tightly coordinated with the nuclear division cycle, which drives DNA replication and segregation (Fig. 4.4D). By contrast, some strains of *Schizosaccharomyces pombe* (Fig. 4.4E), the rod-shaped fission yeast cells elongate through tip growth, and then use fission in the cell middle to separate two symmetrical daughter cells.

Like *S. cerevisiae*, most budding yeast cells of *Aureobasidium* spp. divide by making a single bud, and this single budding is also thought to be key to making precise segregation of the genome and cellular contents when their cells are divided (Fig. 4.4A–C).

In *S. cerevisiae*, *K. lactis*, *C. albicans*, *Ustilago maydis*, *C. neoformans*, and most strains of *Aureobasidium* spp., their cells can only form a single bud in each cell cycle as mentioned above. A conserved cell polarity pathway in their cells guarantees the formation of a single bud in each cell cycle. Then, accurate delivery of one daughter nucleus into the single bud in them is achieved by orienting the mitotic spindle along the mother-bud axis, and the cytoskeletal elements at the mother-bud neck only occur following properly oriented mitosis. So now, it is still poorly understood how a specific nucleus is selected for delivery into each bud, and how each bud receives only a nucleus. It has been reported that in *S. cerevisiae*, feedback mechanisms prevent additional bud emergence once the initial bud has formed. In the single budding of *Aureobasidium* spp., the similar feedback mechanisms may also happen in the cells of *Aureobasidium* spp.

In addition, this may be closely related to the cytoskeletal elements as mentioned below.

It has been known that in *S. cerevisiae*, all of the cytoskeletal elements, including actin, microtubule, and septin networks, are responsible for the accurate segregation of nuclei and other organelles into the bud, as mentioned above. Actin is an essential protein in yeast and functions in the fundamental processes of cell polarity, cytokinesis, and endocytosis. The actin also functions in generation of the polar budding pattern and is required specifically for proper selection of bud sites in the mother MATa/α cells of *S. cerevisiae*. The yeast actin cytoskeleton is organized into three distinct structures: (a) the actin cables that run along the mother-daughter cell axis; (b) a contractile cytokinetic ring at the neck of dividing cells; (c) the actin rings in cooperation with myosin II. All three distinct structures are also responsible for septum formation.

Microtubules (MTs) are the unbranched and rigid tubules composed of tubulin subunits and play a crucial role in cell division, mitosis, cell polarity, intracellular transport, and the structural framework in eukaryotic cells. They are important for the distribution of nuclei or other organelles. Nuclei migrate probably along MTs until they reach a certain position. Thus, they are important for rapid hyphal growth. However, they are not essential for spore germination, but only for the site selection of germination.

Septins are the GTP-binding proteins that form short (~32-nm-long) filaments and higher-order structures on the cell cortex of eukaryotic cells and are associated with actin and microtubule cytoskeletal networks. Septins are important for cytokinesis, cell cycle regulation of the polarized cell growth, and formation of a diffusion barrier in the yeast plasma membrane to prevent free translocation of bud-specific membrane proteins into the mother cell (Takeshita et al. 2014).

However, the functions of actin, microtubule, and septin in the single budding of *Aureobasidium* spp. are still completely unknown and are waiting to be elucidated by their genome editing, protein engineering, and structural biology.

Filamentous growth of *Aureobasidium* spp. occurs with occasional septation and branching, followed by single budding from various sites on the filament (Fig. 4.4C). This mode of growth/budding is dominant when the initial cell density is low. For example, on microcrystalline cellulose, hyphae of *Aureobasidium* sp. TE-18 isolated from soil of sugar cane field in Mexico can grow well and have high cellulase activity while it is grown on a PAD medium, the large numbers of blastospores including the germinating yeast-like cells can be observed and the hyphae is well developed, even a scar (arrow) left on the hyphae can be seen by separation of a blastospore (Fig. 4.5) (Gilbon and Huitron 1986).

The activity of the transcriptional activator Msn2 in *A. melanogenum* is controlled by the cAMP-protein kinase A (PKA) signaling pathway, the nitrogen-sensing TOR signaling pathway, the high osmotic pressure-sensing HOG1 signaling pathway, and the glucose repression SNF pathway (Chi et al. 2022). The disruptant *ΔMsn25*, in which the *MSN2* gene encoding the transcription factor Msn2 was

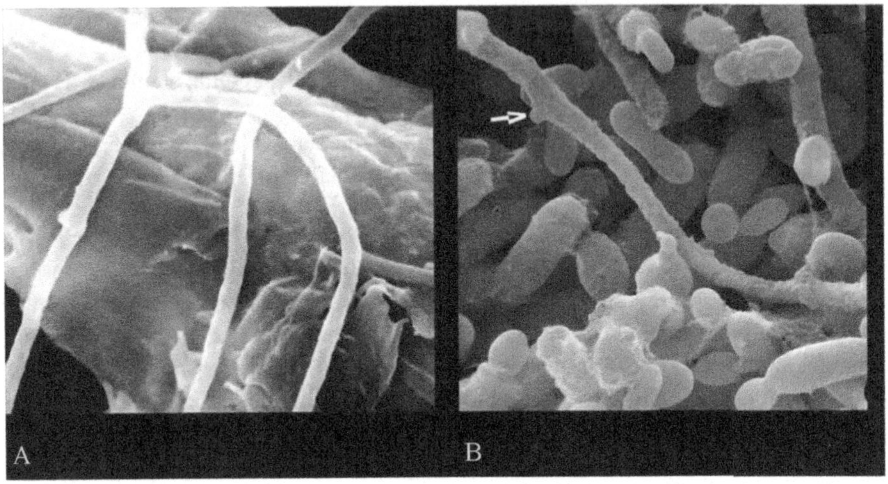

Fig. 4.5 Hyphae grown on microcrystalline cellulose (**A**) and hyphae and blastospores grown on PDA medium (**B**). A scar is indicated by an arrow (Gilbon and Huitron 1986)

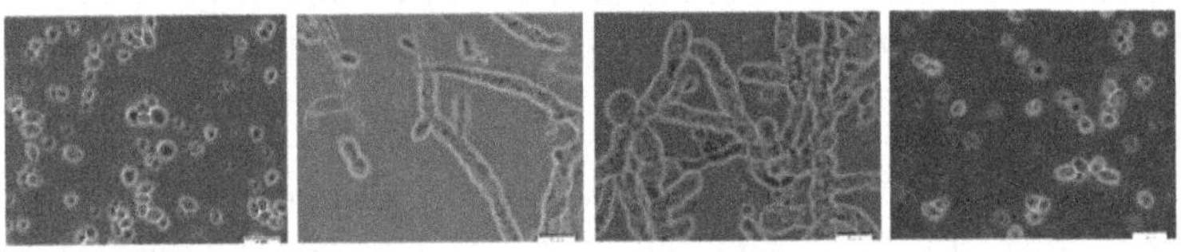

Fig. 4.6 The cell morphologies of the wild-type strain 6-1-2 of *A. melanogenum*, a high liamocin producing yeast, its Δ*Msn16* mutant, and the complementing strain BMsn3 as well as the disruptant ΔMsn25 of *A. melanogenum* P16, a high pullulan-producing yeast

completely removed, loses its ability to form yeast-like cells but forms filamentous cells with single budding, while its wild-type strain 6-1-2 and the complementing strain BMsn3 still keep the yeast-like cells (Zhang et al. 2021a). Similarly, deletion of the *MSN2* gene from the genome of *A. melanogenum* P16, a high pullulan-producing yeast, also makes the *Δmsn16* mutant become filamentous, while the cells of its wild-type strain *A. melanogenum* P16 still keep the yeast-like cells (Yang et al. 2020) (Fig. 4.6). This indicates that deletion of the *MSN2* gene could affect cell morphology and cell growth of *A. melanogenum*. However, the reason for this is still unknown. One of the reasons is that the cell wall division of *A. melanogenum* may be determined by the transcriptional activator Msn2.

Even substrate concentration in the medium can regulate cell morphogenesis in *A. pullulans* isolated from the pelagic open ocean. The multicellular filamentous cell development of *A. pullulans* can be triggered only under high initial substrate conditions, suggesting that hyphal growth could be more advantageous under rich-nutrient conditions and may influence pelagic fungal interactions with particulate organic matter while budding yeast-type unicells grown in the low-nutrient conditions decrease in size and became more elongated in shape, resulting in higher cell surface area to volume ratios, which can affect the yeast dispersal and/or provide a nutrient uptake advantage under such oligotrophic conditions (Diver et al. 2025) while the formation of simple multicellular structures of filamentous fungi can display various ecological benefits (e.g., improved extra-cellular metabolism, competitive ability, and dispersal) that may act as selective drivers of microbial evolution (Heaton et al. 2020; Tong et al. 2022).

4.4.2 *Multiple Budding in a Single Cell Cycle of Aureobasidium spp.*

In recent years, it has been found that some strains (*A. pullulans*, *A. melanogenum*, *A. epipinicola* and *A. insectorum*) of *Aureobasidium* spp. as stated in Chap. 3 can produce multiple buds simultaneously, up to 3 or 10 from the same vicinity of a mother cell of conidium or hyphae (Fig. 4.7). Then, the cells have a synchronous mitosis in which all the nuclei are divided simultaneously so that every bud can receive at least one nucleus and the mother still retains a variable number of nuclei. So, they also have multinucleate yeast cells and multinucleate hyphae with variable morphology. Although the number of buds that a mother produces varies each cell cycle, a single mother cell produces similar numbers of buds in consecutive division cycles, and the sites of bud emergence are often re-used. Now, it has been found that almost 80% of the obtained *Aureobasidium* spp. cells produce >1 bud, and of those, 64% of *Aureobasidium* spp. cell produce multiple rounds of simultaneous buds. Each cell cycle length is about 159.5 min. But the cell cycle duration is not obviously correlated with the number of buds that a mother produces in a given cell cycle. So, firstly a single *Aureobasidium* spp. cell must undergo multiple DNA replications and nuclear divisions to become multinucleated, then the cell simultaneously produces multiple buds from the same site (Fig. 4.7), and finally, the mother itself still keeps a variable number of nuclei in its own cell. The variable numbers of buds might reflect a variable number of nuclei in the mother cell. The mother cells with more nuclei tend to produce higher numbers of buds in a given division cycle. Consequently, each bud generally receives at least one nucleus each, and in some

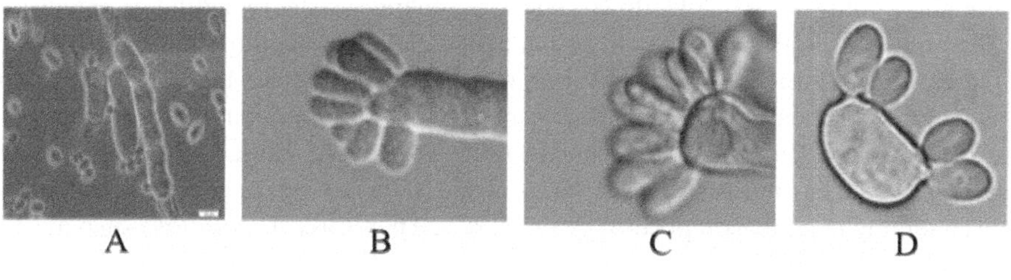

Fig. 4.7 Simultaneous multiple buds within a single cell cycle of *Aureobasidium* spp. (**A**) from *A. melanogenum*; (**B**) from *A. insectorum*; (**C**) from *A. epipinicola*, (**D**) from *A. pullulans*

cases, even the nucleus is deposited when buds are extremely small. Thus, the mother cells appear to produce some number of nuclei in the absence of budding and then distribute nuclei to nascent buds. In this case, each cell ends up with a single nucleus, suggesting that one bud emerges per nucleus, followed by nuclear migration into the new bud (Mitchison-Field et al. 2019). Now, it has been confirmed that the mother cells of *A. pullulans* indeed have highly variable numbers of nuclei during multiple simultaneous budding. Furthermore, each cell has several coexisting polarity sites (Mitchison-Field et al. 2019; Petruccoa et al. 2024), and it is thought that *A. pullulans* can display a semi-open form of mitosis with an intact but permeable nuclear envelope for nuclear migration into the new bud. So this is completely different from the single cell budding as mentioned in the last subsection. In fact in the same culture, some cells have both the single cell budding and the multiple cell budding.

Usually, the larger multinucleate yeast can make multiple buds in a single cell cycle, revealing highly synchronous mitosis in the multinucleate yeast cell that occurs in a semi-open manner with an intact but permeable nuclear envelope for nuclear migration into the new bud. It also has been shown that in the multinucleated cells, mitosis appears very synchronous, with all the nuclei condensing within 30 s of each other. During mitosis, the H2B signal becomes more condensed in a presumed metaphase plate and then undergoes a rapid segregation to the mother and bud. So, it is interesting to note how yeast-like fungal cells decide the number of buds to make and how nuclear segregation is managed so as to distribute nuclei to every bud. This may also be closely linked to the fact that bud formation and nuclear segregation must be achieved by appropriate organization of the cytoskeleton (formation of actin, microtubule, and septin networks) as mentioned above. However, it is still completely unclear why and how some strains of *Aureobasidium* spp. can produce multibudding and multinucleate in one cell cycle. What environmental and genetic factors can determine *Aureobasidium* spp. to produce multibudding and multinucleate in one cell cycle or single budding in one cell cycle? It is also still unknown if multibudding and multinucleate can make their cells multiply faster than single budding. What are the advantages and benefits of the multiple budding compared to those in the single budding? If the multibudding can make their cells multiply faster than single budding, this will promote faster cell growth and metabolite production during fermentation.

It has been well known that animal cells perform an “open” mitosis during which the nuclear envelope breaks down, whereas several yeasts, including *S. cerevisiae*, carry out a “closed” mitosis with an intact nuclear envelope throughout the process. Other fungi exhibit “semi-open” variations of mitosis in which the nuclear envelope is partially broken down, or the nuclear envelope remains intact, but nuclear pore complexes (NPCs) become permeable due to loss of some NPC components. Similarly, during mitosis in *A. pullulans*, the nuclei become permeable due to the disassembly of at least a portion of the NPCs. Therefore, *A. pullulans* displays the form of “semi-open” mitosis where the nuclear pores become permeable, the chromosomes condense, and the nuclear envelope deforms to generate two daughter nuclei before dividing. This may be required for multibudding and multinucleation of *Aureobasidium* spp. as mentioned above.

This also happens in the closely related *A. nidulans*, where the three lobes include two daughter nuclei and a central nucleolar remnant that disperses (Ukil et al. 2009). In *A. nidulans*, a subset of nuclear pore complex proteins disperses during mitosis, leading to a permeable nuclear envelope (De Souza et al. 2004).

These behaviors also raise problems and questions that do not arise in studies of the model yeasts such as *S. cerevisiae*. In contrast, as mentioned above, in *S. cerevisiae*, the feedback mechanisms prevent additional bud emergence once the initial bud has formed. So, during the multiple budding in the single cell cycle, the feedback mechanisms must be stopped. However, it is still completely unknown which signals (which environmental factors, nutrients, physiological state, and genes in the genome) can make the cells stop the feedback mechanisms. Therefore, the *Aureobasidium* spp. super-budders suggest a dramatic rewiring of the conserved signaling pathways in *S. cerevisiae*. The nuclear division process in some strains of *Aureobasidium* spp. also shows that the spatial and temporal links between the cell division cycle and the nuclear division cycle can be uncoupled. This new system raises questions about how the cell decides how many buds to form, how they coordinate the segregation of nuclei and other organelles so that each bud becomes a viable daughter. All the questions still remain to be answered by the multi-omics analysis and genome editing of the yeast-like fungal cells. So, *Aureobasidium* spp. can offer the unique opportunity to pursue many new research topics in yeast cell biology (Mitchison-Field et al. 2019).

It was found that *A. pullulans* genome indeed contains well-conserved cytoskeletal genes encoding actin, microtubule, and septin networks as mentioned above. This includes one each of alpha, beta, and gamma tubulin; a single conventional actin; and the four core septin genes, as well as one additional septin gene. Among them, actin is important for fungal cytokinesis, and it was found that *A. pullulans* can generate actin cables, patches, and rings analogous to those described in other fungi. *A. pullulans* also displays the polarized actin cytoskeleton, microtubule spindle, and cortical septin structures at the neck that are typical of *S. cerevisiae* and other budding yeasts. However, the detailed functions of the actin, microtubule, and septin in the multiple budding at molecular levels in *A. pullulans* remain to be elucidated by genome editing and structural biology.

As mentioned above, many *A. pullulans* cells have an unconventional mode of proliferation in which multinucleate yeast cells make multiple buds within a single cell cycle. That is different from the model budding yeasts that produce only a single bud per cell cycle, including ascomycete such as *S. cerevisiae* and *C. albicans* and basidiomycete such as *C. neoformans*, and *U. maydis*. During the multibudding growth, it is still unclear how mother cells determine the number of buds in a given cycle or whether they deliver similar numbers of nuclei and organelles to each bud. Therefore, *A. pullulans* is a promising emerging model for our further understanding of other fungal strategies for successful proliferation of multinucleate yeast cells.

In fact, it is very common that *Aureobasidium* spp. have such an ability to generate multiple buds simultaneously within a single cell cycle as shown in Figs. 4.2, 4.3 and 4.7 (Czachura and Piatek 2025).

4.5 Possible Mating in *Aureobasidium* spp.

The sexual reproduction of ascomycetous fungi has different arrangements of the mating-type genes that encode the key transcription factors at one or more of the *MAT loci*: the *MAT1-1-1* gene encodes a protein with an alpha1 domain, and the *MAT1-2-1* gene encodes a protein with a high mobility group (HMG) box. For example, it has been well known that the yeast haploid cells have one of two mating types, designated MATa or MATα in *S. cerevisiae* (Fig. 4.8B). Each haploid cell secretes a small peptide pheromone that binds compatible receptors on the surface of cells of the opposite mating type. The pheromone binding causes a characteristic response whereby cells reorientate growth toward the potential mating partner and then fuse to form a diploid cell (Fig. 4.8A, B). The diploid cell formed is no longer capable of mating, but under suitable environmental conditions, it undergoes meiosis and sporulation to form four haploid ascospores (Fig. 4.8B). Indeed, one mating type locus is also identified in each genome of *A. pullulans*, *A. melanogenum*, *A. subglaciale*, and *A. namibiae.* Furthermore, the mating-type locus in all four of these species has a conserved organization and contains one of each of the *MAT1–1-1* and *MAT1–2-1* homologues in opposite orientations. However, although *Aureobasidium* spp. belong to Phylum Ascomycota that are known to reproduce sexually, as mentioned above, the cell hybridization between the haploid cells and meiosis of the diploid cell have not been observed in any native strain of over 55 *Aureobasidium* species obtained so far (Chap. 3). Furthermore, although population genomic analysis of 50 *A. pullulans* genomes shows regular recombination within the species and has a mating locus that occurs in all the analyzed genomes, no sexual reproduction has been detected for them to date, either (Gostinčar et al. 2019). The reasons for this are still completely unknown.

In *S. cerevisiae*, both MAT*α* and MATa contain the same homeodomain proteins, which, in the case of diploid cells, play an essential role in mating regulation by

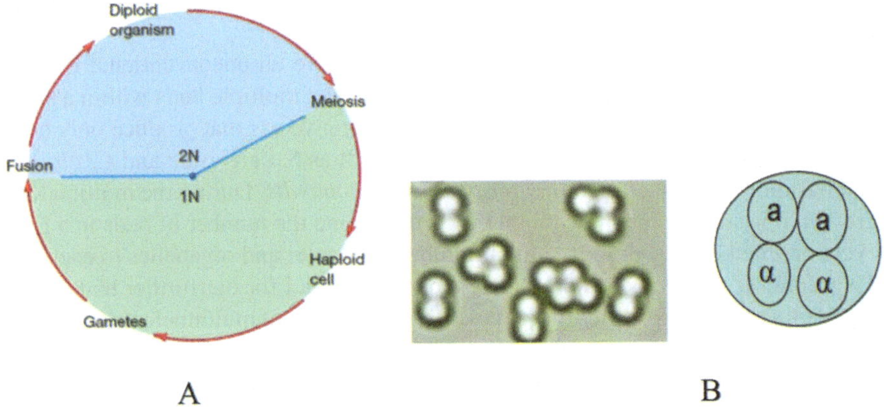

Fig. 4.8 Life cycle (**A**) of *S. cerevisiae* and its ascospores (**B**)

repressing the transcription of cell-specific genes (Fig. 4.8B). This means that the *MAT* locus is conserved throughout *Aureobasidium* spp., and a genomic rearrangement event leads to the two versions of the homothallic *MAT* locus so that no sexual reproduction has been detected for them to date.

However, the high pullulan-producing and osmotic-tolerant yeast *A. melanogenum* TN3-1 strain isolated from natural honey has two subgenomes A and B, and the genomic contents of the two subgenomes are asymmetric with many structural variations (Xue et al. 2019; Chen et al. 2020). Especially, as shown in the next chapter, the genomes (the size is 51.61 Mb) of the TN3–1 strain are regarded as a recent hybrid/fusion between the ancestor of *A. melanogenum* CBS105.22/CBS110374 and the ancestor of another unidentified strain of *A. melanogenum* similar to another high pullulan-producing and osmotic-sensitive yeast *A. melanogenum* P16 strain (its genome size is 25.82 Mb) isolated from the mangrove ecosystem (Ma et al. 2014). We estimated that the two ancient progenitors diverged around 18.38 Mya and merged around 10.66–9.98 Mya (Jia et al. 2023). However, it is still not fully known about the hybrid/fusion process between the ancestor of *A. melanogenum* CBS105.22/CBS110374 and the ancestor of another unidentified strain of *A. melanogenum* similar to *A. melanogenum* P16 strain at the cellular and molecular levels, and there have been no experimental data to support the hypothesis to date.

Gostinčar et al. (2022) also found that different strains of *A. melanogenum* occasionally form stable and highly heterozygous diploid intraspecific hybrids by hybridization, but they did not carry out meiosis or haploidization, either. This may show that during the long-term evolution, the meiosis in *Aureobasidium* spp. was lost. In fact, this also happens in *Candida* spp. and other yeasts.

4.6 Subcellular Organelles and Their Functions

4.6.1 Extracellular Vesicles

The small membrane-bound structures released by prokaryotic and eukaryotic cells are called extracellular vesicles (EVs) which are spherical structures surrounded by a bilayered membrane. It has been reported that *A. pullulans* EXF-150 also can produce 6.1×10^8 nanoparticles (EVs) per milliliter of culture medium. Their sizes range between about 50 and 400 nm, and the average diameter of the nanoparticles was 171.4 nm (Černoša et al. 2022), while the average diameters of the non-melanin, melanin, and total EVs from *A. melanogenum* are 182.6 ± 31.2 nm, 166.7 ± 11.8 nm, and 167.0 ± 15.4 nm, respectively, and concentrations of them are $3.72 \times 10^6 \pm 2.6 \times 10^6$, $9.03 \times 10^7 \pm 11 \times 10^7$, and $6.09 \times 10^7 \pm 3.4 \times 10^7$/mL of the culture (Černoša et al. 2025). However, it is still unknown what functions the EVs have (Černoša et al. 2022). It has been reported that some strains of *A. melanogenum* that can produce a large amount of melanin are potential opportunistic pathogens to humans. However, there is no influence of a non-melanin pool (containing mainly EVs), a melanin pool (containing mainly melanin nanoparticles), and a total pool (containing both EVs and melanin nanoparticles) from *A. melanogenum* on human cell viability. Therefore, the opportunistic potential of *A. melanogenum* remains only partially understood, and the detailed functions of EVs produced by *Aureobasidium* spp. are still completely unknown (Černoša et al. 2025). Furthermore, it is still unknown how the EVs are synthesized and secreted by *A. pullulans* and *A. melanogenum*. It has been known that microvesicles, one kind of EVs, are generated by outward budding from the plasma membrane, followed by pinching off and release to the extracellular space, resulting in EVs ranging from 100 to 1000 nm in size (Silveira Freitas et al. 2019). *A. pullulans* and *A. melanogenum* may also synthesize and secrete their EVs in a similar way.

It has been reported that the EVs secreted by all the cells, including many fungal cells, have many functions. For example, EVs can communicate with neighboring cells in the environment. This communication and signaling can occur between cells of the same or different kingdoms. EVs contain complex cargo that can affect other cells—proteins, nucleic acids, lipids, glycans, polysaccharides, and pigments. Different molecules transported in EVs have diverse effects and functions, including virulence and immunomodulation. The proteins in the EVs may be involved in the methionine biosynthesis process, carbohydrate metabolism, stress response, lipid and protein metabolism, oxidation/reduction reactions, and translation. Fungal EVs are also involved in the synthesis and maintenance of the capsule and cell wall, which are important pathogenic factors for most fungal pathogenic species, as they ensure their survival in the host. EVs secreted by *Candida auris* also can induce resistance to antifungal drugs such as amphotericin B. The EVs from *C. albicans*, *C. neoformans*, *Histoplasma capsulatum*, and *Paracoccidioides brasiliensis* contain virulence factors such as various chaperones, enzymes (catalase, superoxide dismutase, laccase, urease), and sterols. Therefore, the fungal EVs are regarded as virulence bags. Fungal EVs also can carry immunogenic molecules; some of the immunogenic molecules in fungal EVs from *Malassezia sympodialis* can trigger immune reactions and can become allergens. There is also evidence that EVs from *C. neoformans, H. capsulatum,* and *P. brasiliensis* can act on mouse macrophages and induce inflammation. The host also can modulate the pathogenesis of fungal EVs from *H. capsulatum* through the immune system. Therefore, the secretion process, secretion mechanisms of the EVs, and the functions of the EVs secreted by *A. pullulans* and *A. melanogenum* need to be further elucidated.

4.6.2 Functions of Vacuoles in *Aureobasidium* spp.

The fungal vacuole is an extremely complex and versatile organelle, structurally similar to plant vacuoles and mammalian lysosomes (Fig. 4.9). They have a wide variety of functions, such as glycoprotein turnover and hydrolysis, the storage of Ca^{2+}, phosphate, and amino acids, pH regulation, ion homeostasis, cytoplasmic detoxification, and so on. Vacuoles also can be involved in the immediate response

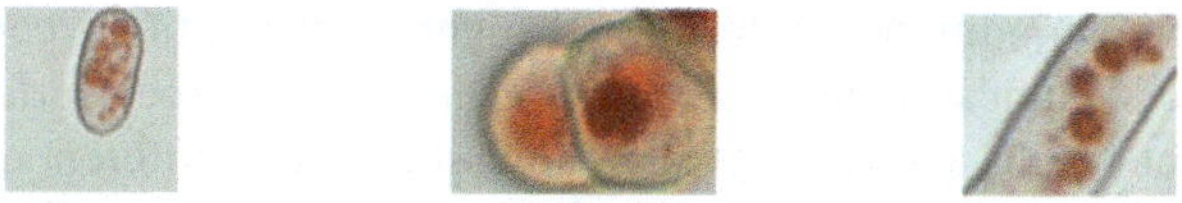

TN3-1 *ΔDV-5* EV-2

Fig. 4.9 The stained vacuoles of the wild-type strain TN3-1, the double mutant *ΔDV-5*, and the complementing strain EV-2

to osmotic shock by osmotic regulation. It can be observed that the numerous small vacuoles of *S. cerevisiae* and the fission yeast *S. pombe* can be fused into big vacuoles quickly in response to low osmotic pressure. This response is beneficial for rapid adaptation of fungi to hypotonic environments. Conversely, vacuolar fission for the formation of a large amount of small vacuoles takes place in response to high concentrations of salt and sugar (Weber 2002). It has been reported that Vsp1, a dynamin-like protein in *S. cerevisiae*, can play a key role in maintaining the fission/fusion balance of vacuoles (Rothlisberger et al. 2009). It has been confirmed that the double mutant *ΔDV-5* in which both the *VSP11* and *VSP12* genes were completely deleted have much bigger vacuoles and less numbers of vacuoles, lower pullulan production (7.4 ± 0.4 g/L), and more sensitivity to high glucose concentration (weak growth at 40% w/v glucose, no growth at 60% w/v glucose) than its complementing strain EV-2 (heavy growth at 60% w/v glucose, 70.0 g/L of pullulan) in which the *VSP11* gene was expressed and its wild type *A. melanogenum* TN3-1 (heavy growth at 60% w/v glucose,70.0 g/L of pullulan) (Chen et al. 2019). Even all the cells of the double mutant *ΔDV-5* are swollen (Fig. 4.9) and produce melanin (Chen et al. 2019). All the results strongly confirm that the vacuoles in *A. melanogenum* can play a crucial role in osmotic tolerance and pullulan biosynthesis and the osmotic tolerance can affect its pullulan biosynthesis. It has been confirmed that the morphological and metabolic changes of one organism, such as cell size, shape, colony formation, growth, metabolism, high pullulan production, and high glucose tolerance, are positively related to the genome duplication (Salmon 1997, Chen et al. 2020). So, the genome duplication, the big cell size, the thick cell wall, and the small and numerous vacuoles are also important for its high osmotic tolerance and high pullulan production by *A. melanogenum* TN3-1.

However, so far, it has been completely unknown about any other subcellular organelles and their functions in the yeast-like fungal cells of *Aureobasidium* spp. Therefore, the cell biology of *Aureobasidium* spp. is still an interesting topic that remains to be uncovered.

4.7 Polyextremotolerance of *Aureobasidium* spp.

Aureobasidium spp. are regarded as the polyextremotolerant yeast-like fungi (Gostinčar et al. 2019) because they can secrete different enzymes, use wide range of substrates and have wide product spectrum, tolerate extreme temperatures and pH, oligotrophic conditions, high salt and sugar concentrations, can live in humid and dry habitats, plant surfaces, coastal and deep seawater, rock surfaces, fuel tanks, wood or in-door areas and show great adaptability to the different environments as stated above. The ability for *Aureobasidium* spp. to tolerate such diverse stresses and wide habitats has been attributed to a high degree of genomic, metabolic, nutritional, and cell polymorphism plasticity via the production of different kinds of enzymes (amylases, cellulases, lipases, proteases, xylanases, mannanases, and killer toxins); microbial antagonism via the production of antimicrobial compounds, such as liamocin, Massoia lactone, siderophores, killer toxin, and antibiotics (Chi et al. 2009; 2012; Price et al. 2013; Moura et al. 2021, Kang et al. 2022); the production of several intracellular compatible solutes, such as trehalose, glycerol, and glycogen (Managbanag and Torzilli 2002; Turk and Gostincar 2018; Jiang et al. 2018a, b; Chen et al. 2020); biofilm formation (Metzger et al. 2006); melanization (Jiang et al. 2016); and exopolysaccharide biosynthesis (Jiang et al. 2019).

Different species of *Aureobasidium* spp. can tolerate extreme environmental conditions; *A. subglaciale* has superior resistance to UV light and heavy metals compared to any other yeasts and bacteria (Liu et al. 2017). *A. pullulans* and *A. mangrovei* are able to resist salt levels up to 17.0% (Gunde-Cimerman et al. 2000; Zalar et al. 2008). Even 20% of the cells of *A. melanogenum* can survive at 200 mM H_2O_2 (Jiang et al. 2016). Especially, the melanized cells of *A. melanogenum* strain XJ5-1 have higher tolerance to UV radiation, oxidation (200.0 mM H_2O_2), heat treatment (40 °C), salt shock (200.0 g/L NaCl), desiccation, and strong acid hydrolysis (6.0 M HCl) at high temperature (80 °C) than the non-melanized cells of the same yeast strain XJ5-1 (Jiang et al. 2016). This means that melanin in the cell walls has many physiological functions in the life of the black yeast *Aureobasidium* spp.

A. melanogenum TN3-1 is a high pullulan-producing and osmotic-tolerant yeast-like fungal strain, as shown in the last subsection. For example，*A. melanogenum* TN3-1 isolated from the natural honey with very high concentrations of glucose and fructose can grow on the plate even with 70% (w/v) glucose, whereas *A. melanogenum* P16 isolated from the mangrove ecosystem with only median salt concentrations only weakly grows on the plate with 20% (w/v) glucose, and more than 20% (w/v) glucose can completely inhibit its growth. This means that *A. melanogenum* P16 is much more sensitive to high concentrations of glucose than *A. melanogenum* TN3-1, and *A. melanogenum* TN3-1 indeed is a hyperosmotic tolerant yeast-like fungus. It was found that the HOG1 signaling pathway controlling glycerol synthesis, glycerol, trehalose, and vacuoles are closely related to high pullulan biosynthesis and high osmotic tolerance of *A. melanogenum* TN3-1 (Chen et al. 2020; Hohmann 2015). So, its double mutant *ΔDH-601* in which both the *HOG11* and *HOG12* genes, the key genes in the HOG1 signaling pathway, were totally removed, cannot grow at glucose concentration more than 50% (w/v) glucose, while their wild- type strain TN3-1 and the complementing strain EH-20 in which the *HOG11* gene was expressed still grow on the plate with 60% (w/v) glucose (Chen et al. 2020). It has been well known that glycerol 3-phosphate dehydrogenase encoded by *GPD1* gene in *S. cerevisiae* is important for yeast survival under osmotic stress, while the yeast glycerol 3-phosphatases Gpp1p and Gpp2p are also essential for glycerol biosynthesis. It also has been well documented that in *S. cerevisiae*, glycerol accumulation during osmo-adaptation is controlled mainly by the high osmolarity glycerol (HOG) signaling pathway. It has been reported that in yeast and filamentous fungi, glycerol is confirmed to be the most prominent solute that regulates cell turgor under high extracellular osmolarity conditions. Similarly, the double mutant *ΔDG-1*, in which both the *GPD1* and *GPD2* genes were totally abolished and glycerol content is zero, cannot grow at glucose concentration more than 40% (w/v) glucose, while their wild- type strain TN3-1 with 18.4 mg glycerol/mg of protein and the complementing strain EG-15 in which the *GPD1* gene was expressed and the glycerol content was 26.3 mg/mg of protein still grow on the plate with 60% (w/v) glucose. Therefore, *A. melanogenum* TN3-1, the hyperosmotic tolerant yeast-like fungus, can be used to commercially produce very high concentration (110.29 ± 2.17 g/L) of pullulan from 140.0 g/L of glucose, while *A. melanogenum* P16 isolated from the mangrove ecosystem transforms 140.0 g/L of glucose into only 45.81 ± 1.7 g/L under the same conditions (Xue et al. 2019). In contrast, when *A. melanogenum* P16 is grown the medium containing 120.0 g/L sucrose, 67.4 g/ L of pullulan is yielded (Ma et al. 2014). However, the price of sucrose is much higher than that of glucose, and glucose released from starch and cellulose is available in every corner of the world.

Trehalose, a non-reducing disaccharide composed of two glucose molecules linked at their 1-carbons can act as a reserve carbohydrate, a highly efficient protectant and a chemical chaperone is to enhance the resistance of cellular components against adverse conditions, such as high temperature, freezing, dehydration, desiccation as it can be involved in the formation of a glassy-matrix in the cytoplasm, which protects proteins and membranes and formation of hydrogen bonds of the polar head groups of phospholipids with the multiple hydroxyl groups of trehalose, resulting in a better stabilizing effect under such adverse physical conditions (Rubio-Texeira et al. 2016). For example, trehalose plays an important role in the adaptation of *A. melanogenum* XJ5-1 isolated from the Taklimakan Desert to heat shock treatment, high oxidation, high desiccation (Jiang et al. 2018a, b), high osmotic pressure, and high ethanol concentration by protecting biological macromolecules and the cell membrane from denaturation (Jiang et al. 2019; Chi et al. 2022). So, the mutant *ΔDT-1* in which both the *TPS11* and *TPS12* genes, the key genes for trehalose biosynthesis, were totally removed and trehalose content is zero, cannot grow at glucose concentration higher than 30% (w/v) while its wild-type strain TN3-1 with 17.1 mg of trehalose/mg of protein and its complementing strain ET-1 in which the *TPS11* gene was complemented and trehalose content is 35.1 mg/mg of protein can grow well at glucose concentration of 60% (w/v).

In addition, the produced pullulan (73.25 ± 2.3 g/L) by *A. melanogenum* 13-2 isolated from the Taklimakan desert also can resist against UV light, high NaCl concentration, heat treatment, strong oxidation of H_2O_2, and desiccation (Jiang et al. 2019). Even some strains of *A. melanogenum* can grow at temperatures up to 37 °C due to the higher amount of black pigment melanin than any other *Aureobasidium* species (Jiang et al. 2020), while *A. subglaciale* strains are able to grow at temperatures ranging from 0 to 30 °C, demonstrating that they prefer colder environments (Zajc et al. 2022). In addition, they also resist high salt concentrations, heavy-metal stress, and radiation, making them extremely robust against harsh surroundings.

The high phenotypic plasticity refers to the high ability of any organism to respond to environmental signals by altering cell morphology, cell metabolism, physiological state, or behavior. This ability can play an important role for the organism to adapt, survive, propagate, and evolve in diverse environments. In Sects. 4.1 and 4.2, *Aureobasidium* spp. have been proposed as a model for the investigation of fungal phenotypic plasticity because of the unique colony morphologies, the frequent changes in appearance and growth-temperature regimes, the different nutrient requirements, and cell morphological responses to environmental signals among its strains and varieties (Slepecky and Starmer 2009). As stated above, *Aureobasidium* spp. have yeast-like cells, swollen cells, septate swollen cells, black thick-walled chlamydospores, septate chlamydospores, and hyaline and black thick-walled hyphae. All the phenotypic plasticity and polymorphism of the fungal cells are also closely related to the polyextremotolerance of *Aureobasidium* spp. For example, the chlamydospores and the melaninized thick-walled hyphae are considered to be resistant to desiccation and ultraviolet irradiation (Figs. 4.2 and 4.3). It is believed that the resistant nature of chlamydospores and the melaninized thick-walled hyphae (Wu et al. 2023) is due to the highly accumulated melanin in their cell wall as well as other molecules (known as electron-dense granular material) that are present in the outermost wall layer and cross walls.

A. melanogenum W12 isolated from the polymer-coated surface of an aircraft has a dikaryon. It was found that its genome contains fatty acid transporters, cutinases, hydroxylases, and lipases potentially used for survival on the polymer coatings on aircraft (Wagner et al. 2024).

After the genomes of *A. pullulans*, *A. melanogenum*, *A. subglaciale*, *A. namibiae*, and *A. aubasidani* were sequenced and annotated, it was found that their genomes indeed contain all the genes that are responsible for stress tolerance, including genes encoding aquaporins, aquaglyceroporins, and alkali-metal cation transporters; enzymes for the synthesis of compatible solutes, exopolysaccharides, melanin, liamocins, Massoia lactone, killer toxin, antibiotics; and bacteriorhodopsin-like proteins (Xiao et al. 2025). In addition, different signaling pathways sensing and responding to various harsh environmental signals, such as Ca^{2+} signaling pathway, the cell wall integrity (CWI) signaling pathway, the target of rapamycin complex 1 (TORC1) signaling pathway, the high glycerol osmotic 1 (HOG1) signaling pathway, cAMP-PKA signaling pathway, Snf1/Mig1 signaling pathway, pH signaling pathway, and the heat shock factor 1 (HSF1) signaling pathway, are widely present in different strains of *Aureobasidium* spp. (Chi et al. 2022, Zhang et al. 2023).

One of the most important traits of *A. pullulans* is its halotolerance, as mentioned above. Such halotolerant fungi should have intracellular compatible solutes and a high and stable intracellular K^+ content in order to eliminate toxic Na^+. The sequenced genomes of 50 strains of *A. pullulans* indeed were found to encode a spectrum of different types of alkali metal cation transporters for uptake of K^+, and the P-type ATPase Na^+ exporter Ena is duplicated in most of the genomes.

During the experimental evolution by harsh stresses, the halotolerant black yeasts *A. pullulans* and *A. subglaciale* were grown at high NaCl or glycerol concentrations for 12 cultivation cycles, and it was found that all the evolved *Aureobasidium* spp. accelerate their growth at low water activity. Moreover, in the whole genomes of the evolved strains, no aneuploidies were detected in any of the genomes. However, several hundred single-nucleotide polymorphisms were identified compared with the genomes of the progenitor strains. Finally, it was found that most of the evolved *Aureobasidium* spp. strains have increased intracellular and decreased extracellular glycerol concentrations at high salinity, suggesting that the evolved strains have modified glycerol biosynthesis. Therefore, the experimental evolution can make *Aureobasidium* spp. from the extreme environments further improve their growth rates under the constant harsh conditions in a short time, but cannot make them have large-scale genomic rearrangements (Hariri Akbari et al. 2024).

4.8 Conclusions and Perspectives

As shown above, although the high cellular morphological and phenotypic plasticity and complex life cycle are considered to be closely related to their polyextremotolerances, the detailed molecular mechanisms of their polyextremotolerances are still unclear. Similarly, although many species of *Aureobasidium* spp. can have the unique multiple budding within one single cell cycle, the detailed molecular mechanisms of multiple budding within one single cell cycle and the physiological advantages of the multiple budding remain to be elucidated.

It is also completely unknown why *Aureobasidium* spp. lost meiosis and only asexual reproduction is kept during their evolution. If the possibilities for their genetic recombination and diversification by sexual reproduction could be verified, this would provide a bank of recombinant progeny, new recombinant products, and new recombinant enzymes suitable for various practical applications and basic biological investigation.

Therefore, *Aureobasidium* spp. are considered to be the non-model fungi for investigating cell biology, evolution, and synthetic biotechnology.

References

Bermejo JM, Dominguez JB, Goñi FM, Uruburu F (1981) Influence of carbon and nitrogen sources on the transition from yeast-like cells to chlamydospores in *Aureobasidium pullulans*. Antonie Van Leeuwenhoek 47(2):107–119

Brown RG, Hanic LA, Hsiao M (1973) Structure and chemical composition of yeast chlamydospores of *Aureobasidium pullulans*. Can J Microbiol 19(2):163–168

Campbell BS, Siddique ABM, McDougall BM, Seviour RJ (2004) Which morphological forms of the fungus *Aureobasidium pullulans* are responsible for pullulan production? FEMS Microbiol Lett 232(2):225–228

Černoša A, Gostinčar C, Lavrin T, Kostanjšek R, Lenassi M, Gunde-Cimerman N (2022) Isolation and characterization of extracellular vesicles from biotechnologically important fungus *Aureobasidium pullulans*. Fung Biol Biotechnol 9:1–17

Černoša A, Gostinčar C, Holcar M, Kostanjšek R, Lenassi M, Gunde-Cimerman N (2025) The impact of *Aureobasidium melanogenum* cells and extracellular vesicles on human cell lines. Scient Rep 15:1413

Chen L, Chi Z, Liu GL, Xue SJ, Wang ZP, Hu Z, Chi ZM (2019) Improved pullulan production by a mutant of *Aureobasidium melanogenum* TN3-1 from a natural honey and capsule shell preparation. Int J Biol Macromol 141:268–277

Chen L, Wei X, Liu GL, Hu Z, Chi ZM, Chi Z (2020) Glycerol, trehalose and vacuoles had relations to pullulan synthesis and osmotic tolerance by the whole genome duplicated strain *Aureobasidium melanogenum* TN3–1 isolated from natural honey. Int J Biol Macromol 165:131–140

Chi ZM, Wang F, Chi Z, Yue LX, Liu GL,Tong Zhang T (2009) Bioproducts from *Aureobasidium pullulans*, a biotechnologically important yeast. Appl Microbiol Biotechnol 82:793–804

Chi Z, Wang XX, Ma ZC, Buzdar MA, Chi ZM (2012) The unique role of siderophore in marine-derived *Aureobasidium pullulans* HN6.2. Biomet 25:219–230

Chi Z, Kong CC, Wang ZZ, Wang Z, Liu GL, Hu Z, Chi ZM (2022) The signaling pathways involved in metabolic regulation and stress responses of the yeast-like fungi *Aureobasidium* spp. Biotechnol Adv 55:107–898

Czachura P, Piątek M (2025) The genus *Aureobasidium* from sooty mould communities in Poland, Including *A. epipinicola* sp. nov. and *A. insectorum*, new to Europe. Plant Pathol 74:1593–1611

De Souza C, Osmani A, Hashmi S, Osmani S (2004) Partial nuclear pore complex disassembly during closed mitosis in *Aspergillus nidulans*. Curr Biol 14:1973–1984

Deshpande MS, Rale VB, Lynch JM (1992) *Aureobasidium pullulans* in applied microbiology: a status report. Enzy Microb Technol 14:514–527

Diver P, Ward BA, Cunliffe M (2025) Cell morphological plasticity in response to substrate availability of a cosmopolitan polymorphic yeast from the open ocean. Mycol 117(1):95–109

Gilbon A, Huitron C (1986) Scanning electron microscopy of a true cellulolytic strain of *Aureobasidium* grown on crystalline cellulose. Mycol 78(5):804–809

Goshima G (2022) Growth and division mode plasticity is dependent on cell density in marine-derived black yeasts. Gen Cell 27:124–137

Gostinčar C, Turk M, Zajc J, Gunde-Cimerman N (2019) Fifty *Aureobasidium pullulans* genomes reveal a recombining polyextremotolerant generalist. Environ Microbiol 21(10):3638–3652

Gostinčar C, Sun X, Černoša A, Fang C, Gunde-Cimerman N, Song Z (2022) Clonality, inbreeding, and hybridization in two extremotolerant black yeasts. GigaSci 11:giac095

Gunde-Cimerman N, Zalar P, de Hoog S, Plemenitaš A (2000) Hypersaline waters in salterns-natural ecological niches for halophilic black yeasts. FEMS Microbiol Ecol 32: 235–240

Hansali K, Wang P, Wei X, Liu GL, Chi Z, Chi ZM (2026) Dual role of the transcriptional factor NsdD in regulation of fumarate and melanin biosynthesis in *Aureobasidium aubasidani* GC1 strain. Fung Biol 130:101793

Hariri Akbari F, Song Z, Turk M, Gunde-Cimerman N, Gostinčar C (2024) Experimental evolution of extremotolerant and extremophilic fungi under osmotic stress. IUBMB Life 76(9):617–631

Heaton LL, Jones NS, Fricker MD (2020) A mechanistic explanation of the transition to simple multicellularity in fungi. Nat Commun 11:1–9

Hohmann S (2015) An integrated view on a eukaryotic osmoregulation system. Curr Genet 61:373–382

Humphries Z, Seifert KA, Hirooka Y, Visagie CM (2017) A new family and genus in *Dothideales* for *Aureobasidium*-like species isolated from house dust. IMA Fung 8(2):299–315

Jia SL, Zhang M, Liu GL, Chi ZM, Chi Z (2023) Novel chromosomes and genomes provide new insights into evolution and adaptation of the whole genome duplicated yeast-like fungus TN3–1 isolated from natural honey. Funct Integr Genom 23(3):206

Jiang H, Liu NN, Liu GL, Chi Z, Wang JM, Zhang LL, Chi ZM (2016) Melanin production by a yeast strain XJ5-1 of *Aureobasidium melanogenum* isolated from the Taklimakan desert and its role in the yeast survival in stress environments. Extrem 20:567–577

Jiang H, Liu GL, Chi Z, Hu Z, Chi ZM (2018a) Genetics of trehalose biosynthesis in desert-derived *Aureobasidium melanogenum* and role of trehalose in the adaptation of the yeast to extreme environments. Curr Genet 64:479–491

Jiang H, Xue SJ, Li YF, Liu GL, Chi ZM, Hu Z, Chi Z (2018b) Efficient transformation of sucrose into high pullulan concentrations by *Aureobasidium melanogenum* TN1–2 isolated from a natural honey. Food Chem 257:29–35

Jiang H, Chen TJ, Chi Z, Hu Z, Liu GL, Chi ZM (2019) Macromolecular pullulan produced by *Aureobasidium melanogenum* 13–2 isolated from the Taklimakan desert and its crucial roles in resistance to the stress treatments. Int J Biol Macromol 132:429–436

Jiang H, Chi Z, Liu GL, Hu Z, Zhao SZ, Chi ZM (2020) Melanin biosynthesis in the desert-derived *Aureobasidium melanogenum* XJ5-1 is controlled mainly by the CWI signal pathway via a transcriptional activator Cmr1. Curr Genet 66:173–185

Kang XX, Jia SL, Wei X, Zhang M, Liu GL, Hu Z, Chi ZM (2022) Liamocins biosynthesis, its regulation in *Aureobasidium* spp., and their bioactivities. Crit Rev Biotechnol 42:93–105

Klement T, Büchs J (2013) Itaconic acid–a biotechnological process in change. Biores Technol 135:422–431

Kockova-Kratochvilova A, Černakova AM, Slavikova E (1980) Morphological changes during the life cycle of *Aureobasidium pullulans* (de Bary) Arnaud. Folia Microbiol 25(1):56–67

Liu T, Zhu L, Zhang Z, Huang J, Zhang L, Jiang L (2017) Protective role of trehalose during radiation and heavy metal stress in *Aureobasidium subglacial* F134. Sci Rep **7**:17586

Lu Y, Wang H, Wang Z, Cong Y, Zhang P, Liu GL, Liu CG, Chi ZM, Chi Z (2019) Metabolic rewiring improves the production of the fungal active targeting molecule fusarinine C. ACS Synth Biol 8:1755–1765

Ma ZC, Fu WJ, Liu GL, Wang ZP, Chi ZM (2014) High-level pullulan production by *Aureobasidium pullulans* var *melanogenium* P16 isolated from mangrove ecosystem. Appl Microbiol Biotechnol 98:4865–4873

Managbanag JR, Torzilli AP (2002) An analysis of trehalose, glycerol,and mannitol accumulation during heat and salt stress in a saltmarsh isolate of *Aureobasidium pullulans*. Mycol 94:384

McNeil B, Kristiansen B, Seviour RJ (1989) Polysaccharide production and morphology of *Aureobasidium pullulans* in continuous culture. Biotechnol Bioeng 33(9):1210–1212

Metzger U, Lankes U, Hardy EH, Gordalla BC, Frimmel FH (2006) Monitoring the formation of an *Aureobasidium pullulans* biofilm in a bead-packed reactor via flow-weighted magnetic resonance imaging. Biotechnol Lett 28(16):1305–1311

Mitchison Field LMY, Vargas Muniz JM, Stormo BM, Lew DJ, Field CM, Gladfelter AS (2019) Unconventional cell division cycles from marine derived yeasts. Curr Biol 29:3439–3456

Moura VS, Pollettini FL, Ferraz LP, Mazzi MV, Kupper KC (2021) Purification of a killer toxin from *Aureobasidium pullulans* for the biocontrol of phytopathogens. J Basic Microbiol 61(2):77–87

Petruccoa CA, Crockerb AW, D'Alessandroa A, Medinac EM, Gormana O, McNeill J, Gladfelterd AS, Lew DJ (2024) Tools for live-cell imaging of cytoskeletal and nuclear behavior in the unconventional yeast, *Aureobasidium pullulans*. Mol Biol Cell 35:br10

Price NP, Manitchotpisit P, Vermillion KE, Bowman MJ, Leathers TD (2013) Structural characterization of novel extracellular liamocins (mannitol oils) produced by *Aureobasidium pullulans* strain NRRL 50380. Carbohyd Res 370:24–32

Rensink S, van Nieuwenhuijzen EJ, Sailer MF, Struck C, Wösten HAB (2024) Use of *Aureobasidium* in a sustainable economy. Appl Microbiol Biotechnol 108:202

Rothlisberger S, Jourdain I, Johnson C, Takegawa K, Hyams JS (2009) The dynamin related protein Vps1 regulates vacuole fission, fusion and tubulation in the fission yeast, *Schizosaccharomyces pombe*. Fung Genet Biol 46:927–935

Rubio-Texeira M, Van Zeebroeck MG, Thevelein JM (2016) Trehalose metabolism: enzymatic pathways and physiological functions. In: Hoffmeister D (ed) The Mycota III, Biochemistry and molecular biology, 3rd edn. Springer, Cham, pp 192–220

Salmon JM (1997) Enological fermentation kinetics of an isogenic ploidy series derived from an industrial *Saccharomyces cerevisiae* strain. J Biosci Bioeng 83:253–260

Samson RA, Houbraken J, Thrane U, Frisvad JC, Andersen B (2019) Food and indoor fungi. Westerdijk Fungal Biodiversity Institute, Utrecht, p 34

Silveira Freitas M, Deperon Bonato VL, Moreira Pessoni A, Rodrigues ML (2019) Fungal extracellular vesicles as potential targets for immune interventions. mSphere 4(6):e00747-19

Slepecky RA, Starmer WT (2009) Phenotypic plasticity in fungi: a review with observations on *Aureobasidium pullulans*. Mycol 101:823–832

Takeshita N, Manck R, Grun N, de Vega SH, Fischer R (2014) Interdependence of the actin and the microtubule cytoskeleton during fungal growth. Curr Opin Microbiol 20:34–41

Tong K, Bozdag GO, Ratcliff WC (2022) Selective drivers of simple multicellularity. Curr Opin Microbiol 67:102141

Turk M, Gostincar C (2018) Glycerol metabolism genes in *Aureobasidium pullulans* and *Aureobasidium subglaciale*. Fung Biol 122:63–73

Ukil L, De Souza C, Liu H, Osmani S (2009) Nucleolar separation from chromosomes during *Aspergillus nidulans* mitosis can occur without spindle forces. Mol Biol Cell 20:2132–2145

Wagner DN, Varaljay VA, Lyon WJ, Crouch AL, Allex-Buckner C, Bifffnger JC, Crookes-Goodson WJ, Kelley-Loughnane NK, Stamps BW (2024) Draft genome sequence of potentially dikaryotic black fungus *Aureobasidium melanogenum* isolated from aircraft. Environ. Microbiol 13(3):1–3

Wang K, Chi Z, Liu GL, Qi CY, Jiang H, Hu Z, Chi ZM (2020) A novel PMA synthetase is the key enzyme for polymalate biosynthesis and its gene is regulated by a calcium signaling pathway in *Aureobasidium melanogenum* ATCC62921. Intern J Biol Macromole 156:1053–1063

Weber RWS (2002) Vacuoles and the fungal lifestyle. Mycol 16:10–20

Wei X, Zhang M, Wang GY, Liu GL, Chi ZM, Chi Z (2023) The ornithine-urea cycle involves fumaric acid biosynthesis in *Aureobasidium pullulans* var. *aubasidani*, a green and eco-friendly process for fumaric acid production. Synth Syst Biotechnol 8:33–45

Wu F, Feng Z, Wang M, Wang Q (2023) Proposal of four new *Aureobasidium* species for exopolysaccharide production. J Fung 9(4):447

Xiao D, Driller M, Dielentheis-Frenken M, Haala F, Kohl P, Stein K, Lars M, Blank LM, Tiso T (2024) Advances in *Aureobasidium* research: Paving the path to industrial utilization. Microb Biotechnol 17:e14535

Xiao D, Driller, Stein K, Blank LM, Tiso T (2025) Genome mining the black-yeast *Aureobasidium pullulans* NRRL 62031 for biotechnological traits. BMC Genom 26(1):244

Xue SJ, Chen L, Jiang H, Liu GL, Chi ZM, Hu Z, Chi ZM (2019) High pullulan biosynthesis from high concentration of glucose by a hyperosmotic resistant, yeast-like fungal strain isolated from a natural comb-honey. Food Chem 286:123–128

Yang G, Liu GL, Wang SJ, Chi ZM, Chi Z (2020) Pullulan biosynthesis in yeast-like fungal cells is regulated by the transcriptional activator Msn2 and cAMP-PKA signaling pathway. Int J Biol Macromol 157:591–603

Zajc J, Černoša A, Sun X, Fang C, Gunde-Cimerman N, Song Z, Gostinčar C (2022) From glaciers to refrigerators: the population genomics and biocontrol potential of the black yeast *Aureobasidium subglaciale*. Microbiol Spect 10:e0145522

Zalar P, Gostinčar C, De Hoog GS, Uršič V, Sudhadham M, Gunde-Cimerman N (2008) Redefinition of *Aureobasidium pullulans* and its varieties. Stud Mycol 61:21–38

Zeng N, Zhang N, Wang D, Long J, Wang Y, Zhang Y, Pu F, Li Z, Babar Baloch F，Li B (2023) Regulation of cell differentiation to promote pullulan synthesis in *Aureobasidium pullulans* NG. Appl Microbiol Biotechnol 107: 6761–6773

Zhang M, Gao ZC, Chi Z, Liu GL, Hu Z, Chi ZM (2021a) cAMP-PKA and HOG1 signaling pathways regulate liamocin production by different ways via the transcriptional activator Msn2 in *Aureobasidium melanogenum*. Enzy Microb Technol 143:109705

Zhang M, Gao ZC, Chi Z, Wang Z, Liu GL, Li XF, Hu Z, Chi ZM (2021b) Massoia lactone displays strong antifungal property against many crop pathogens and its potential application. Microb Ecol 84:376–390

Zhang M, Wei X, Wang P, Chi Z, Liu GL, Chi ZM (2023) Liamocin biosynthesis is induced by an autogenous host acid activation in *Aureobasidium melanogenum*. Biotechnol J 23:e2200440

Chapter 5
Genomes and Genome-Editing Techniques of *Aureobasidium* spp.

Abstract The genomes of many species of *Aureobasidium* spp. have been sequenced, characterized, and annotated. The size of the genomes of haploid *Aureobasidium* spp. is about 26.0 Mbp large, while the size of the genomes of diploid *A. melanogenum* is about 51.0 Mbp. So far, the genomes of only some *A. melanogenum* strains have been found to be diploid. Although their genomes contain the genes encoding MATa and MATα, their sexual reproduction has not been discovered. All the haploid strains have 12 chromosomes, while the wholly duplicated genome of *A. melanogenum* contains 24 chromosomes. The GC contents of all the genomic DNAs of *Aureobasidium* spp. are around 50%. Three efficient genome editing techniques for *Aureobasidium* spp. have been well developed. All these are crucial for overproduction of industrial bioproducts by metabolic engineering and synthetic biology, elucidation and molecular editing of synthetic pathways, and clarification of regulation mechanisms of their multiple fermentation products by genome editing.

Keywords Genomic DNA · Genome editing · Transformation · Haploid · Diploid

5.1 Introduction

Many strains of *Aureobasidium* spp. have been defined as a polyextremotolerant generalists because they are able to colonize vastly different ecological niches around the world, especially harsh environments. They also have high adaptive capacity: biofilm formation, secretome composition, extraordinary morphology, phenotypic and genotypical plasticity and variability, production of pullulan, polymalate, liamocin, fumarate, intracellular lipid, β-glucan, siderophores and antibiotics, and competition for nutrients as stated in the last chapter. Therefore, the sequencing and annotation of more wild-type *Aureobasidium* spp. genomes can lead to important discoveries about their origin, adaptation, and evolution and increase in the repertoire of phenotypic and genetic diversity of the unconventional yeasts. *A. pullulans* and *A. melanogenum* among *Aureobasidium* spp. are the most

Z. Chi et al., *Biology and Synthetic Biotechnology of Aureobasidium spp.*,
https://doi.org/10.1007/978-981-92-2607-8_5

studied species with currently over 80 published genomes and are the best known for producing pullulan. In this chapter, the available genomes of *Aureobasidium* spp. and their genome editing techniques are summarized and discussed.

5.2 General Characteristics of the Genome of *Aureobasidium* spp.

Usually, the size of the genomes of *A. melanogenum* is about 26.0 Mbp in the case of a haploid strain and about 51.0 Mbp in the case of a diploid strain. Now, it is known that the genomes of only some *A. melanogenum* strains are diploid. It was found that the diploid strains are highly heterozygous hybrids of phylogenetically relatively distant haploids and a consequence of occasional hybridization between relatively heterozygous haploids. But this did not happen to *A. pullulans*, which is closely related to *A. melanogenum.* For example, the genome sizes of *A. pullulans*, haploid *A. melanogenum*, diploid *A. melanogenum*, *A. subglaciale*, *A. namibiae*, *A. aubasidani*, *A. hainanensis*, *A. mustum*, *A. vineae*, and *A. uvarum* are 28.1 Mb, 26.1, 50.1 Mbp, 29.6 Mbp, 25.4 Mbp, 30.8 Mbp, 24.4 Mbp, 27.2 Mbp, 25.9 Mbp, and 25.6 Mbp, respectively (Li et al. 2025; Onetto et al. 2020), suggesting that the mean genome size of the diploid strain of *A. melanogenum* strains is the highest while some strains of *A. melanogenum*, *A. pullulans*, *A. subglaciale*, *A. namibiae*, *A. aubasidani*, *A. hainanensi*, *A. mustum*, *A. vineae*, and *A. uvarum* are haploid. However, among the haploid strains of *Aureobasidium* spp., the genome sizes of *A. pullulans*, *A. subglaciale*, and *A. aubasidani* are the highest. This may be related to the most halotolerance (osmotic tolerance) of *A. pullulans*, the low temperature tolerance of *A. subglaciale*, and the biosynthesis of β-glucan and fumaric acid of *A. aubasidani* because some strains of *S. cerevisiae*, the extremely halotolerant black yeast *Hortaea werneckii* and *A. melanogenum* TN3-1 isolated from honey which have experienced recent whole-genome duplication also increase their genome sizes when they respond to high concentrations of salt and high concentration of sugar (Lenassi et al. 2013; Jia et al. 2023). It was found that all the haploid strains always are limited to clonal reproduction (Černoša et al. 2021), while some diploid hybrids have always been very stable, allowing them to disperse over long distances and constitute a large proportion of the species in some harsh habitats (Gostincar et al. 2022). Why do only some strains of *A. melanogenum* have a wholly duplicated genome and are the diploids formed by hybridization of haploid? One reason for this is that they are obtained from harsh environments, such as natural honey and aircraft. Another reason for this is that most strains of *Aureobasidium* spp. obtained from natural environments so far have been *A. melanogenum.* Therefore, after more genomes from *A. pullulans* and any other species of *Aureobasidium* spp. are isolated, sequenced, and annotated, the diploid strains of any other species may also be found. But why do not the diploid strains whose genomes contain the genes encoding MATa and MATα undergo meiosis to form haploid cells by sexual reproduction, as stated in the last chapter, and why are not the haploid strains hybridized to form diploid strains, as shown in Fig. 4.8 in the last chapter? These two questions must be answered and explained in the near future, regarding genome editing and synthetic biotechnology. It has been known that such a reproductive strategy also happens to *Candida parapsilosis, Candida orthopsilosis*, and *Candida metapsilosis* because of their absence of meiosis. But it is still completely unknown about meiosis in *Aureobasidium* spp. Maybe meiosis in *Aureobasidium* spp. was also lost during evolution, as mentioned in the last chapter. This must be elucidated by mining all the genes in the genomes related to meiosis and by editing the genomes by synthetic biology. The fact that most of the haploid strains cannot be hybridized to form diploid strains may be due to that the genes cannot encode the active MATa and MATα must be clarified by analyzing, codon-optimizing and expressing the active *MATa* and *MATα* genes.

The gene numbers in the genomes of *A. pullulans*, haploid *A. melanogenum*, diploid *A. melanogenum, A. subglaciale, A. namibiae, A. aubasidani, A. hainanensis, A. mustum, A. vineae*, and *A. uvarum* are 10,646, 9789, 17,863, 10,809, 10,266, 10,596, 9089, 10,353, 9742, and 9831 (Xiao et al. 2025; Jia et al. 2023; Onetto et al. 2020), and GC content (%) of almost all the genomes of *Aureobasidium* spp. is around 50%.

All the haploid strains of *Aureobasidium* spp. such as high pullulan-producing *A. melanogenum* P16 isolated from the mangrove ecosystem, *A. mustum* isolated from grape juice, and *A. pullulans* NBB 7.2.1 have 12 chromosomes. But the wholly duplicated genome of *A. melanogenum* TN3-1 isolated from natural honey and other diploid strains of *Aureobasidium* spp. contains 24 chromosomes.

The genome (accession number: QNUU00000000) of *A. pullulans* strain P25 producing high level of gluconic acid has a total assembly size of 30.97 Mb, 459 scaffolds, 495 contigs, 50.28% of G-C content, and 10,922 genes (Zhao et al. 2019). The analysis of the genomic sequence indicates that the gluconic acid biosynthesis pathway of the strain P25 is the same as that in any other fungi.

The genome of *A. aubasidani* CBS 100524 (the accession no. JADGIM000000000) consists of 30.3 Mbp, which GC content is 50.50% and has 10,978 genes (Vignolle et al. 2021). The size of the genome (the GenBank accession number is JADDKP000000000) of the high fumarate-producing *A. aubasidani* DH177 is 31.34 Mbp, and its GC content is 50.3% (Wei et al. 2023). The whole genome of the DH177 strain contains 10,215 encoding genes, and the average length of each gene is 1513 bp. However, there is only one copy of the *FUM* gene encoding fumarase in its genome. Therefore, it is thought that there is no cytoplasmic reductive TCA pathway for fumaric acid biosynthesis in the DH177 strain. This is one of the reasons why fumaric acid is biosynthesized through OUC pathway in *A. aubasidani* as stated in Chap. 8.

The whole genome size (the accession number: RZIQ01000000) of *A. hainanensis* sp. nov. P6 isolated from the mangrove ecosystem is 24.0 Mbp, and the G + C content of its genome is 50.69%. Totally, 6098 protein-coding genes and 264 non-coding RNA genes were predicted. The yeast-like fungal strain P6 has the complete metabolic pathways of TCA cycle, EMP pathway, pentose phosphate pathway, glyoxylic acid cycle, and other central metabolic pathways as mentioned in Chap. 3. It can be clearly observed that the genome size (24.0 Mb) of *A. hainanensis* sp. nov. P6 is the smallest among those of any other haploid strain genomes of *Aureobasidium* spp. with the published genome sequences of *A. pullulans* P25 (30.97 Mbp) (Zhao et al. 2019), *A. pullulans* (29.62 Mbp), *A. melanogenum* (26.2 Mbp), *A. subglaciale* (25.8 Mbp), *A. namibiae* (25.43 Mbp) (Gostinčar et al. 2014), and *A. pullulans* AY4 (26.72 Mbp) (Chan et al. 2012), but they have the similar G + C content of around 50%. This means that the strain P6 draft genome is different from that of any other strains of *Aureobasidium* spp. and is identified to be a new species (Jia et al. 2019).

A. pullulans strain NBB 7.2.1 is strongly antagonistic against several plant pathogens and highly competitive against other yeasts on apple fruits. It can inhibit *F. oxysporum* NRRL 26381/CL57 more strongly than *A. melanogenum* CBS 110374, *A. namibiae* CBS 147.97, and *A. subglaciale* EXF-2481. *A. pullulans* NBB 7.2.1 genome has a size of 28.41 Mbp with 10,925 genes and consists of 12 chromosomes, and a circular, complete mitogenome (37563351128 bp (Rueda-Mejia et al. 2021).

The genome size (accession number is JAULBJ000000000) of *A. melanogenum* W12 isolated from the surface of an aircraft is 53.2 Mbp with 50.13% G + C content. This strain is a dikaryon (Fig. 5.1) and may also have 24 chromosomes, resulting in the large genome size compared to those of any other *Aureobasidium* species and the large number of 90.8% duplicated orthologs (Wagner et al. 2024).

Because the genomes (the size is about 50 Mb, over 95.4% genes are duplicated; and only over 3.6% of the genes is one copy gene) of *A. melanogenum* FSL-S8-0006 and *A. melanogenum* NRRL YB-395 isolated from caramel sauce and pineapple are also duplicated, they are diploid. In contrast, because the genomes (the sizes are 29.4 Mbp, 27.1 Mbp, and 26.8 Mbp, and 98.8%, 99.1%, and 92.3% of the genes are one copy gene) of *A. pullulans* NRRL 62816, *A. melanogenum* NRRL Y-12974, and *A. melanogenum* NRRL Y-9624 isolated from plum seaweed plant and seawater have only one copy of each gene, these strains are haploid. All four *A. melanogenum* strains produce melanin, and their cultures become more pigmented as the cultures

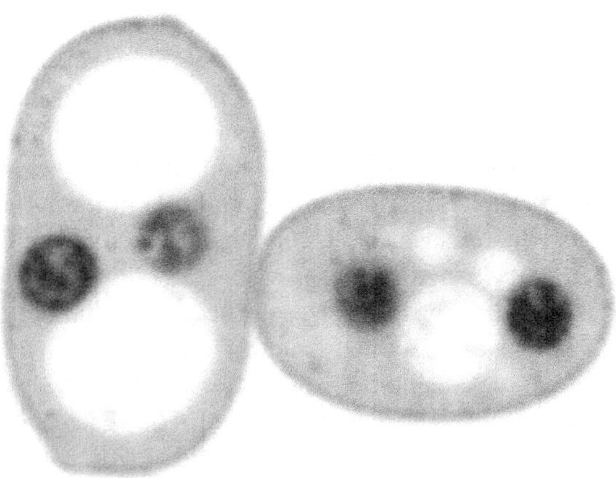

Fig. 5.1 The dikaryon cells of *A. melanogenum* W12 (Wagner et al. 2024)

become old. By contrast, *A. pullulans* NRRL 62816 has pinkish cultures and remains pink under the same conditions. However, the copy number of stress tolerance genes was found not to be correlated with survival of food processing treatments among the *Aureobasidium* strains (Cai and Snyder 2022).

Usually, many genes, such as melanin synthesis-related genes, polymalate synthesis-related genes, pullulan biosynthesis-related genes, and Liamocin synthesis-related genes in different strains of *Aureobasidium* spp., are clustered together because the clustered genes have many advantages, such as co-evolution and efficient expression under the control of the specific transcriptional activators. For example, the Pks for melanin biosynthesis in *A. melanogenum* XJ5–1 isolated from the Taklimakan desert belongs to non-reducing PKS (NR-Pks) and has many domains of a β-ketoacyl synthase (KS), an acetyl transferase (AT), the two acyl carrier proteins (ACP), a thioesterase (TE), and a cyclase (CYC). Many genes responsible for the DHN (a dihydroxynaphthalene)-melanin biosynthesis) are physically linked together, which is called the *PKS* gene cluster. Expression of some of the *PKS* gene clusters is regulated by a melanin-specific transcriptional activator Cmr1 (Jiang et al. 2020), and expression of some other *PKS* gene clusters is regulated by other transcriptional factors. For example, expression of the *HR-PKS* gene cluster for liamocin biosynthesis is controlled by a liamocin-specific transcriptional activator Gal1 (Xue et al. 2019). The *PMAs* gene encoding NRPS for polymalate biosynthesis is governed by the transcriptional activator Crz1 in the Ca^{2+} signaling pathway (Wang et al. 2020).

The repetitive DNA sequences in the genome are classified into two categories according to their complexity and multitude: a: interspersed repeats, which are scattered throughout the genome, and b: tandem repeats, which are repeated in an organized pattern in one localized region. The tandem repeats mainly include microsatellites, simple sequence repeats (2–6 bases as a repeating unit), minisatellites (10–100 bases of long sequences as a repeating unit), and others. The genome of *A. pullulans* contains 54 small RNAs, 2615 simple repeats, and 387 sequences with low complexity, but has no satellite RNA sequences, short interspersed nuclear elements and interspersed repeats, long interspersed nuclear elements, and long terminal repeats and DNA transposon (Xiao et al. 2025), whereas the whole genome duplicated chromosomes of *A. melanogenum* TN3-1 isolated from honey harbor a high level of transposable elements as shown below (Jia et al. 2023).

The genomes of different *Aureobasidium* species harbor a large diversity of genes, including abundance of different families of carbohydrate-active enzymes, sugar transporters, and genes associated with polyextremotolerance (Gostincar et al. 2014). Most of the genes in the genomes of *Aureobasidium* species are responsible for starch, cellulose and sucrose, and other carbohydrate transport and metabolic process, nitrogen compound transport and metabolic process, lipid transport and metabolic process, cellular metabolism, amino sugar and nucleotide sugar metabolisms, glycolysis/gluconeogenesis, butanoate metabolism, and the central intracellular metabolism such as TCA cycle, glyoxylate shunt, and pentose phosphate pathway and amino acid transport and metabolism. Furthermore, a large number of the genes in the genomes of *Aureobasidium* species are involved in signal transduction. This again demonstrates that *Aureobasidium* spp. indeed have evolved to feature various signaling systems for the utilization of many substrates and the adaptation to harsh environments as mentioned above.

So far, it has been found that *A. fermentans* and *A. essambei* weakly ferment D-glucose. *A. peruviana* ferments glucose, sucrose, and inulin, while *A. fermentans*, *A. essambei A. vanuatuense*, and *A. albui* can ferment sucrose, with only *A. albui* capable of fermenting melibiose. In contrast, *A. pullulans* and *A. melanogenum* and any other species cannot ferment glucose. This means that the genomes of *A. pullulans* and *A. melanogenum* do not contain any genes for fermentation, while those of *A. fermentans*, *A. essambei*, *A. vanuatuense*, *A. albui*, and *A. peruviana* carry the genes for fermentation. However, it is still completely unclear which genes in their genomes are responsible for sugar fermentation and what fermentation products can be produced by such fermenters of *Aureobasidium* spp.

The genome of *A. pullulans* CRUB 182 isolated from an area exposed to high UV radiation contains the genes for photoprotection and antioxidant activity, such as the mycosporine-glutaminol-glycoside gene cluster responsible for the synthesis of UV sunscreen compounds, the genes for carotenoid synthesis and some ROS scavenging enzymes, the core genes for the circadian clock, eight light-induced genes, and five sporulation genes (Parra et al. 2023).

The genome of *A. pullulans* NRRL 62031 carries the genes for 17 β-glucosidases, 11 endoglucanases, 3 cellobiohydrolases, 8 xylanases, 2 β-xylosidases, 5 α-amylases, 11 glucoamylases, 1 polygalacturonase, 8 pectinesterases, 1 pectin lyase, 2 cutinases, lipase and laccase biosynthesis and melanin, pullulan, polymalate, liamocin, yanuthone D, burnettramic acid A, and choline biosynthesis and their transcriptional regulation (Xiao et al. 2025). This means that *A. pullulans* has a full of biosynthetic pathways. *Aureobasidium* sp. isolated from the Elizabel river sediment has polycyclic aromatic hydrocarbon (PAH) degradation ability. Its genome size is 27, 536, 729 bp with 23 contigs, N_{50} of 2,191, 637 bp and a GC content of 50.05% and has 9,629 genes encoding the common PAH degradation enzymes (Babusci and Gunsch 2026). The genome size of *A. melanogenum* DAP1, which can degrade aflatoxin FB1 (AFB1) and inhibit *A. falvus* growth, is 24 Mbp with a GC content of 50.1%. Its genome contains putative AFB1 detoxifying genes including the genes encoding cytochrome P450 monooxygenase (CYP531), glutathione transferase (GST), epoxize hydrolase (EPHX1) and aflatoxin B1 aldehyde reductase (AKR7) (Harahap et al. 2026).

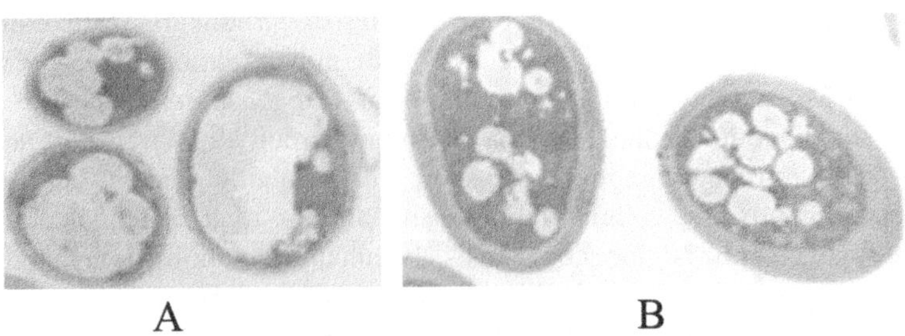

Fig. 5.2 The microstructures of the yeast-like fungal cells of *A. melanogenum* P16 (**A**) and *A. melanogenum* TN3-1 (**B**) observed under a transmission electron microscope

5.3 The Novel Characteristics of the Duplicated Genome of *A. melanogenum* TN3-1

As mentioned above, *A. melanogenum* TN3-1 strain isolated from the natural honey produces over 110 g/L pullulan from 140 g/L glucose while *A. melanogenum* P16 strain isolated from the mangrove ecosystem only can produce 45.81 ± 1.7 g/L pullulan from 140.0 g/L of glucose because *A. melanogenum* TN3-1 strain can tolerate higher osmotic pressure (high glucose concentration) due to its bigger cells, thicker cell wall, and smaller vacuoles than *A. melanogenum* P16 strain (Fig. 5.2) (Xue et al. 2019). So, in this subsection, the genomes of *A. melanogenum* TN3-1 and *A. melanogenum* P16 being used as a representative of the osmotic tolerant strain and a representative of the osmotic sensitive strain, respectively, are compared and discussed here.

The genome size (the accession number: CP061980-CP062003) of *A. melanogenum* TN3-1 is 51.61 Mbp with the contig N50 of 2.19 Mbp and 24 haploid chromosomes, while that (the accession number: CP061917-CP061928) of *A. melanogenum* P16 is 25.82 Mbp with the contig N50 of 2.26 Mbp and with 12 haploid chromosomes. The genomes of *A. melanogenum* TN3-1 are highly duplicated with 95.9% being duplicated genes, while the genome of *A. melanogenum* P16 is not highly duplicated, with 99.3% being only single-copy genes and cannot tolerate high osmotic pressure (Chen et al. 2019). Therefore, the genome size of *A. melanogenum* TN3-1 is biggest among those of the 43 representative Dothideomycetes genomes (average genome size was 36.75 Mbp), 50 *A. pullulans* genomes (the average size was 28.04 Mbp) (Gostinčar et al. 2019), and any other strains of *Aureobasidium* spp. Furthermore, the gene numbers and the DNA repeats in the TN3-1 strain are also the highest among all the strains of *Aureobasidium* spp.

In contrast, *A. melanogenum* OUC strain, whose genome (the accession number: JACXJB000000000-J) is also highly duplicated, is a non-pullulan producer with low stress resistance (Chen et al. 2020b), but can produce high-molecular-weight (> 4.9×10^5 Da) polymalate with the titer 34.6 g/L (Qi et al. 2020), while *A. melanogenum* ATCC62921, whose genome (the accession number: GCA_016807825.1) is not duplicated, produces low-molecular-weight (3.8×10^4 Da) polymalate with the titer 53.3 g/L (Wang et al. 2020). Thus, *A. melanogenum* TN3-1 must be a highly stress-tolerant strain with the highest pullulan-producing ability, whereas *A. melanogenum* P16 strain and ATCC62921 strain are not. Although *A. melanogenum* OUC strain and ATCC62921 strain are similar, they have completely different phenotypic characteristics, which may also be caused by the genome duplication of *A. melanogenum* OUC strain (Qi et al. 2020; Wang et al. 2020). It was found that all the genomes of *A. melanogenum* TN3-1, *A. melanogenum* P16, *A. melanogenum* OUC strain, and ATCC62921 strain contain the *AmAGS2* gene encoding the multidomain AmAgs2 (α-glucan synthase 2) for cell wall α-1,3 glucan biosynthesis (Chen et al. 2019). All the multidomain AmAgs2 contain the amylase domain (Amy_D), the Big_5 domain, the glycogen synthetase domain (Gys_D), and the transmembrane regions in which the exopolysaccharide transporter domain

(EPST_D) is embedded (Chen et al. 2019). However, because only one copy of the two genomes of *A. melanogenum* TN3-1 and the genome of *A. melanogenum* P16 contains the *AmAGS2* gene encoding the Big_5 domains with the conserved amino acids sequence LQS, these two strains can produce pullulan, whereas in the genomes of *A. melanogenum* OUC strain and ATCC62921 strain, the *AmAGS2* gene encoding the Big_5 domains has the conserved amino acids sequence LET so that they cannot produce any pullulan, but polymalate (Qi et al. 2020; Wang et al. 2020). Like carbohydrate binding module (CBM), the importance of the Big_5 domain is to enhance the binding affinity of an enzyme to its substrate; it is implicated in the attachment and positioning of the catalytic module of the enzyme on the surface of the carbohydrate substrate residues, and it is essential for maintaining the normal function of the enzyme (Han et al. 2013). Furthermore, the binding affinity, attachment, and positioning represent a novel strategy for polysaccharide degradation. So, the conserved amino acids LQS of the Big_5 domain was thought to be required for pullulan biosynthesis, probably due to its enhancing the AmAgs2 specifically to be bound to the pullulan precursor and promoting the Amy_D of the AmAgs2 to cut the maltotriose from the pullulan precursor and transfer the released maltotriose to be connected together to form pullulan molecules on Lph-glucose according to the pullulan synthetic pathway, which will be discussed in the Chap. 7 (Chen et al. 2020a).

Because a recent cell fusion or possible hybridization of the different genomes happened in the ancestor of the TN3-1 strain, the genomes (24 chromosomes) of the TN3-1 strain are divided into two subgenomes A and B based on the average nucleotide identity (ANI) as mentioned in Chap. 3. The genome size, gene content, and gene density of the subgenome A are slightly higher than those of the subgenome B, while the transposable element (TE) density in the subgenome A is lower than that of the subgenome B, indicating there is an asymmetry of the genome content of the TN3-1 strain. Finally, the ancestor of the subgenome A in the TN3-1 strain was thought to come from *A. melanogenum* OUC strain or *A. melanogenum* CBS110374 strain, while the ancestor of the subgenome B was regarded to come from the ancestor of an unknown strain similar to *A. melanogenum* P16 (Jia et al. 2023). This is due to the fact that the subgenome A of the TN3-1 strain is closely related to the genomes of the OUC strain or the CBS110374 strain, whereas the subgenome B of the TN3-1 strain is far from their genomes. Furthermore, there is no obvious interchromosomal translocation that happens between the P16 genome and the subgenome A of the TN3-1 strain. In contrast, the chromosome pairs between the P16 genome and the subgenome B of the TN3-1 strain have conserved synteny. Moreover, the active α-1, 3-glucan synthase (AmAgs2) encoded by both the genome of the P16 strain and the subgenome B of TN3-1 strain contains the same Big_5 domain with the conserved amino acids LQS as stated above. So, the whole-genome duplication from different sources in the TN3-1 strain genomes does enable its rapid adaptation to new environments, such as bee honey, where it was isolated. Furthermore, the whole-genome duplication from different sources in the TN3-1 strain genomes promotes its genome and gene diversity and evolution during the long-term post-WGD (whole genome duplication) evolution. This means that the genomes from different sources in one cell are helpful for its genome evolution. This may also be an example of heterosis in organisms.

In contrast, the two subgenomes of the OUC strain mentioned above have a higher genome identity (95.84%) than that (82.17%) of the TN3-1 strain. This may be due to the fact that the two subgenomes in the OUC strain were formed by direct genome doubling without separation after its cell division at ancient time. Therefore, the rapid adaptation does not happen to the OUC strain due to its direct doubling of the same genome. In addition, the rapid adaptation does not happen to the P16 strain which genome is not duplicated, either. This again confirms that the genomes from the same sources in one cell are not helpful for its genome evolution and adaptivity.

Centromeres are a complex of centromere DNA (CEN DNA) and the specific factors that help mediate microtubule-dependent movement of chromosomes during mitosis. The centromere is usually composed of three conserved DNA sequences, but only CDE I and CDE III occur in most of the fungi (Chaitanya 2019). It was found that there are four centromeres which are localized close to the central regions on chr2, chr19, chr23, and chr24 of the TN3-1 strain, while seven centromeres seem to have deviated from the central locations. In contrast, there are three centromeres localized close to the central regions on chr4, chr5, and chr6 of the P16 strain. This means not all the chromosomes in the TN3-1 strain have centromeres.

Telomere, which has the conserved repeat sequences "TTA GGG /CCC TAA," is importantly used as a protective structure at the end of the chromosome (O'Sullivan and Karlseder 2010). To mitigate telomere sequence loss during cell division, the telomerase can prevent any detrimental effects on genome stability triggered by excessive telomere shortening. In the genomic DNAs of the TN3-1 strain, both ends of 13 chromosomes and one end of 8 chromosomes of the TN3-1 strain have the telomeric repeats "TTA GGG /CCC TAA," while the ends of only chr5, chr7, and chr8 of the P16 strain have the telomeric repeat "TTA GGG" sequence. Indeed, each chromosome from *Aspergillus* spp., most of any other fungi, and humans also contains such conserved core "TTA GGG /CCC TAA" as the telomeric repeat sequences. Although some chromosomes of the TN3-1 strain have no "TTA GGG /CCC TAA" repeats, each chromosome contains at least one or more long-interspersed nuclear elements (LINEs). LINE is a retrotransposon that self-replicates, and its uncontrolled expression has the potential to induce genomic instability. Besides, there are more LINEs distributed on the chromosome ends of the TN3-1 strain than those distributed on the chromosome ends of the P16 strain, *S. cerevisiae* S288C, *S. pastorianus* CBS1483, and *Z. rouxii* ATCC42981 (Jia et al. 2023). In addition, only one type of telomerase encoding gene exists on the chr5 and chr13 of the TN3-1 strain, while the telomerase encoding gene was lost at the ends of most of its chromosomes. This may be related to gene diversity, rapid genome evolution, and high stress tolerance of the TN3-1 strain to high osmotic pressure (over 60% glucose) as mentioned above.

Because of such specific chromosomes as stated above, different strains of *Aureobasidium* spp., even different strains of the same species of *Aureobasidium* spp., have an extreme variability and genetic instability and could be easily adapted to various harsh environments by rapid genome evolution.

Transposon is a transposable segment of DNA that is capable of inserting copies of itself into other DNA sites along a chromosome within the same cell. It is also found that the number of transposons scattered on different chromosomes of the TN3-1 strain is more than that of the transposons scattered on different chromosomes of the P16 strain. There are many intrachromosomal rearrangements, including some translocations, and a large number of inversions between the two subgenomes of the TN3-1 strain because of the high number of transposons on different chromosomes of the TN3-1 strain. This may lead to the high numbers of transposons and LINEs in the genome of the TN3-1 strain, as mentioned above.

The transposable elements (TEs) have been reported to be DNA sequences that can change their position within a genome, sometimes creating or reversing mutations and altering the cell's genetic identity and genome size. In the genomes of the TN3-1 strain, there are more transposable elements (TEs) than those in the genomes of the P16 strain, the OUC strain, *A. melanogenum* CBS110374, *A. pullulans* EXF-150, and most of any other strains of *Aureobasidium* spp. These could be regarded as the source of new genes, gene variation, and ultimately biological evolution by insertion of a cDNA into the genome. Therefore, more TEs in the genomes of the TN3-1 strain might also cause rapid genome evolution, precise regulation of the differential gene expression, and excellent adaptation of the TN3-1 strain to the harsh environments, such as honey for high levels of pullulan production.

All the chromosomal translocations of the TN3-1 strain mentioned above might result from natural selection under the high osmotic pressure in the honey and might lead to its genetic diversity and high osmotic tolerance. The hypertonic environment such as the natural honey might also cause polyploidization and hybridization of its genomes. The same stories also happen to the genomes of the extremely halotolerant *H. werneckii* isolated from the extreme environments and the genomes of *Coccidioides immitis*. Therefore, the genomes of *A. melanogenum* TN3-1 indeed have many unique characteristics that make it have high tolerance to harsh environments and high pullulan production.

5.4 Mitochondrial Genome of *A. pullulans*

Mitochondria are important eukaryotic organelles for ATP production by oxidative phosphorylation, ion homeostasis, cell senescence, and apoptosis. The typical fungal mitogenomes (Fig. 5.3) consist of 14 core protein-coding genes that are related to the electron transport chain which are ATP synthase subunits and apocytochrome b, NADH dehydrogenase subunits, cytochrome oxidase subunits and a large subunit, a small subunit of the ribosomal RNA gene, different tRNA genes, ribosomal protein S3 and the mitochondrial RNA subunit RNase P which is involved in tRNA processing.

The mitochondrial genome of *A. pullulans* isolated from the leaves of the endemic tree *N. pumilio* in Otto Hill on the southern side of Lake Nahuel Huapi in the city of San Carlos de Bariloche, Rio Negro, Patagonia, Argentina, is a 40.59-Kb

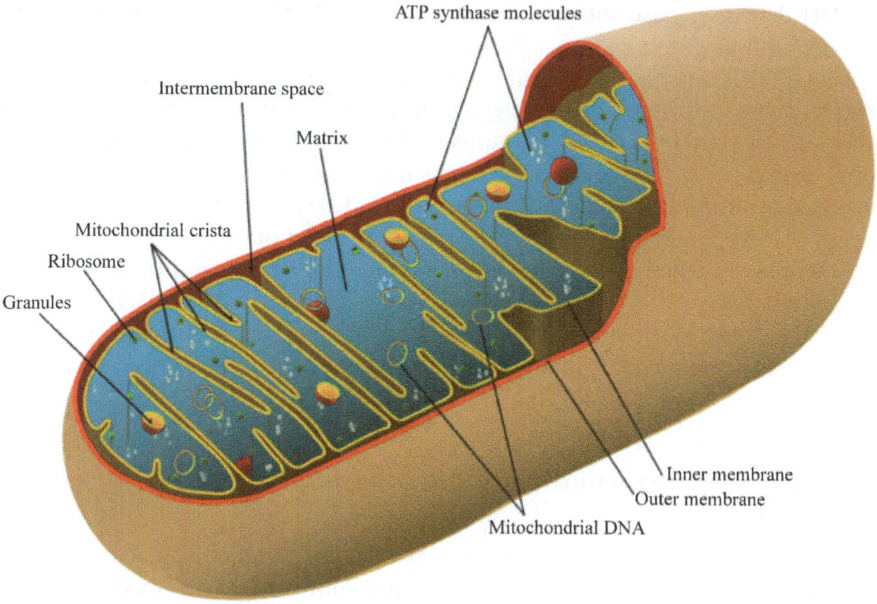

Fig. 5.3 Map of fungal mitochondrion and its circular genomic DNA

complete circular genome with a GC content of 30.50%. The genes include the genes encoding complex I (NADH dehydrogenase subunit 1 encoded by the *NAD1* gene, NADH dehydrogenase subunit 2 encoded by the *NAD2* gene, NADH dehydrogenase subunit 3 encoded by the *NAD3* gene, NADH dehydrogenase subunit 4 encoded by the *NAD4* gene, NADH dehydrogenase large subunit 4 encoded by the *NAD4L* gene, NADH dehydrogenase subunit 5 encoded by the *NAD5* gene, and NADH dehydrogenase subunit 6 encoded by the *NAD6* gene), complex III (Apocytochrome b), complex IV (cytochrome c oxidase subunit 1 encoded the *COX1* gene, cytochrome c oxidase subunit 2 encoded by the *COX2* gene, and cytochrome c oxidase subunit 3 encoded by the *COX3* gene), and complex V (ATP synthase F0 subunit 6 encoded by the *ATP6* gene), ATP synthase F0 subunit 8 encoded by the *ATP8* gene, and ATP synthase F0 subunit 9 encoded by the *ATP9* gene); two ribosomal genes (Large subunit RNA encoded by the *RNL* gene) and small subunit RNA encoded by the *RNS* gene, the gene encoding ribosomal protein S3 encoded by the *RPS3* gene; 28 tRNA genes; and evidence for homing endonuclease genes belonging to families LAGLI-DADG and GIY-YIG in the *NAD1* and *NAD* genes. Ten genes present translational exceptions; at both start and stop codons (ATP synthase F0 subunit 6 encoded by the *ATP6* gene, cytochrome c oxidase subunit 1 encoded by the *COX1* gene, GIY-YIG endonuclease, LAGLI-DADG endonuclease, NADH dehydrogenase subunit 1 encoded by the *NAD1* gene, NADH dehydrogenase subunit 4 encoded by the *NAD4* gene, and ribosomal protein S3 encoded by the *RPS3* gene or at the stop codon (ATP synthase F0 subunit 9 encoded by the *ATP9* gene, cytochrome c oxidase subunit 2 encoded by the *COX2* gene, and

NADH dehydrogenase subunit 3 encoded by the *NAD3* gene, while three genes present in-frame stop codons (ATP synthase F0 subunit 6 encoded by the *ATP6* gene, GIY-YIG endonuclease, and NADH dehydrogenase subunit 4 encoded by the *NAD4* gene). Beyond the complete set of core genes, two cytochrome c oxidase subunit 2 encoded by the *COX2* gene-like genes are also identified. In the mitochondrial genome of *A. aubasidani*; it contains fragments of ATP synthase F0 subunit 6 encoded by the *ATP6* gene and cytochrome c oxidase subunit encoded by the *COX1* genes (<200 bp) and a putative novel gene of unknown function.

The mitochondrial genomes could be used to improve the production of secondary metabolites and ATP and enhance electron transport (Parra et al. 2023). So, sequencing and elucidating more mitochondrial genomes in different species of *Aureobasidium* spp. are of signicance in biotechnology.

5.5 DNA Transformation of *Aureobasidium* spp.

After the whole genomes of various strains of *Aureobasidium* spp. are sequenced, annotated, and analyzed, in order to elucidate physiology, metabolic pathways, genetic regulation, and evolution in *Aureobasidium* spp., the development of fast and efficient genome-editing techniques is crucial for basic and applied research on *Aureobasidium* spp. So far, the techniques for transformation of foreign genes into the yeast-like fungal cells include chemical transformation, spheroplast transformation, and electroporation transformation.

5.5.1 *Chemical Transformation of Aureobasidium spp.*

So far, the most popular yeast model systems, *S. cerevisiae* and *S. pombe*, have been typically transformed using lithium acetate (LiAc)/single-stranded DNA (ssDNA)/polyethylene glycol (PEG) to generate chemically competent cells. Due to its use of inexpensive and readily available reagents, production of transformants appears in just 3–4 days, and applicability to intact cells requiring no spheroplasting or electroporation, this transformation technique has been very popular, effective and convenient, enabling targeted homology-mediated manipulation of the genome in *A. pullulans*.

It has been well known that most strains of *Aureobasidium* spp. are sensitive to hygromycin B, nourseothricin, and geneticin. *HPT* gene encodes a hygromycin phosphotransferase that can phosphorylate hygromycin B and inactivate it, while *NAT* gene encodes an acyltransferase that can acylate nourseothricin and inactivate its activity. In addition, geneticin is an aminoglycoside antibiotic that can irreversibly bind 70S or 80S ribosomes, inhibiting protein extension in procaryotic and eucaryotic cells. The *G418* gene encodes a geneticin phosphotransferase, which detoxifies the aminoglycoside antibiotic geneticin. So, the *HPT* gene, the *NAT* gene, and the *G418* gene can be used as the genetic and selectable markers during the transformation of *Aureobasidium* spp. (Wang et al. 2025; Colarusso et al. 2025). In addition, in order to avoid using antibiotics which can cause high costs, environmental concerns and spread of the antibiotics resistance gene in organisms, a novel glucose-dependent selection system (the phosphofructokinase gene (*PFK*) is expressed in a glucose-deficient mutant of *Δpfk*), a selectable genetic marker and is also used to achieve customizable and stable multilocus chromosomal integration of the target genes in *Aureobasidium* spp. (Zhang et al. 2024). This unique system can achieve a 100% positive rate of transformation and stable multilocus chromesomal integration of the target genes.

5.5.1.1 Preparation of the Competent Cells

The competent *A. pullulans* cells are prepared based on the methods described by Wirshing et al. (2024).

5.5.1.2 Transformation

Before the transformation, 10.0 mg/mL single-stranded fish sperm DNA in a solution containing 10.0 mM Tris-HCl, 10.0 mM NaCl, and 1.0 mM EDTA, pH 8.0, is boiled for 5 min. The suspension is briefly vortexed and cooled on ice; 10.0 μL carrier DNA and 10.0 μg linear DNA fragments carrying *HPT* gene or *NAT* gene are added to an aliquot of the competent cells obtained above. 2.5 μL 40% glucose in the competence buffer mentioned above is also added to the competent cells. The whole cell suspension is gently mixed. Next, 600.0 μL transformation buffer (10.0 mM Tris pH 8, 1.0 mM EDTA, 40% [w/v] PEG 3350) is added, and the cell suspension is inverted quickly several times to mix. The cell suspension is incubated at 24 °C for 1 h with mixing and then heat-shocked at 37 °C or 42 °C in a heat block for 15 or 30 min. Following the heat shock, the treated cell suspension is centrifuged at 9391 × g for 10 s, and the PEG solution is removed.

5.5.1.3 Isolation of Transformants

The transformed cells are resuspended in 1.0 mL of 1.0 M sorbitol, and the suspension is added to 1.0 mL of YPD medium. The cells are incubated and recovered for the indicated time at 24 °C with gentle shaking at 100 rpm before plating on YPD plates supplemented with hygromycin B or nourseothricin at a final concentration of ~50.0 μg/mL. The plates are incubated at 24 °C, and the colonies typically appear after 2–3 days. Finally, the positive colonies are examined based on the required.

The detailed methods refer to those described by Wirshing et al. (2024).

5.5.2 *Spheroplast Transformation*

5.5.2.1 Preparation of Competent Spheroplast Cells

Several freshly grown colonies of *Aureobasidium* spp. are inoculated into the liquid HC medium, and the culture is aerobically cultivated at 28 °C and 180 rpm for 10 h until OD_{600nm} = 1.0–1.2 is reached. The culture is examined under a microscope, and 25.0–35.0 mL of the culture with most of the yeast-like fungal cells is centrifuged at 8000 × *g* and room temperature for 3 min and the supernatant obtained is removed. The pellets are washed with 30.0 mL of the sorbitol-citrate buffer (pH 5.8) by centrifugation at 5000 × *g* and room temperature for 3 min and the supernatants are removed. The pellets are resuspended in 3.0 mL of sorbitol-citrate buffer (pH 5.8). 500.0 μL of zymolyase and 1.0 mL of driselase are added to the resuspension, and the mixture is incubated at 28 °C and 110 rpm for 6 h to remove the cell walls. The treated cell suspension is centrifuged 3000 × *g* and 4 °C for 5 min, and the supernatants are removed. The pellets are mildly suspended in 30.0 mL of a STC buffer, the cell suspension is centrifuged at 3000 × *g* and 4 °C for 5 min, and the supernatants are removed. The new pellets are mildly suspended in 6.0 mL of the STC buffer (pH 7.5), the cell suspension is centrifuged at 3000 × *g* and 4 °C for 5 min, and the supernatants are removed.

5.5.2.2 Transformation

The competent cells are mildly suspended in 3.0 mL of the STC buffer (pH 7.5). 200.0 μL of the competent cells, 40.0 μL of the linear DNA fragments carrying the *HPT* gene or the *NAT* gene, and 50.0 μL of the STC buffer (pH 7.5) are mixed well in an Eppendorf tube, and the tube is bathed at 22 °C for 10 min. The blank without the linear DNA fragments is used as a control. 200.0 μL of the PTC buffer and 800.0 μL of the PTC buffer are in order added to the sample and the blank and mildly mixed, respectively. The mixtures are bathed at 22 °C for 60 min. Both the sample and blank are centrifuged at 3000 × *g* and room temperature for 15 min, and the supernatants are removed.

The pellets are washed with 1.0 mL of 2 × HCS buffer, the washed pellets are centrifuged at 2000 × *g* and room temperature for 15 min, and the supernatants are removed. The pellets are mildly suspended in 0.5 mL 2 × HCS buffer, and the cell suspension is incubated at 28 °C and 100 rpm for 90 min to regenerate cell walls.

5.5.2.3 Isolation of the Positive Transformants

The regenerated cells are mildly spread on the double HCS plates with the lower plate containing 50.0 μg/mL hygromycin B or nourseothricin, and the double HCS plates are incubated at 28 °C for 48 h. Each colony appeared on the double HCS plates is transferred to YPD plate with 30.0 μg/mL of hygromycin B or 30.0 μg/mL of nourseothricin. After the plates are incubated at 28 °C for 48 h, the positive transformants are screened and examined.

The buffers and media used here are listed below:

STC buffer: Tris base 0.30 g, $CaCl_2$ 0.28 g, and sorbitol 18.22 g are dissolved in 90 mL of distilled water, and pH of solution is adjusted to be 7.5. The total volume of solution is adjusted to be 100 mL. The solution is sterilized at 115 °C for 30 min.

PTC buffer: 50 g of PEG3350 is completely dissolved in the STC buffer at 70 °C. 4.0 mL of β-mercaptoethanol is added to the PEG solution. The total volume of the solution is adjusted to be 100 mL. The solution must be freshly prepared and used.

HC medium: 1.0 g of glucose, 0.1 g of beef extract, 0.3 g of yeast extract, 1.0 g of polypeptone, 0.3 g of malt extract are dissolved in 100 mL of distilled water. The medium is sterilized at 115 °C for 30 min.

HCS medium: Based on 100 mL of the HC medium, 1.0 M sorbitol and 2.0 g of agar are added. The medium is sterilized at 115 °C for 30 min. After cooling, 5 mg of 50.0 μg/mL hygromycin B or nourseothricin is added to 100 mL of the medium.

5.5.3 Electroporation Transformation

5.5.3.1 Preparation of the Competent Cells

Several freshly grown colonies of *Aureobasidium* spp. are inoculated into 50.0 mL of the liquid YPD medium, and the culture is aerobically grown at 28 °C and 180 rpm for 8 h. The culture is examined under microscope to make sure that most of the cells are yeast-like fungal cells in the culture. 300.0 mL of the culture with most of yeast-like cells is centrifuged at 5000 × *g* and room temperature for 5 min and the supernatant obtained is removed. The pellets are resuspended in 40.0 mL of the pretreatment solution. After the resuspension is left statically for 30 min, it is centrifuged at 3500 × *g* for 5 min and the supernatant obtained is removed. After the new pellets are washed three times with 5.0 mL of the precooled 1.0 M sorbitol solution, the pellets obtained are resuspended in a suitable amount of 1.0 M sorbitol solution, and the cell density of resuspension is adjusted to be around 3.5×10^7 cells/mL.

5.5.3.2 Transformation

200.0 μL of the cell resuspension is mixed with a suitable amount of the linear DNA carrying the *HPT* gene, the *NAT* gene, or *G418* gene in the electroporation cup, and the cup left in the ice is electrically shocked for 5 min under the conditions of voltage of 1.5 kV, electric resistance of 200.0 Ω, and electric capacity of 25.0 μF. After the electric shock, 1.0 mL of YPD-1.0 M sorbitol solution is added to the treated cell resuspension and the new mixture is left statically for 2 h.

5.5.3.3 Isolation of the Positive Transformants

A suitable amount of the treated cell resuspension is mildly spread on the double plates of the YPD-1.0 M sorbitol, with the lower plate containing 50.0 μg/mL of hygromycin B or nourseothricin or geneticin. The plates are incubated at 28 °C for 72 h, and the transformants appear on the plates in 2–3 days. The positive transformants are screened and obtained.

The pretreatment solution: 10.92 g of sorbitol,1.02 g of Li acetate, and 1.0 mL of 1.0 M Tris-HCL are dissolved in 100 mL of distilled water.

5.6 The Genome-Editing Techniques in *Aureobasidium* spp.

So far, an efficient Cre/loxp site-specific recombination system, fast and efficient CRISPR-mediated genome editing, and *Agrobacterium tumefaciens*-mediated transformation (AMT) for random integration of an endogenous or exogenous gene have been successfully developed for genome editing in *Aureobasidium* spp.

5.6.1 The Efficient Cre/Loxp Site-Specific Recombination System

A combination of these tools with the Cre/loxP system, where the specific loxP sequences are recognized and recombined by a Cre recombinase, has been established for model organisms, non-conventional yeasts, and filamentous fungi (Li et al. 2015). Recombination of the loxP sites flanking the selection marker after genetic modification facilitates the generation of marker-less strains, enabling repeated use of this system for combined modifications. The laboratory of the authors in this book has successfully developed such a Cre/loxP system (pAMGDlox-1, Fig. 5.4, pAMEXlox-1, Fig. 5.5, and pAMCRE1, Fig. 5.6) based on homologous recombination for genome editing in *Aureobasidium* spp., and performed several gene deletions as well as gene expression in various strains of *A. pullulans*, *A. melanogenum*, and *A. aubasidani* (Zhang et al. 2019a, b; Zhao et al. 2019; Wei et al. 2023; Driller et al. 2026).

Both the knock-out plasmid (pAMGDlox-1, Fig. 5.4) and the knock-in plasmid (pAMEXlox-1, Fig. 5.5) are composed of a *PGK* promoter, nourseothricin gene (*NAT* gene), a specific terminator (polyA), and a loxP sequence (ATGTATGC).

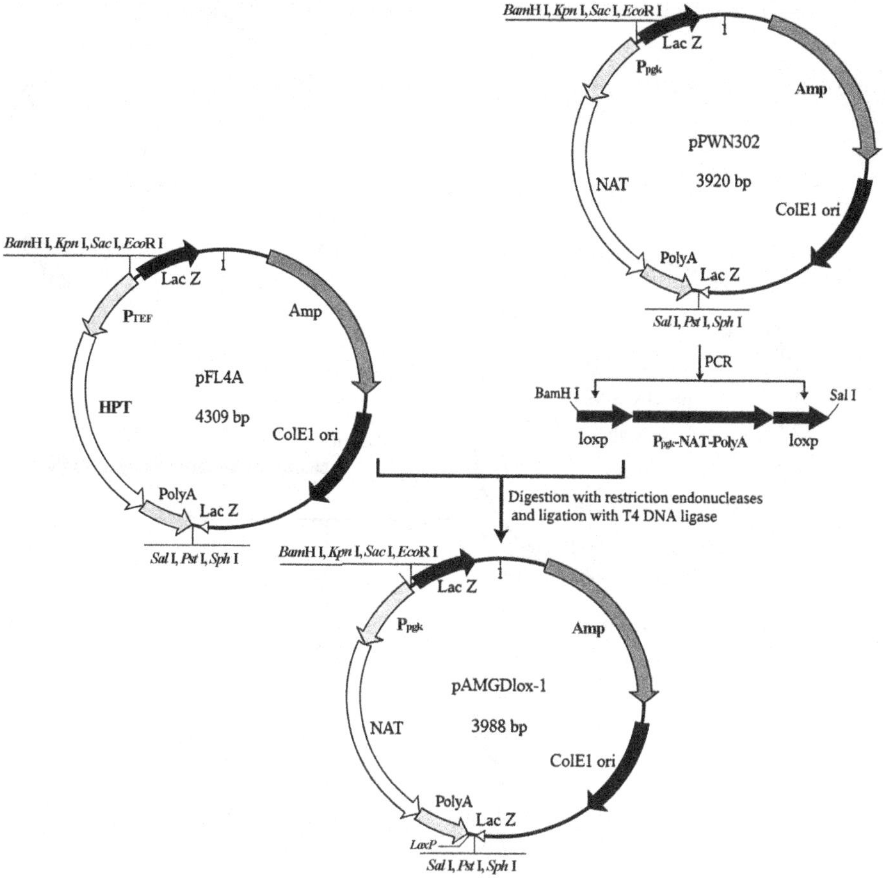

Fig. 5.4 Construction of the knock-out plasmid pAMGDlox-1 containing loxp sequence 5′-CGTATGT↓A-3′; ↓ means cutting site by the Cre recombinase

Before the transformations as described above, the plasmids must be linearized using the DNA restriction enzyme digestion shown on the plasmids (Figs. 5.4 and 5.5) or PCR techniques to release linear DNA fragments (5′-arm of the target gene-LoxP-P_{gk}-NAT-PolyA-LoxP-arm-3′ of the target gene which must be linked into pAMGDlox-1 and 26S-rDNA-P_{gk}-NAT-PolyA-Target gene-P_{TEF}-18S rDNA which must be linked into pAMEXlox-1) from the recombinant knock-out plasmid (pAMGDlox-1) and the recombinnat knock-in plasmid (pAMEXlox-1), respectively. Then, the linear DNA fragments obtained are transformed into the competent cells, which are sensitive to hygromycin B or nourseothricin as described above. Finally, the positive transformants are selected on the plates with hygromycin B or nourseothricin.

In order to continue to delete and express several genes simultaneously, after each transformation, the *NAT* gene in the transformants must be removed by using

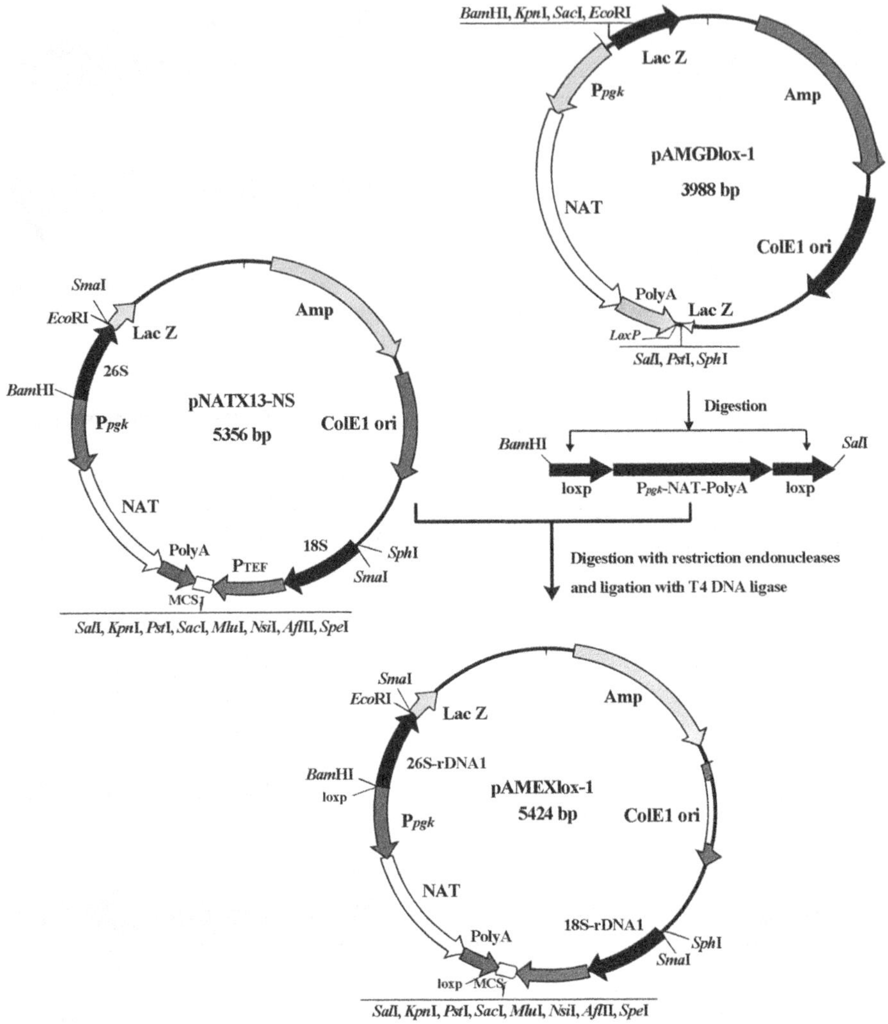

Fig. 5.5 Construction of the knock-in plasmid pAMEXlox-1 containing loxp sequence 5'-CGTATGT↓A-3'; ↓ means cutting site by the Cre recombinase

pAMCRE1 (Fig. 5.6). Furthermore, now, the antibiotics resistance gene and antibiotics resistance bacteria are regarded as the dangerous pollutants. Therefore, any antibiotics resistance gene in the transformants must be cleared. So, the pAMCRE1 (Fig. 5.6) for removal of antibiotics resistance gene constructed in this study consists of the autonomously replicating sequence (ARS), the DNA fragments (TEF-HPT-polyA) from the plasmid pFL4A, and the codon-optimized P_{gap}-Cre-Cyc1 fragment amplified from the plasmid pUC57-Cre. The pAMCRE1 in the genome editing system has many advantages over any other plasmids. In this system, a constitutive promoter (*PGD*) of 3-phosphate glycerol dehydrogenase gene in *A.*

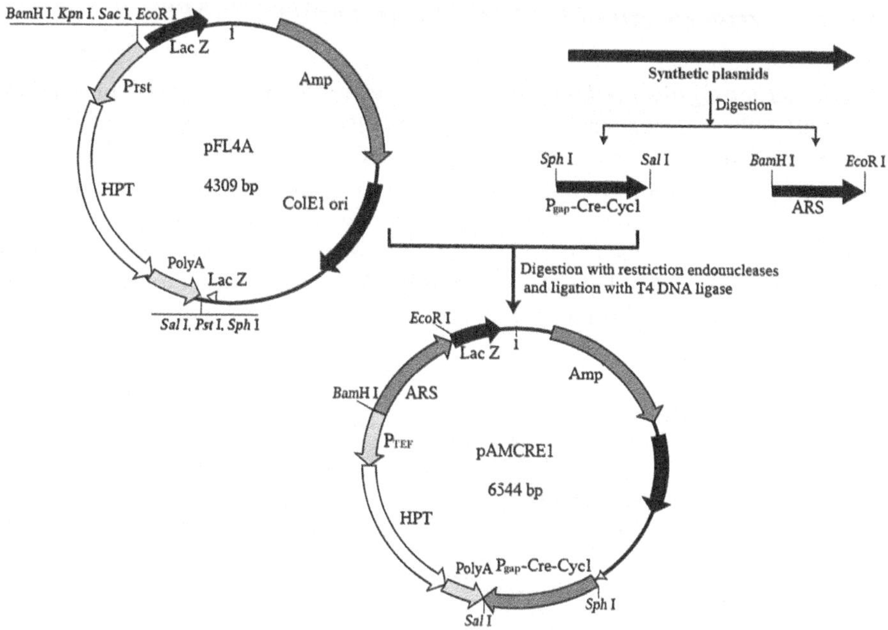

Fig. 5.6 Construction of the resistance gene removal plasmid pAMCRE1 containing Cre recombinase gene (CRE and ARS)

melanogenum is used to drive the transient expression of the Cre recombinase and a terminator of cytochrome C (CYC) gene in *S. cerevisiae* is used to terminate expression of the Cre recombinase in the pAMCRE1. At the same time, the autonomously replicating element from *Astilago maydis* is also used to construct the pAMCRE1 to keep its temporary stability in the transformant cells in order to make the produced Cre recombinase effectively cut and remove the antibiotics resistance gene (nourseothricin resistance gene (the *NAT* gene) and hygromycin B resistance gene (the *HPT* gene) and afterwards to allow the plasmid pAMCRE1 (the *HPT* gene) to be lost under none selection pressure (without any antibiotics) by curing. Therefore, the whole plasmid pAMCRE1 must be transformed into the transformants by using the transformations as described above. After the produced Cre recombinase effectively cuts and removes the antibiotic resistance gene (nourseothricin resistance gene (*NAT* gene) and hygromycin B resistance gene (*HPT* gene) in the transformants, the plasmid pAMCRE1 will be automatically lost by curing.

The Cre/loxp site-specific recombination system has many advantages, such as the high efficiency, high specificity, accuracy, extensiveness, simplicity of its use, and its adaptability to address diverse biological questions. It can be used as a tool for generating a recyclable marker system and multigene manipulation. Furthermore, the Cre/loxP-based system has been successfully used for elucidation of the pullulan, polymalate, liamocin, siderophore, melanin, fumarate, and intracellular lipid biosynthesis pathways and regulation in *Aureobasidium* spp. in this laboratory (Chi et al. 2022; Wei et al. 2023).

5.6.2 *Agrobacterium-Mediated Gene Transfer (AMT)*

The bacterium *Agrobacterium tumefaciens* contains a tumor-inducing plasmid (Ti plasmid) with a segment known as the transfer DNA (T-DNA) which contains right border and left border repeat and can be transferred into target cells and integrated into their genomes (Fig. 5.7). The T-DNA can be modified to replace the plant tumor induction genes with a desired target gene and the modified T-DNA is transformed into the competent *A. tumefaciens* cells with the electroporation technique as mentioned above. So, when *A. tumefaciens* carrying the binary plasmid vector containing the right border (*R/B*) and left border (*L/B*) sequences of *A. tumefaciens* T-DNA,

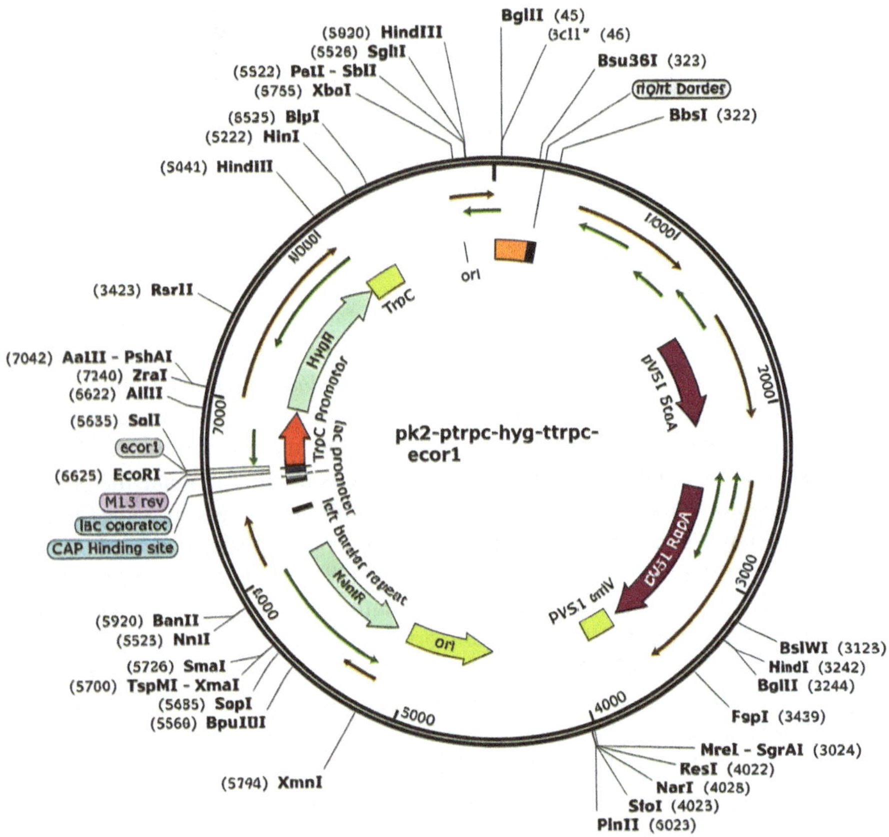

Fig. 5.7 The map of the plasmid carrying T-DNA with right border and left border repeat and the inserted *HPT* gene (Hyg^R)

the *HPT* gene or the *NAT* gene or the *G418* gene and the desired target genes located between the border sequences and controlled by the promoter (P_{TrpC}) of tryptophan synthetase C gene and the terminator of tryptophan synthetase C gene from *A. pullulans* is mixed with the competent *Aureobasidium* spp. cells. Then, the mixture is incubated for 3 days at room temperature, during which period the binary plasmid vector can be transferred to the yeast-like fungal and fungal cells and the target genes can be integrated into their genome at random sites. After that, 50.0 μg/mL carbenicillin and 50 μg/mL tetracycline are added to kill *A. tumefaciens*, and 50.0 μg/mL hygromycin B or 50 μg/mL nourseothricin is added to select the positive *A. pullulans* transformants. The positive *A. pullulans* transformants are screened and examined on the YPD plate with 50.0 μg/mL hygromycin B or 50.0 μg/mL nourseothricin.

Such *Agrobacterium tumefaciens*-mediated transformation system can yield stable transformants that can grow in the medium containing hygromycin B. However, the optimal transformation conditions for the *Agrobacterium tumefaciens*-mediated transformation system must be determined. For example, the number of fungal cells and the recombinant *Agrobacterium* cells during the co-cultivation must be optimized to obtain a higher number of resulting transformed cells, and the time of the co-cultivation also must be optimized.

The *Agrobacterium*-mediated gene transfer (AMT) has been adapted for use in diverse fungi including Ascomycetes, Basidiomycetes, and Mucoromycetes, yeasts, plants and human cells and integrate the desired target gene into their genomes. Recently, the AMT technique also has been used to obtain *A. pullulans* transformants resistant to hygromycin B (HPT) or nourseothricin (NAT) or geneticin (G418) as mentioned above.

For example, such a plasmid containing the *HPT* gene was used to improve polymarate production by *A. pullulans* CCTCC M2012223. The *HPT* gene could be stable in the 80–120 transformants per 10^7 single cells during asexual proliferation, indicating that the plasmid has been integrated into the *A. pullulans* genome. A high-titer PMA mutant H27 increases PMA by 24.5% compared with PMA produced by the control by the disruption of phosphoglycerate mutase gene (Tu et al. 2015).

This AMT technique offers numerous advantages, such as a wider range of potential transformation recipients, greater stability of transformants, increased efficiency of transformation, easily producing transformed fungi just simply by mixing the recombinant *Agrobacterium* cells with fungal cells and mycelia, and an efficient tool for insertional mutagenesis of various fungi.

Although the AMT technique has been successfully used to transform the recombinant plasmids expressing fluorescent probes for nuclear and cytoskeletal elements in *A. pullulans* (Petrucco et al. 2024), several commonly used fluorescent proteins were not detectable in *A. pullulans*, possibly due to the presence of TTA Leucine codons that are very rare in that system (Petrucco et al. 2024). Because AMT frequently leads to non-homologous integration of DNA into the genome, many transformants affect gene expression in unintended ways.

5.6.3 CRISPR-Mediated Genome Editing Techniques

In recent years, CRISPR (the clustered regularly interspaced short palindromic repeat system from *Streptococcus pyogenes*)-mediated genome editing quickly has become a widely used method for genome editing in many different species because of the ease of target programming, modification efficiency, and multiplexing capacity. Generally, the CRISPR/Cas9 system consists of a single multifunctional Cas nuclease (Cas9) and a single guide RNA (sgRNA) which directs the expressed Cas9 nuclease to cause a double-strand break (DSB) in a 20 nt-target sequence upstream of a protospacer adjacent motif (PAM, 5′-NGG-3′) and a downstream chimeric scaffold sequence (gRNA scaffold) (Fig. 5.8). The sgRNA can guide the expressed Cas9 nuclease to generate a specific DNA double-strand break (DSB) in the target gene, followed by a short Protospacer Adjacent Motif (PAM). After the DSB, the DNA double-strand breaks (DSBs) are repaired through error-prone nonhomologous end

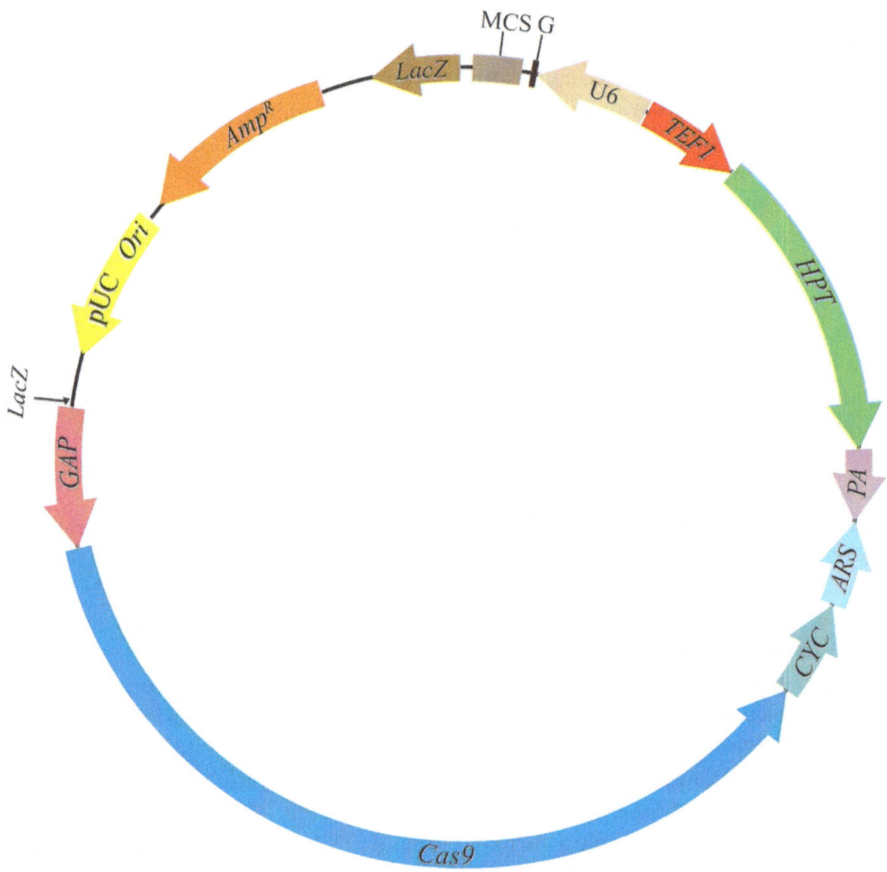

Fig. 5.8 The map of the plasmid used for the CRISPR-mediated genome-editing techniques

joining (NHEJ), causing insertion/deletion mutations at the target gene and the homology-directed repair (HDR), causing insertion of a defined sequence at the target gene. The CRISPR/Cas9 system has been successfully used to edit many genomes of several fungi, including *A. pullulans* (Zhang et al. 2019a). As shown in Fig. 5.8, the expression of the Cas nuclease gene (Cas9 gene) is controlled by the *GAP* promoter and terminated by *CYC* terminator. The autonomously replicating sequence (ARS) is cloned from the genomes of *Aureobasidium* spp., and expression of the *HPT* gene is controlled by the promoter of the *TEF* (translation elongation factor 1) gene and the terminator of polyA. The artificially synthesized guide RNA (gRNA) which directs the Cas9 nuclease to introduce a double-strand break (DSB) in a 20 nt-target sequence upstream of a protospacer adjacent motif (PAM, 5′-NGG-3′) is inserted into the multiple cloning site (MCS). After the whole plasmid is transformed into the competent cells of *Aureobasidium* spp., the gRNA gene is expressed under the control of the promoter of the *TEF* gene, and the expressed gRNA is bound to Cas9, specifically cutting the target gene, producing a double-strand break (DSB) and resulting in gene disruption through NHEJ and HDR as mentioned above. After cultivation of the positive transformants for several generations, the plasmid is automatically lost by curing, yielding the sensitive strain again for deletion of next gene.

However, the commonly used CRISPR/Cas9 systems are protospacer adjacent motif (PAM) dependent, have high off-target effects due to the prolonged presence of the Cas9 and/or guide RNA, need time-laborious construction of the plasmids, do not allow multiplex genome editing, and the plasmid DNA might be inserted at unwanted sites in the genome (Zhang et al. 2019a; Kreuter et al. 2022). Therefore, a CRISPR/Cas9-based genome-editing method using the pre-assembled Cas9-RNA-complexes (RNPs) for editing the genomes in *A. pullulans* and *A. melanogenum* is developed. The ribonucleoproteins (RNPs) consisting of the Cas9 protein and the sgRNA can be assembled in vitro and inserted into the target cell. This is a fast and easy delivery technique of the CRISPR components that does not require cloning or in vivo transcription and translation. The RNPs enable immediate transient gene editing with reduced off-target effects. It was found that this method can be used for single and multiplex genome editing using only RNPs by targeting *URA3* (encoding for orotidine-5′-phosphate decarboxylase), *ADE2* (encoding for phosphoribosylaminoimidazole carboxylase), and *ARG4* (encoding for argininosuccinate lyase) in *Aureobasidium* spp. Furthermore, it is sufficient to use the RNPs to improve the homologous recombination rate and 20 bp long homologous flanks. Therefore, the repair cassettes can be constructed by a single PCR, abolishing the need for laborious and time-consuming cloning as mentioned above. Therefore, the RNP-based method can be used for safe, fast, and efficient genome editing in *Aureobasidium* spp. (Kreuter et al. 2022). Furthermore, an AMA1 (Autonomously Maintained in *Aspergillus*) plasmid for gRNA and CRISPR/Cas9 expression is also used for efficient genomic deletions in *A. pullulans*. AMA1 is a DNA sequence with an inverted repeat promoting extrachromosomal replication. The CRISPR/Cas9 self-replicating plasmid carries a gRNA, a Cas9 gene, the half-shortened AMA1 sequence, and the *HPT* gene and can cause deletion mutations in the genomic DNA at the locus targeted by the gRNA without the *HPT* gene integration in host genome (Masi et al. 2024). Therefore, it is not necessary to remove the *HPT* gene after each transformation.

5.7 Conclusions and Perspectives

So far, only genomes of *A. pullulans*, *A. melanogium*, *A. aubasidani*, A. *hainanensis, A. subglaciale*, *A. namibiae, A. mustum*, *A. vineae,* and *A. uvarum* have been sequenced, annotated, and characterized; any other genomes of most of the species of *Aureobasidium* mentioned in Chap. 3 are still unknown. So, as more and more genomes of *Aureobasidium* spp., especially the fermentation strains, are uncovered, more functional genes and fermentation products will be mined and produced.

All the efficient gene-editing tools mentioned above can also be easily used to carry out the top-down approach (genome reduction) and the bottom-up approach (genome synthesis) in *Aureobasidium* spp. in the future. In addition, all the genome-editing tools available must be further improved for more efficient genome editing.

References

Babusci E, Gunsch CK (2026) Draft genome sequence of Elizabeth River PAH-degrading fungal isolates Trichoderma deliquescens and Aureobasidium sp. Microbiol Res Announc 15(5):1–2

Cai S, Snyder AB (2022) Genomic characterization of polyextremotolerant black yeasts isolated from food and food production environments. Front Fung Biol 3:928622

Černoša A, Sun X, Gostinčar C, Fang C, Gunde-Cimerman N, Song Z (2021) Virulence traits and population genomics of the black yeast *Aureobasidium melanogenum*. J Fung 7(8):665

Chaitanya KV (2019) Eukaryotic genome organization, regulation, evolution and control. In: Genome and genomics. Springer, Singapore, pp 200–220

Chan GF, Bamadhaj HM, Gan HM, Rashid NAA (2012) Genome sequence of *Aureobasidium pullulans* AY4, an emerging opportunistic fungal pathogen with diverse biotechnological potential. Eukaryot Cell 11:1419–1420

Chen L, Chi Z, Liu GL, Xue SJ, Wang ZP, Hu Z, Chi ZM (2019) Improved pullulan production by a mutant of *Aureobasidium melanogenum* TN3-1 from a natural honey and capsule shell preparation. Int J Biol Macromol 141:268–277

Chen L, Xin Wei X, Liu GL, Hu Z, Chi ZM, Chi Z (2020a) Glycerol, trehalose and vacuoles had relations to pullulan synthesis and osmotic tolerance by the whole genome duplicated strain *Aureobasidium melanogenum* TN3-1 isolated from natural honey. Int J Biol Macromole 165:131–140

Chen TJ, Liu GL, Wei X, Wang K, Hu Z, Chi Z, Chi ZM (2020b) A multidomain α-glucan synthetase 2 (AmAgs2) is the key enzyme for pullulan biosynthesis in *Aureobasidium melanogenum* P16. Int J Biol Macromol 150:1037–1045

Chi Z, Kong CC, Wang ZZ, Wang Z, Liu GL, Hu Z, Chi ZM (2022) The signaling pathways involved in metabolic regulation and stress responses of the yeast-like fungi *Aureobasidium* spp. Biotechnol Adv 55:107–898

Colarusso A, Williams A, Gladfelter AS, Wirshing ACE, Lew DL (2025) Optimized vectors for genetic engineering of *Aureobasidium pullulans*. bioRxiv 27:634885

Driller M, Grad M, Dielentheis-Frenken MRE, Stein K, Kohlsted M, Wittmann C, Lars M. Blank LM, Tiso T (2026) Aureobasidium pullulans NRRL 62042 as robust platform for efficient lignin valorization and sustainable cis,cis-muconic acid production. Biores Technol 443:133851

Gostinčar C, Ohm RA, Kogej T, Sonjak S, Turk M, Zajc J, Zalar P, Grube M, Sun H, Han J, Sharma A, Chiniquy J, Yee Ngan C, Lipzen A, Barry K, Grigoriev IV, Gunde-Cimerman N (2014) Genome sequencing of four *Aureobasidium pullulans* varieties: biotechnological potential, stress tolerance, and description of new species. BMC Genom 15:549

Gostinčar C, Turk M, Zajc J, Gunde-Cimerman N (2019) Fifty *Aureobasidium pullulans* genomes reveal a recombining polyextremotolerant generalist. Environ Microbiol 21:3638–3652

Gostinčar C, Sun X, Černoša A, Fang C, Gunde-Cimerman N, Song Z (2022) Clonality, inbreeding, and hybridization in two extremotolerant black yeasts. GigaSci 11:giac095

Han Q, Liu N, Robinson H, Cao L, Qian C, Wang Q, Xie L, Ding H, Wang Q, Huang Y, Li J, Zhou Z (2013) Structure of a GH10 xylanase from termite gut bacteria reveal a novel structural feature and significance of its bacterial Ig-like domain. Biotechnol Bioeng 110:3093–3103

Harahap I, Maryam R, Idwan Sudirman L, Indri Astuti R (2026) Whole-genome analysis reveals the genetic potential of yeast Aureobasidium melanogenum DAP1 for aflatoxin B1 decontamination and Aspergillus flavus inhibition in nutmeg seeds. Intern J Food Microbiol 453:111728

Jia SL, Ma Y, Chi Z, Liu GL, Hu Z, Chi ZM (2019) Genome sequencing of a yeast-like fungal strain P6, a novel species of *Aureobasidium* spp: insights into its taxonomy, evolution, and biotechnological potentials. Ann Microbiol 69:1475–1488

Jia SL, Zhang M, Liu GL, Chi ZM, Chi Z (2023) Novel chromosomes and genomes provide new insights into evolution and adaptation of the whole genome duplicated yeast-like fungus TN3-1 isolated from natural honey. Funct Integr Genom 23(3):206

Jiang H, Chi Z, Liu GL, Hu Z, Zhao SF, Chi ZM (2020) Melanin biosynthesis in the desert-derived *Aureobasidium melanogenum* XJ5-1 is controlled mainly by the CWI signal pathway via a transcriptional activator Cmr1. Curr Genet 66:173–185

Kreuter J, Stark G, Mach RL, Mach-Aigner AR, Zimmermann C (2022) Fast and effcient CRISPR-mediated genome editing in *Aureobasidium* using Cas9 ribonucleoproteins. J Biotechnol 350:11–16

Lenassi M, Gostinčar C, Jackman S, Turk M, Sadowski I, Nislow C, Jones S, Birol I, Cimerman NG, Plemenitaš A (2013) Whole genome duplication and enrichment of metal cation transporters revealed by de novo genome sequencing of extremely halotolerant black yeast *Hortaea werneckii*. PLoS One 8:e71328

Li C, Lei J, Zhao Y, Xu X, Li S (2015) Effect of saline water irrigation on soil development and plant growth in the Taklimakan Desert Highway shelterbelt. Soil Tillage Res 146:99–107

Li S, Wu X, Meng J (2025) Draft genome sequence of *Aureobasidium pullulans* ATCC15233. Microbiol. Res Announc 14(2):e0075624

Masi A, Wögerbauer K, Mach RL, Mach-Aigner RM (2024) Genomic deletions in *Aureobasidium pullulans* by an AMA1 plasmid for gRNA and CRISPR/Cas9 expression. Fung Biol Biotechnol 11:6

Onetto CA, Schmidt SA, Roach MJ, Borneman AR (2020) Comparative genome analysis proposes three new *Aureobasidium* species isolated from grape juice. FEMS Yeast Res 20:foaa052

O'Sullivan RJ, Karlseder J (2010) Telomeres: protecting chromosomes against genome instability. Nat Rev Mol Cell Biol 11:171–181

Parra M, Libkind D, Todd Hittinger C, Álvarez L, Bellora N (2023) Assembly and comparative genome analysis of a Patagonian *Aureobasidium pullulans* isolate reveals unexpected intraspecific variation. Yeast 40:197–213

Petrucco CA, Crocker AW, D'Alessandro A, Medinac EM, Gormana McNeill OJ, Gladfelter AS, Lew DJ (2024) Tools for live-cell imaging of cytoskeletal and nuclear behavior in the unconventional yeast, *Aureobasidium pullulans*. Mole Biol Cell 35: br10, 1–12

Qi CY, Jia SL, Wei X, Yang G, Chi Z, Liu GL, Hu Z, Chi ZM (2020) The differences between fungal α-glucan synthase determining pullulan synthesis and that controlling cell wall α-1,3 glucan synthesis. Int J Biol Macromol 162:436–444

Rueda-Mejia MP, Nageli L, Lutz S, Hayes RD, Varadarajan AR, Grigoriev IV, Ahrens CH, Freimoser FM (2021) Genome, transcriptome and secretome analyses of the antagonistic, yeast-like fungus *Aureobasidium pullulans* to identify potential biocontrol genes. Microb Cell 8(8):184–202

Tu G, Wang Y, Feng J, Li X, Guo M, Zou X (2015) *Agrobacterium* tumefaciens-mediated transformation of *Aureobasidium pullulans* and high-efficient screening for polymalic acid producing strain. Sheng Wu Gong Cheng Xue Bao 31:1063–1072

Vignolle GA, Mach RL, Mach-Aigner AR, Derntl C (2021) Genome sequence of the black yeast-like strain *Aureobasidium pullulans* var. *aubasidani* CBS 100524. Microbiol Res Announ 10:e01293-20

Wagner DN, Varaljay VA, Lyon WJ, Crouch AL, Allex-Buckner C, Bifffnger JC, Crookes-Goodson WJ, Kelley-Loughnane NK, Stamps BW (2024) Draft genome sequence of potentially dikaryotic black fungus *Aureobasidium melanogenum* isolated from aircraft. Environ Microbiol 13(3):1–3

Wang K, Chi Z, Liu GL, Qi CY, Jiang H, Hu Z, Chi ZM (2020) A novel PMA synthetase is the key enzyme for polymalate biosynthesis and its gene is regulated by a calcium signaling pathway in *Aureobasidium melanogenum* ATCC62921. Int J Biol Macromole 156:1053–1063

Wang P, Tiso T, Chi Z, Chi ZM (2025) Efficient transformation of inulin into liamocins by a mutant of *Aureobasidium melanogenum* HN6-1 and antimicrobial activity of the produced Massoia lactone. Int J Biol Macromole 318:145245

Wei X, Zhang M, Wang GY, Liu GL, Chi ZM, Chi Z (2023) The ornithine-urea cycle involves fumaric acid biosynthesis in *Aureobasidium pullulans* var. *aubasidani*, a green and eco-friendly process for fumaric acid production. Synth Syst Biotechnol 8:33–45

Wirshing ACE, Petrucco CA, Daniel J, Lew DJ (2024) Chemical transformation of the multibudding yeast, *Aureobasidium pullulans*. J Cell Biol 223(10):e202402114

Xiao D, Driller M, Stein K, Lars M, Blank LM, Tiso T (2025) Genome mining the black-yeast *Aureobasidium pullulans* NRRL 62031 for biotechnological traits. BMC Genom 26:244

Xue SJ, Chen L, Jiang H, Liu GL, Chi ZM, Hu Z, Chi Z (2019) High pullulan biosynthesis from high concentration of glucose by a hyperosmotic resistant, yeast-like fungal strain isolated from a natural comb-honey. Food Chem 286: 123-128

Zhang Z, Lu Y, Chi Z, Liu GL, Jiang H, Hu Z, Chi ZM (2019a) Genome editing of different strains of *Aureobasidium melanogenum* using an efficient Cre/loxp site-specific recombination system. Fung Biol 123:723–731

Zhang Y, Feng J, Wang P, Xia J, Li X, Zou X (2019b) CRISPR/Cas9-mediated efficient genome editing via protoplast-based transformation in yeast-like fungus *Aureobasidium pullulans*. Gene 709:8–16

Zhang S, Ma T, Zheng FH, Aslam M, Wang YJ, Chi ZM, Liu GL (2024) Customizable and stable multilocus chromosomal integration: a novel glucose-dependent selection system in *Aureobasidium* spp. Biotechnol Biofuel Bioprod 17:81

Zhao SF, Jiang H, Chi Z, Liu GL, Chi ZM, Chen TJ, Yang G, Hu Z (2019) Genome sequencing of *Aureobasidium pullulans* P25 and overexpression of a glucose oxidase gene for hyper-production of Ca^{2+}-gluconic acid. Antonie Van Leeuwenhoek 112(5):669–678

Chapter 6
Enzymes from *Aureobasidium* spp. and Their Applications

Abstract The enzymes produced by *Aureobasidium* spp. include amylase, cellulase, xylanase, inulinase, pectinase, mannanase, esterases, lipase, protease, laccases, and tannase. As the genomic DNAs of most strains of *Aureobasidium* spp. have been sequenced and characterized and their efficient genomic editing techniques have been developed well, all these enzymes can be easily overproduced by metabolic engineering and synthetic biology. All the produced enzymes are closely related to their wide distributions, nice adaptation to different environments including harsh surroundings and multiple bioproducts (pullulan, polymalate, liamocin, lipids, melanin, fumarate, siderophore, β-1,3-1,6 glucan, gluconic acid) by assimilating sucrose, glucose, starch, cellulose, xylose, xylan, inulin, pectin, mannan, lipid, protein, lignin, synthetic chemicals, tannins, waste raw materials, and so on. All the produced enzymes also have multiple applications in various sectors of biotechnology, such as agriculture, food, pharmaceutical, medical, and biofuel industries. Furthermore, *Aureobasidium* spp. and the enzymes they produce also can be used for recycling waste compounds in forestry, agriculture, and industries.

Keywords Enzymes · Applications · Xylooligosaccharides · Inulooligosaccharides

6.1 Introduction

The ascomycete fungi *Aureobasidium* species, characterized by their yeast-like cells and polymorphic nature, are highly valued in biotechnology for their ability to produce various biodegradable compounds. These include extracellular poly-α-1,6-maltotriose (pullulan), polymalate (PMA), liamocin as a type of polyol lipid, intracellular lipids, fumarate, melanin, siderophore, gluconic acid, and β-1,3–1,6 glucan (Wang et al. 2022). These compounds find diverse applications across the food, pharmaceutical, chemical, environmental, and nanomaterial industries. These fungi

Z. Chi et al., *Biology and Synthetic Biotechnology of Aureobasidium spp.*,
https://doi.org/10.1007/978-981-92-2607-8_6

are extensively present in soil, water, and the atmosphere globally, with notable prevalence in extreme environments, including phyllospheres, deserts, outer space, honey, seawater, deep-sea habitats, hypersaline waters found in salterns, rocks, monuments, polar regions, subglacial ice, and glacial ice, as stated in Chap. 2. Thus, these fungi play crucial ecological roles by decomposing deceased organisms, recycling organic carbon and nitrogen into their inorganic forms in the environment, and by establishing ecological relationships with various organisms like plants, pathogens, and symbionts. Throughout their life cycle, these fungi generate blastoconidia, swollen cells, septate and non-septate hyphae, arthroconidia, as well as melanin-rich, thick-walled chlamydospores, exhibiting adaptability to diverse extreme environments as stated in Chap. 4. To date, more than 55 species of *Aureobasidium* have been identified and characterized through traditional and phylogenetic analyses (Rensink et al. 2024). The complete genomic DNAs of more than 70 distinct strains of *Aureobasidium* species have been sequenced, annotated, and revealed to possess duplicated genomes originating from diverse ancestors (Gostinčar et al. 2014, 2019; Vignolle et al. 2021; Jia et al. 2023). This resilience is also attributed to their exceptional ability to withstand harsh environmental conditions. In this scenario, researching the genetics and evolution of this genus is highly feasible. Therefore, many researchers refer to them as polyextremotolerant generalists as stated in Chap. 4. Several genome editing techniques have been successfully developed for *Aureobasidium* species, including an efficient Cre/loxP site-specific recombination system, swift and effective CRISPR-mediated genome editing, and *Agrobacterium tumefaciens*-mediated transformation (ATMT) for the random integration of endogenous and exogenous genes, as stated in Chap. 5. Therefore, editing their entire genomes through gene deletion, expression of target genes, and the removal of genetic marker genes is highly feasible. The functions of all the genes within the cells and synthetic pathways leading to the production of various valuable products have been successfully elucidated (Chi et al. 2022; Wei et al. 2023). These fungi, which can exhibit extremophilic, mesophilic, thermophilic, or even psychrophilic characteristics, have been utilized to produce enzymes that can thrive in extreme environments. Additionally, their ease of genomic editing has positioned them as significant sources of various industrially important enzymes. Their ability to utilize a wide range of diverse natural substrates stems from their capacity to synthesize and excrete a variety of hydrolytic enzymes tailored for the breakdown and utilization of various natural substrates found in nature. The hydrolytic enzymes produced by these fungi encompass a range of enzymes such as amylase, cellulase, xylanase, inulinase, pectinase, protease, esterase, lipase, mannanase, laccases, and tannase. The enzymes obtained, along with the functional genes responsible for their production, offer numerous potential applications. Thus, this chapter primarily focuses on summarizing, discussing, and reviewing the enzymes produced by *Aureobasidium* species and their respective applications.

6.2 Enzymes from *Aureobasidium* spp. and Their Applications

6.2.1 Amylase

Starch stands out as a prevalent and crucial form of renewable biomass derived from various crop sources. It serves as an optimal substrate for the large-scale production of microbial cells and their fermentation products, owing to its ease of hydrolysis, cost-effectiveness, and widespread availability as a raw material across many regions worldwide. Starch can be categorized into two main components: amylose and amylopectin.

Amylases catalyze the hydrolysis of starch molecules, breaking them down into glucose, maltose, and dextrin. Amylases can be classified into three main types: α-amylase (EC 3.2.1.1), β-amylase (EC 3.2.1.2), and glucoamylase (EC 3.2.1.3). Amylases find numerous applications across various industries, including the bread and baking industry, starch liquefaction and saccharification processes, textile desizing, paper industry, detergent industry, analysis in medical and clinical chemistry, pharmaceutical industries, and the production of digestive aids for clinical use.

The strain *A. pullulans* N13d, derived from deep-sea sediments in the Pacific Ocean, exhibits the capability to yield 10.0 U/ml of extracellular glucoamylase with debranching activity. This crude glucoamylase effectively hydrolyzes potato starch granules, converting 68.5% of them into glucose when in the presence of 20.0 U/ml of the enzyme within a short time (Li et al. 2007a). The crude glucoamylase possesses the ability to hydrolyze both α-1,4 and α-1,6 linkages within the starch molecules of potato starch (Li et al. 2007b). The purified enzyme from *A. pullulans* N13d has a molecular weight of 98.0 kDa and exhibits optimal activity at a pH of 4.5 and a temperature of 60 °C. The purified enzyme can effectively absorb and digest raw potato starch, resulting in the hydrolysis of 15.8% of raw potato starch (1.0%, w/v) and 21.1% of cooked potato starch (1.0%, w/v) within 30 min when employing 0.5 U of glucoamylase activity per milligram of starch (Li et al. 2007c). Glucoamylase enzyme, with its ability to efficiently degrade raw starch, plays a crucial role in various industries such as fermentation and food processing. By eliminating the need for energy-intensive cooking at high temperature and liquefaction processes required for breaking down starch, such a glucoamylase offers several advantages in terms of cost-effectiveness and resource efficiency. Its wide range of applications includes the production of glucose, ethanol, high-fructose corn syrup, and various other food products from raw starch.

The availability of sequenced and characterized genomic DNAs of various strains of *A. pullulans* has significantly simplified the process of accessing and studying functional genes, including those encoding glucoamylase and amylase. Researchers and biotechnologists can now easily download these gene sequences from databases or repositories, enabling them to conduct further studies on the enzyme's properties, structure, and expression patterns. This accessibility to genetic information has greatly expedited research and development efforts focused on utilizing glucoamylase and amylase for various industrial applications, such as in fermentation and food processing.

The amylase enzyme from *A. pullulans* Cau 19, isolated from the leaves of spinach, cauliflower, and coriander, has the K_m value of 6.3 mg/mL, indicating the enzyme's affinity for its substrate, and has a turnover value of 5.0×10^2/min, reflecting the enzyme's catalytic efficiency. Bioconjugation of this amylase enzyme from *A. pullulans* Cau 19 to the gold nanoparticles synthesized using the biomass of its own yeast-like fungus offers a promising approach for enzyme immobilization. By attaching the enzyme to the surface of the nanoparticles, enhanced stability and reusability of the enzyme can be achieved, making it suitable for various applications in biocatalysis and biotechnology. The bioconjugation of the amylase enzyme from *A. pullulans* Cau 19 to the gold nanoparticles, carried out by Mulay and Deopurkar (2018), has resulted in an enzyme with improved properties. The conjugated amylase shows enhanced tolerance to a broader pH range of 3.0–9.0, allowing it to function effectively under varying environmental conditions. Additionally, the bioconjugated enzyme exhibits increased stability at higher temperatures than the native enzyme, making it more robust and suitable for applications requiring elevated temperatures. Indeed, the bioconjugation of the enzyme to the gold nanoparticles can significantly enhance its stability under extreme conditions such as elevated temperatures and a broad pH range. This increased stability is attributed to the protective effect offered by the nanoparticle platform, which shields the enzyme from denaturation and degradation. As a result, the bioconjugated enzyme is able to maintain its catalytic activity and structural integrity under harsh environmental conditions, making it a valuable biocatalyst for various industrial applications. The use of the gold nanoparticles for molecular immobilization offers a versatile platform with diverse applications in various fields, as confirmed by Mulay and Deopurkar (2018). These nanoparticles can serve as effective supports for immobilizing enzymes, enabling the construction of biosensors and complex multienzyme systems.

The production of thermostable α-amylase by *A. pullulans* further highlights the potential of this yeast-like fungus for industrial enzyme production. The purified α-amylase from *A. pullulans* demonstrates enhanced thermostability, particularly at temperatures ranging from 50 to 60 °C, in contrast to the crude α-amylase. This increased thermostability makes the purified enzyme a valuable candidate for applications that require high-temperature conditions (Ademakinwa et al. 2019). The α-amylase gene from *A. pullulans* NRRL Y-12974, with the accession number HM590632, is composed of 2247 base pairs. The putative mRNA transcribed from this gene is 1878 base pairs long and encodes an α-amylase consisting of 625 amino acid residues. The detailed genetic information provides valuable insights into the structure and composition of the α-amylase enzyme produced by this particular strain of *A. pullulans*. The transcription of the α-amylase gene from *A. pullulans* NRRL Y-12974 occurs in the cultures grown in both sucrose and starch media, as reported by Manitchotpisit et al. (2011). This suggests that the expression of the gene encoding the α-amylase enzyme is not limited to a specific carbon source, indicating the versatility of the enzyme production process in *A. pullulans* under different growth conditions.

The studies by Elyasi Far et al. (2020), Wang et al. (2011), and Pedersen et al. (2000) have demonstrated that α-amylase produced by industrial *Bacillus* species and glucoamylase produced by industrial *Aspergillus* species and *Saccharomycopsis fibuligera* exhibit significantly higher enzymatic activity against starch than the α-amylase from *A. pullulans*. These findings suggest that the enzymes from *Bacillus* spp. and other fungal species may have distinct properties that make them more efficient in starch degradation, potentially offering advantages for various industrial applications. Indeed, as mentioned above, the α-amylase and glucoamylase produced by *A. pullulans* represent valuable bioresources with unique properties that contribute to the widespread distribution of this fungus across different environments on Earth. These enzymes may also play key roles in the metabolism and adaptation of *A. pullulans* to various ecological niches, enabling the fungus to thrive in diverse conditions.

6.2.2 Cellulase

Lignocellulosic biomasses are abundantly produced annually and serve as a significant low-cost and renewable feedstock in fermentation industries. To achieve the complete hydrolysis of lignocellulosic biomasses into sugar monomers, the cellulase preparation requires high activity levels of exoglucanase (EG) or cellobiohydrolases (CBH) (filter paperase [FPase]), endoglucanase (carboxymethyl cellulase [CMCase]), β-glycosidases (BGLs), xylanase, and other essential accessory enzymes.

Cellulose consists of glucose units linked together by β-1,4-glycosidic bonds. Hence, there is significant economic interest in developing processes for the efficient treatment and utilization of cellulosic wastes as cost-effective carbon sources, such as glucose.

Cellulases are the enzymes responsible for breaking down crystalline cellulose into glucose. Three types of cellulases—endoglucanases (EGs) (EC 3.2.1.4, endo-1,4-β-D-glucanases), exo-cellobiohydrolase (CBH) (EC 3.2.1.91), and β-glucosidases (BGLs) (EC 3.2.1.21) or cellobiase (EC 3.2.1.21)—are essential for the in vivo degradation of crystalline cellulose to glucose, acting synergistically. Typically, the first two enzymes are inhibited by cellobiose, whereas β-glucosidase activity is significantly inhibited by glucose. In addition, β-glucosidase has the ability to hydrolyze cellobiose and other cello-dextrins into glucose, thus reducing the inhibitory effect of cellobiose on the cellulases. Cellulases find diverse applications in the environmental, fermentation, food, agricultural, and textile industries. Among these applications, β-glucosidase can also enhance the bioavailability of isoflavones in the human intestine and improve the aromatic composition of juices and wines. It is now crucial to identify β-glucosidases that are resistant to product (glucose) inhibition, can enhance the saccharification process of lignocellulosic materials, and exhibit high thermal stability to maintain their catalytic activity at elevated temperatures.

Currently, fungi such as *Trichoderma reesei* and *Aspergillus niger* are the primary producers of cellulases (Pant et al. 2022). For instance, strains of *A. niger* (Brito Santos et al. 2022) can only achieve activities of CMCase (0.4 U/mL) and FPase (0.2 U/mL), while strains of *T. reesei* (Hirasawa et al. 2019) can only produce activities of cellobiohydrolase (1.5 U/mL), endoglucanase (50.0 U/mL), and β-glucosidase (0.5 U/mL). However, neither of these fungi can produce all the essential enzymes required for the complete degradation of lignocellulosic biomasses into sugar monomers. Furthermore, editing the entire genomic DNA in *T. reesei* and *A. niger* is not a straightforward task. Recently, it has been discovered that various strains of *Aureobasidium* spp. are also effective cellulase producers.

A. pullulans strain 98, isolated from a marine environment, demonstrates the ability to produce CMCase (4.51 U/mg of protein) and FPAase (4.75 U/mg of protein). The crude enzyme can hydrolyze carboxymethyl cellulose (CMC) and filter paper, yielding a substantial quantity of monosaccharides and a minimal amount of disaccharides (Zhang and Chi 2007). The molecular mass of the endo-1,4-β-D-glucanase is 67.0 kDa, and the purified enzyme exhibits the optimal pH and temperature of 5.6 and 40 °C, respectively. After hydrolysis with the purified CMCase, only oligosaccharides are released from carboxymethyl cellulose (CMC), and no monosaccharides are detected. The structural gene of CMCase (*CMC1* gene) consists of 954 base pairs, and its sequence shows significant similarity to the CMCase genes found in *Aspergillus terreus* and *Aspergillus fumigatus*. The N-terminal amino acid sequence of the purified cellulase (CMCase) matches the sequence deduced from the cloned gene, as reported by Rong et al. (2015). Such sequence is found to be M-A-P-H-A-E-P-Q-S-Q-T-T-E-Q-T-S-S-G-Q-F.

Under the optimal conditions, *A. pullulans* LB83 cultivated on sugarcane bagasse can produce maximum cellulase activity, measured at 3.1 U/mL and 8.7 U/mL. When the cellulases produced by *A. pullulans* LB83 (5.0 FPU/g) are combined with the commercial CTec2 (5.0 FPU/g) from Novozyme Inc. in Curitiba, Brazil, and the mixtures are incubated at 50 °C and 200 rpm in a shaking incubator, 92.4% of sugarcane bagasse can be converted. This suggests that *A. pullulans* can be utilized for on-site cellulase production in lignocellulose biorefineries (Maitan Vieira et al. 2020).

When the thermophilic fungus *Thermoascus aurantiacus* CBMAI-756 and the mesophilic yeast-like fungus *A. pullulans* ER-16 were cultivated in a medium containing wheat bran, *T. aurantiacus* produced 7.0 U/mL or 70 U/g of β-glucosidase within 48–72 h, whereas *A. pullulans* was able to produce 1.3 U/mL or 13 U/g of β-glucosidase at 120 h. The β-glucosidase produced by *A. pullulans* exhibits higher pH stability (pH 4.5–10.0 compared to pH 4.5–8.0) and greater thermostability (maintaining 90% activity after 1 h at 75 °C as opposed to 85% activity after 1 h at 70 °C) than the corresponding enzyme from *T. aurantiacus* CBMAI-756 (Ribeiro Leite et al. 2008). This interestingly indicates that the β-glucosidase produced by the mesophilic *A. pullulans* possesses distinct properties compared to that produced by the thermophilic *T. aurantiacus*, suggesting that the mesophilic fungi also can produce unique enzymes.

It was determined that wheat bran was the most suitable substrate for β-glucosidase production by *A. pullulans* ER-16, isolated from orange juice industrial residues in Catanduva, Sao Paulo State, Brazil. The β-glucosidase activity can reach 13.0 U/g when *A. pullulans* ER-16 is cultivated on the solid-state medium for 120 h. The β-glucosidase produced exhibits stability at pH 4.5 and 75 °C, demonstrating its high thermal stability. However, similarly to any other fungal β-glucosidases, it is also significantly inhibited by glucose (Ribeiro Leite et al. 2008). When *A. pullulans* from temperate regions is cultivated on a synthetic medium incorporating cell walls of the Msasa tree, it can generate endoglucanase activity ranging from 2.4 to 12.9 μmol glucose/mg protein/h and exoglucanase activity ranging from 0.3 to 22.4 μmol glucose/mg protein/day. It is found that enzyme production is stimulated by both carboxymethyl cellulose (CMC) and cell walls of the Msasa tree (Kudanga and Mwenje 2005). The yeast-like fungal *A. leucospermi* LB86 grown in wheat bran as inducer of hydrolytic enzyme production was found to produce cellulase, xylanase, β-glucosidase and β-xylosidase. The produced cellulase and xylanase have optimal activity at pH 5.0–6.5 °C and at temperatures 55.0–55 °C. However, the produced β-glucosidase and β-xylosidase show the maximum activity at 40 °C and exhibit moderate tolerance to glucose, xylose and arabinose, maintaning residual activities above 60% at the concentrations of up to 0.5 M. Furthermore, β-glucosidase activity produced by *A. leucospermi* LB86 is not affected by the presence of 10% (v/v) ethanol (Roberto Ribeiro et al. 2026).

All results indicate that some strains of *Aureobasidium* spp. have the capability to produce all three types of cellulases: endoglucanases, exo-cellobiohydrolase, and β-glucosidases, or cellobiase, for the comprehensive degradation of cellulose. This demonstrates that some strains of *Aureobasidium* spp. are also good bioresources of cellulases and better than the filamentous fungi.

Upon analyzing the sequenced genomic DNAs from various strains of *Aureobasidium* spp. mentioned in this book, the results unequivocally confirm the presence of all the genes encoding enzymes required for cellulose degradation in *Aureobasidium* spp. (Tables 6.1 and 6.2). In contrast, *Scheffersomyces stipitis*, a proficient xylose-utilizing yeast, is found to lack most enzymes essential for cellulose degradation and lytic polysaccharide monooxygenase (Wei et al. 2015) (Tables 6.1 and 6.2). However, *T. reesei*, a renowned fungus as a cellulase producer (Shen and Sun 2021), exhibits fewer copies of β-glucosidases from the GH3 family (Fig. 6.1). Hence, the synergistic collaboration between cellulases and accessory hemicellulases, as elaborated below with regards to *Aureobasidium* spp., could further enhance the degradation of cellulose.

6.2.3 Xylanase

Hemicelluloses, typically found in plant cell walls alongside cellulose and lignin, constitute approximately one-third of the renewable organic carbon on Earth, primarily in the form of xylan. Xylan, the predominant polysaccharide component of

Table 6.1 The genes encoding endoglucanases (EGs) in *Aureobasidium* spp., *T. reesei* and *S. stipitis*

Strain		Enzyme (accession No)	Family	Reference
Aureobasidium spp.	110374	EG1 (KEQ64414.1)	GH7	EG1 P07981
		EG2 (KEQ64359.1, KEQ63578.1)	GH5	EG2 P07982, O74706
	A62921	EG1 (KEQ63408.1-D2)	GH7	EG1 P07981
		EG2 (KEQ64359.1, KEQ63578.1)	GH5	EG2 P07982, O74706
	CBS105.22	EG1 (A04798, A05175)	GH5	EG1 P07981
		EG2 (A05125, A04904, A09569, A15996)	GH5	EG2 P07982, O74706
	HN6.2	EG1 (g4902)	GH7	EG1 P07981
		EG2 (g4786, g5203)	GH5	EG2 P07982, O74706
	P5	EG1 (A3478)	GH5	EG1 P07981
		EG1 (A2875, A4241)	GH5	EG2 P07982, O74706
	P6	EG1 (A5683)	GH5	EG1 P07981
		EG2 (A4494, A0111)	GH5	EG2 P07982, O74706
	P16	EG1 (A9213)	GH7	EG1 P07981
		EG2 (A9268, A1214)	GH5	EG2 P07982, O74706
	P25	EG1 (A6331)	GH5	EG1 P07981
		EG2 (A0122, A1214)	GH5	EG2 P07982, O74706
	DH177	EG1 (A04225)	GH7	EG1 P07981
		EG1 (A03944, A07250)	GH5	EG2 P07982, O74706
	TN3-1	EG1 (A09782, A14780)	GH5	EG1 P07981
		EG2 (A04578, A09677, A15573, A16813)	GH5	EG2 P07982, O74706
	XJ5-1	EG1 (M437DRAFT_47104-D4)	GH7	EG1 P07981
		EG2 (M437DRAFT_64904-D2, M437DRAFT_83954-D2)	GH5	EG2 P07982, O74706
T. reesei	QM6a	EG1 (XP_006965674.1)	GH5	EG1 P07981
		EG2 (XP_006962383.1, XP_006969897.1)	GH5	EG2 P07982, O74706
S. stipitis	CB S6054	/	/	EG1 P07981
		/	/	EG2 P07982, O74706

Table 6.2 The genes encoding cellobiohydrolases (CBHs) in *S. stipitis, T. reesei* QM6a, and different strains of *Aureobasidium* spp. mentioned in this book

Strain		Enzyme (accession no)	Family	Reference
Aureobasidium spp.	110,374	CBH1 (KEQ63408.1) CBH2 (KEQ63409.1)	GH7 GH6	CBH1 P62694 CBH2 P07987
	A62921	CBH1 (KEQ63408.1) CBH2 (KEQ63409.1)	GH7 GH6	CBH1 P62694 CBH2 P07987
	CBS105.22	CBH1 (A15447, A09461) CBH2 (A09460, A15448)	GH7 GH6	CBH1 P62694 CBH2 P07987
	HN6.2	CBH1 (g5124) CBH2 (g5127)	GH7 GH6	CBH1 P62694 CBH2 P07987
	P5	CBH1 (P5GA4058) CBH2 (P5GA4057)	GH7 GH6	CBH1 P62694 CBH2 P07987
	P6	CBH1 (P6GA5683) CBH2 (P6GA5684)	GH7 GH6	CBH1 P62694 CBH2 P07987
	P16	CBH1 (A4883) CBH2 (A4882)	GH7 GH6	CBH1 P62694 CBH2 P07987
	P25	CBH1 (P25GA6331) CBH2 (P25GA6330)	GH7 GH6	CBH1 P62694 CBH2 P07987
	DH177	CBH1 (A04225) CBH2 (A04224)	GH7 GH6	CBH1 P62694 CBH2 P07987
	TN3-1	CBH1 (A09782, A14780) CBH2 (A14781, A10667)	GH7 GH6	CBH1 P62694 CBH2 P07987
	XJ5-1	CBH1 (M437DRAFT_47104-D2) CBH2 (M437DRAFT_46665-D2)	GH7 GH6	CBH1 P62694 CBH2 P07987
T. reesei	QM6a	CBH1 (XP_006969224.1) CBH2 (XP_006962580.1)	GH7 GH6	CBH1 P62694 CBH2 P07987
S. stipitis	CBS6054	/ /	/ /	CBH1 P62694 CBH2 P07987

hemicelluloses, is composed of D-xylose units interconnected via β-1,4-glycosidic bonds in the main chain, with various carbohydrates (such as uronic acids, L-rhamnose, and oligosaccharides) branching off as the side chains. Due to its intricate composition, the comprehensive breakdown of xylan necessitates the coordinated activity of multiple hydrolytic enzymes, including endoxylanase (EC 3.2.1.8) and β-xylosidase (EC 3.2.1.37). The endo-β-1,4-xylanase plays a crucial role in hydrolyzing the internal glycosidic bonds of the main chain, making it one of the primary xylanolytic enzymes. In the absence of β-xylosidase, the breakdown process can result in the release of smaller and oligomer xylooligosaccharides. β-xylosidase then enzymatically cleaves individual unsubstituted xylose units from the non-reducing ends of xylo-oligosaccharides to release xylose. The thorough conversion of hemicellulose necessitates the involvement of various enzymes that cleave both the main chain and side chains, including endoxylanase, β-xylosidase, α-glucuronidase, α-arabinofuranosidase, and acetylxylan esterase. All the xylanases find diverse applications in industries such as papermaking, fermentation, food processing, juice clarification, xylooligosaccharide production, and waste treatment. To date, xylan-rich substrates derived from agricultural by-products and wastes, including corn cobs, wheat bran, rice straw, sugarcane bagasse, hazelnut shells, soybean residue, and sorghum biomass, have been intensively investigated for xylanase production by various fungal strains.

In particular, the xylo-oligosaccharides liberated from xylan exhibit a sweet and edible taste. These non-digestible oligomers, composed of two to ten xylose units connected by β-1,4 bonds, possess the unique ability to traverse the upper gastrointestinal tract of humans and animals without undergoing degradation. This property stimulates the proliferation of beneficial probiotic bacteria, such as *Bifidobacterium* spp. and *Lactobacilli* spp., while inhibiting the growth of harmful microorganisms in the human intestine. Therefore, in the food industry, these compounds can also serve as low-calorie sweeteners. Notably, xylobiose and xylotriose, among others, demonstrate the greatest prebiotic potential. Moreover, it has been discovered that fast-growing invasive tropical weeds, including grasses and aquatic plants, represent ideal sources of xylans because of their plentiful availability, minimal economic worth, and the necessity for their eradication. Such obtained xylans are suitable substrates for the production of xylo-oligosaccharides. Recent findings also have highlighted the superior qualities of xylo-oligosaccharides compared to any other non-digestible sugars, such as fructooligosaccharides from inulin and sucrose, as stated in Chap. 7. Xylo-oligosaccharides are known for their stability in high temperatures and acidic environments, positive impacts on human health at low

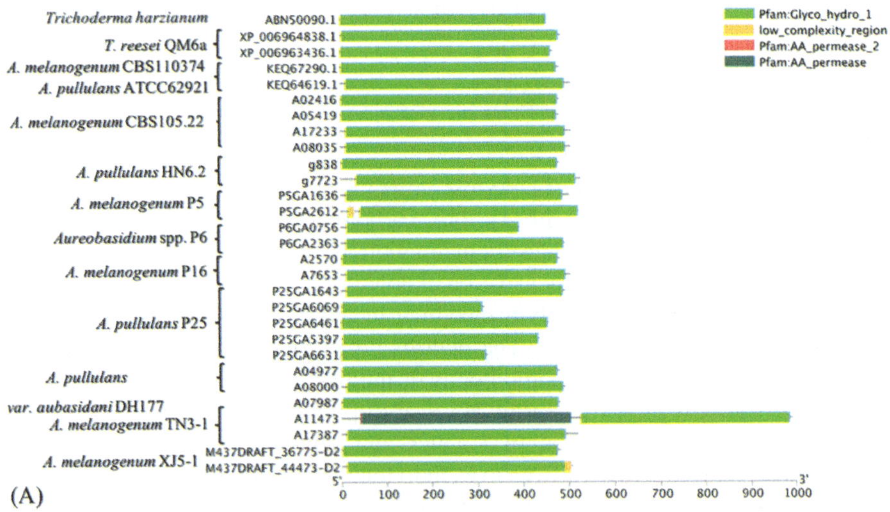

Fig. 6.1 Alignments of the domains of β-glucosidase 2 (BGL2) (**A**) and β-glucosidase 1 BGL1 (**B**) from *T. reesei* QM6a. *T. harzinum, S. stipitis* CBS6054, and all the strains of *Aureobasidium* spp. mentioned in this book

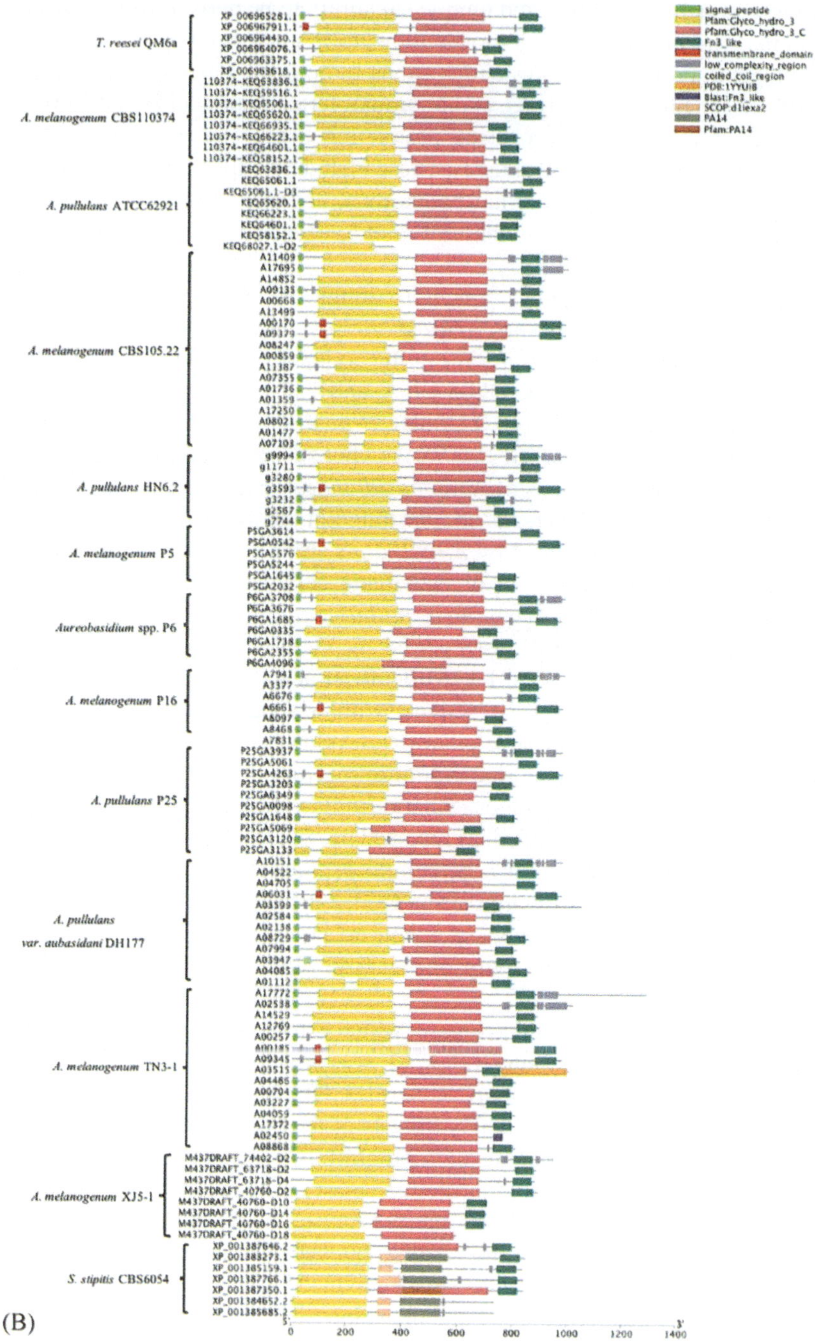

(B)

Fig. 6.1 (continued)

daily doses, competitive pricing, and unique capability to be derived from lignocellulosic biomass as stated above. Moreover, they are characterized by non-toxic and non-carcinogenic properties, along with various biological benefits. In contrast to autohydrolysis, chemical hydrolysis, and microbial fermentation, enzymatic hydrolysis of xylan for xylo-oligosaccharide production using endoxylanase without β-xylosidase offers numerous advantages. These include the specificity of xylanases and other auxiliary enzymes in xylan breakdown, absence of toxic by-products, eco-friendly processing, minimal release of monosaccharides, and operation under the mild reaction conditions.

Prominent xylanase-producing filamentous fungi encompass *Aspergillus* spp., *Disporotrichum* spp., *Penicillium* spp., *Neurospora* spp., *Fusarium* spp., *Chaetomium* spp., and *Trichoderma* spp. For instance, the xylanase activities generated by *A. niger* and *A. fumigatus* can reach 89.5 U/mL and 125.1 U/mL, surpassing those produced by *A. nidulans* KK-99 (40.0 U/mL), *Penicillium oxalicum* (14.5 U/mL), and *Chaetomium thermophilum* (61.0 U/mL) (Khonzue et al. 2011). However, in recent discoveries, various strains of *Aureobasidium* spp. also have been identified as exceptional xylanase producers, as demonstrated in the following.

The xylanase enzyme produced by *A. pullulans* strain CCT 1261 cultivated on rice bran exhibits the highest level of enzymatic activity. When cultivated in a medium containing 61.9 g/L of rice bran, *A. pullulans* strain CCT 1261 can achieve a xylanase activity of 82.2 U/mL, as reported by Victoria Gauterio et al. (2020). Additionally, *A. pullulans* strain CCT 1261 demonstrates the capability to generate 74.9 U/mL of endo-β-1,4-xylanase activity with a specific activity measuring 158.7 U/mg protein. When the crude endo-β-1,4-xylanase is utilized to hydrolyze untreated rice bran obtained from rice grain milling for a duration of 12 h, a total concentration of 1.7 mg of xylo-oligosaccharides/mL, primarily consisting of xylobiose and xylotriose, can be obtained, as reported by Victoria Gauterio et al. (2018).

Oat hulls are known to consist of 7.0% cellulose, 28.0–35.0% hemicellulose, and 19.0–23.0% lignin. When *A. pullulans* strain CCT 1261 is cultured in oat hulls pretreated with alkaline and ultrasound, it is able to produce a xylanase activity of 30.2 U/mL and achieve a productivity rate of 0.42 U/mL/h (Blumberg Machado et al. 2021).

The crude and partially purified xylanase, which exhibits low β-xylosidase activity, produced by *A. pullulans* strain CCT 1261, was also employed for hydrolyzing beechwood xylan to obtain xylooligosaccharides. The xylanase enzymes exhibit optimal activity at 50 °C and pH 4.5. Hydrolysis of the beechwood xylan can yield the total xylooligosaccharide contents of 7.7 mg/mL or 7.9 mg/mL when subjected to catalysis of the crude or partially purified xylanases. The resulting xylooligosaccharides are mainly composed of xylobiose (50.9% and 53.8%) and xylotriose (30.9% and 32.8%), according to the results obtained by Victoria Gauterio et al. (2021). Additionally, the endo-β-1,4-xylanase produced by *A. pullulans* strain CCT 1261 exhibits the optimal temperature and pH values of 50 °C and 4.5, respectively. The K_m and V_{max} values of the purified xylanase, when utilizing beechwood xylan, are 74.9 mg/mL and 476 U/mL, respectively. The presence of salts, such as $CaCl_2$, in the reaction system can enhance the xylanase activity. The xylanase obtained, when incubated at 45 °C and pH 4.5, exhibits superior thermal stability (half-life = 169 h) compared to any other conditions. Under these specified conditions and in the presence of Ca^{2+} (10 mM/L), the enzyme exhibits even greater stability, with a half-life of 231 h. Under the mentioned optimal conditions, it is possible to achieve a total xylooligosaccharide content of 6.7 mg/mL, and a xylan conversion to xylooligosaccharides is 22.3% within a 2-h incubation period. Major components found in the hydrolysates include xylobiose, xylotriose, and xylose. Furthermore, the addition of Ca^{2+} ions enhances the concentration of xylobiose in the hydrolysate while reducing the content of xylose (Patipong et al. 2019).

The xylanase activity produced by *A. melanogenum* strain PBUAP46 grown in the production medium containing 3.9% (w/v) rice straw reaches 5.2 ± 0.1 U/mL. The crude enzyme functions effectively and exhibits relative stability at pH 7.0 and room temperature. The produced xylanase actively hydrolyzes xylans from cogon grass, napier grass, and vetiver grass, releasing 20.4 ± 0.8, 17.5 ± 0.3, and 19.4 ± 0.4 mg of reducing sugar per 100 mg of xylan, respectively. The digestion of vetiver grass xylan at 27.9 U of xylanase per gram of xylan over 92 h yields the highest amount of reducing sugars (23.7 ± 1.3 mg per 100 mg of xylan). The resulting hydrolysate predominantly contains xylobiose and xylotriose, which are known to promote the growth of *L. brevis* and *L. casei* (Patipong et al. 2019).

When cultivated in a xylanase production medium with corncob (1.0% w/v) as the sole carbon source, the strain *A. pullulans* NRRL 58523 can produce 22.4 ± 0.3 U/mL of endoxylanase activity. The crude endoxylanase from *A. pullulans* strain NRRL 58523 effectively hydrolyzes the extracted arabinoxylan from sugarcane leaf at 50 °C and pH 4.0, demonstrating relative stability under these conditions (87.4 ± 1.3% of the activity remains after 60 h). Under the optimum conditions (50 °C, pH 4.0, 65.31 U/g of xylan, 53 h), the endoxylanase hydrolysis can yield 237.5 ± 17.7 mg of xylobiose and xylotriose per gram of xylan. The xylooligosaccharides exhibit species-specific prebiotic activity toward *L. brevis* TISTR 868, *L. casei* TISTR 390, and *L. casei* subspecies *rhamnosus* TISTR 047 (Na Nongkhai et al. 2023). This enzymatic hydrolysis is considered more favorable than chemical acid hydrolysis due to its specificity and absence of toxic byproducts like furfural.

A. pullulans strain Y-2311-1 is another yeast-like fungus known for xylanase production. When *A. pullulans* Y-2311-1 is cultured in a medium containing 1.0% oat spelt xylan or xylose for a duration of 4 days, it generates four primary xylanases. The enzyme, characterized by a pI of 9.4 and a molecular weight of approximately 25 kDa, displays a purified xylanase with a K_m of 7.6 mg/mL and a V_{max} of 2650 μM/min/mg towards birchwood xylan at 28 °C and pH 4.5. Its optimal enzymatic activity is observed at pH 4.8 and 54 °C (Li et al. 1993). The *XYNA* gene, isolated from *A. pullulans* NRRL Y-2311-1, comprises 895 base pairs, encoding a protein of 221 amino acids with an estimated mass of 23.5 kDa, and harbors a signal peptide composed of 34 amino acids. The produced xylanase is classified within the glycoside hydrolase family G. Nevertheless, the gene expression is inhibited by glucose, as reported by Li and Ljungdahl (1994). Similarly, the cultivation of *A. pullulans* NRRL Y-2311-1 on globe artichoke stem yields xylanase activity levels of 66.5 U/mL. The optimal conditions for the xylanase action are at pH 4.0 and a temperature range of 30–50 °C. Additionally, the enzyme exhibits remarkable stability across a broad pH spectrum of 3.0–8.0. The crude xylanase exhibits significant activity in saccharifying corn cob pretreated with aqueous ammonia, yielding 357 mg of reducing sugar per gram of dry substrate. Hence, the crude xylanase sourced from *A. pullulans* presents itself as a promising alternative for the saccharification of lignocellulosic biomass in the context of biological fuel production. Furthermore, the globe artichoke stem represents an affordable and renewable lignocellulosic biomass, ideal as a substrate for the economical production of xylanases (Yegin 2016).

An extremophilic xylanase derived from *A. pullulans* NRRL Y-2311-1 cultured on wheat bran has demonstrated potential for its large-scale application in the bread-making industry. The utilization of such novel xylanase can enhance water absorption, dough development time, and stability, while reducing the dough softening degree and mixing tolerance index at a dosage of 100 U/100 g flour. The xylanase at a concentration of 125 U/100 g flour shows a significant 30% enhancement in bread specific volume compared to the available commercial xylanases. The moisture content of all the bread samples falls within the optimal range of 35–40%. Moreover, the xylanase has been identified as the most efficient enzyme in reducing crumb firmness (Yegin et al. 2018). Incorporation of the xylanase (100 U/100 g flour) from *A. pullulans* NRRL Y-2311-1 into rice flour-based cookies yields soft doughs with reduced absolute stickiness. Compared to wheat flour-based cookies, the hardness and absolute stickiness of the rice dough are notably decreased (Yegin et al. 2024). Therefore, the extremophilic xylanase derived from *A. pullulans* NRRL Y-2311-1 offers numerous advantages compared to any other xylanases and has high potential for large-scale application in the bread-making industry.

The melanin-free strain of *A. pullulans* 2012 cultivated in a medium containing 60.0 g/L wheat bran yields 545 mM/L ferulic oligosaccharides, 64.1 U/mL xylanase, and 0.14 U/mL ferulic acid esterase. Such hemicellulases from *A. pullulans* are primarily induced by xylan and wheat bran, but their activities are also inhibited by glucose and sucrose. The xylanase and ferulic acid esterase enzymes are secreted into the culture broth, whereas β-xylosidase is localized in the hyphal filaments, and arabinofuranosidase is a membrane-bound enzyme (Yu and Gu 2014). The released ferulic oligosaccharides can function as a natural antioxidant, guarding against hemolysis in mouse red blood cells and exhibiting positive effects against Fe^{2+}, H_2O_2, and hydroxyl radicals. Simultaneously, ferulic oligosaccharides can promote the growth of *Lactobacillus* spp. and Bifidobacteria while inhibiting the growth of gram-positive bacteria in the human body and lymphocytes in leukemia patients. Hence, they can serve as a component of health-promoting food for humans and as a component of animal feeds.

Furthermore, certain strains of *A. melanogenum* demonstrate the ability to produce an acidophilic endo-1,4-β-xylanase. For instance, the acidophilic endo-1,4-β-xylanase generated by *A. melanogenum* (ATCC 20524) when it was cultivated on oat-spelt xylan exhibits an apparent molecular weight of 24 kDa, an isoelectric point of 6.7, and its optimal activity at pH 2.0 and 50 °C. The xylanase gene (*XYN1* gene) has a length of 663 base pairs, encoding a mature protein consisting of 187 amino acids along with a signal peptide of 34 amino acids. Upon expressing the

XYN1 gene extracted from cDNA of *A. melanogenum* (ATCC 20524) in *S. cerevisiae*, the resultant recombinant xylanase exhibits an optimal pH of 4.8, akin to the xylanase derived from the color variant strain NRRL Y-2311-1 of *A. pullulans* (Ohta et al. 2001).

Under the specific conditions, certain strains of *A. pullulans* cultivated on xylose demonstrate the production of hemicellulolytic enzymes, predominantly exhibiting xylanase and β-xylosidase activities while lacking cellulase activity. Such enzymes from *A. pullulans* display the capability to selectively hydrolyze the hemicellulose component of unbleached sulfite pulps, facilitating the production of dissolving pulp. The pulp treated with the *A. pullulans* enzymes shows lower levels of xylan (48%) and glucomannan (15%) compared to the untreated control of dissolving pulp (Christov and Prior 1996). So, such xylanase and β-xylosidase can be used in paper pulp making.

Additionally, certain strains of *Aureobasidium* spp. are capable of producing a cell-associated β-xylosidase. For instance, *A. pullulans* strain ATCC 20524 grown on oat spelt xylan produces a cell-associated β-xylosidase belonging to glycoside hydrolase family 3. The β-xylosidase exhibits an apparent molecular weight of 88.5 kDa and demonstrates a specific activity of 288 U/mg of protein for pNP-β-D-xylopyranoside. The optimal pH and temperature for the β-xylosidase action are 3.5 and 70 °C, respectively. It exhibits K_m and V_{max} values of 3.5 mM and 263 μM/mg/min for pNP-β-D-xylopyranoside. Additionally, 35.8% of pNP-β-D-arabinofuranoside can be hydrolyzed under these optimal conditions. The β-xylosidase gene (*XYLI* gene) derived from the genome of *A. pullulans* strain ATCC 20524 spans 2415 base pairs, encoding a mature protein consisting of 785 amino acid residues. The protein includes a 20-residue signal peptide and exhibits a calculated molecular weight of 85.0 kDa. Within this β-xylosidase enzyme, Asp-294 and Glu-498 serve as the nucleophile and acid/base, respectively, pivotal for catalyzing the enzymatic reaction. Upon expression of the cloned *XYLI* gene from the cDNA of *A. pullulans* strain ATCC 20524 in the yeast *Pichia pastoris*, the resulting recombinant β-xylosidase activity reaches 14.9 U/ml as reported by Ohta et al. (2001).

The presence of α-L-arabinofuranoside substitutions on arabinoxylan has been noted to impede the activity of endoxylanases and β-xylosidases. α-L-Arabinofuranosidases (EC 3.2.1.55), also known as α-L-arabinofuranoside arabinofuranohydrolases, are exoglycosidases responsible for cleaving the terminal non-reducing L-arabinofuranoside residues found in hemicelluloses like arabinan, arabinoxylan, and arabinogalactan. In arabinan, the backbone comprises α-(1 → 5)-linked L-arabinofuranose units, accompanied by α-(1 → 2)- and α-(1 → 3)-linked side chains, whereas arabinogalactan features α-(1 → 3)- and α-(1 → 6)-linked side chains of L-arabinofuranose. The α-L-arabinofuranosidase gene (*ABFB* gene) cloned from *A. pullulans* ATCC 20524 spans 2097 base pairs, encoding a mature protein weighing 74.2 kDa with a pI of 4.95, along with a 17-residue signal peptide. The pivotal amino acids Glu-362 and Glu-440 within the protein function as the acid/base and nucleophile, respectively, in catalytic reactions. Following the expression of the *ABFB* gene in *P. pastoris*, the resulting recombinant AbfB exhibits an apparent molecular weight of 110 kDa and demonstrates an enzyme activity of 56.6 U/mg of protein under the optimal conditions of pH 4.5 and 75 °C. The AbfB enzyme, when acting on pNP α-L-arabinofuranoside, displays K_m and V_{max} values of 6.27 mM and 78.1 mM/mg/min, respectively. As reported by Ohta et al. (2013), the recombinant enzyme displays significant activity toward rye arabinoxylan and pNPα-L-arabinofuranoside. This highlights the capability of the recombinant AbfB to effectively hydrolyze L-arabinofuranoside residues present in hemicelluloses.

The collective results outlined above affirm that various strains of *Aureobasidium* spp. are capable of producing a diverse array of xylanases, facilitating the effective degradation of varying hemicellulosic compositions. Therefore, these xylanases generated by *Aureobasidium* spp. can serve as valuable bioresources in enzymatic applications.

After analyzing the sequenced genomic DNAs from various strains of *Aureobasidium* spp. mentioned in this book, the results unequivocally reveal the presence of the genes encoding enzymes essential for hemicellulose degradation and xylose utilization within the genomes of *Aureobasidium* spp. (Figs. 6.2, and 6.3). However, *S. stipitis* lacks many of the enzymes necessary for hemicellulose degradation (Figs. 6.2, and 6.3). On the contrary, *T. reesei* lacks certain accessory hemicellulases and possesses a lower number of copies of β-glucosidases from the GH3 family, endo-1,4-β-xylanases, and other accessory hemicellulases (Figs. 6.2, and 6.3). These findings align with the earlier observations shown in this chapter. Thus, based on the genomic DNA analysis, it is evident that various strains of *Aureobasidium* spp. possess significant potential for hemicellulose degradation, particularly xylan and the different kinds of xylanases from *Aureobasidium* spp., and their functional genes have many uses in biotechnology.

Very recently, it has been found that the marine-derived yeast-like fungus, *A melanogenum* TN2-1-2, can efficiently co-utilize glucose and xylose for pullulan production. The AmXlnR encoded by the *AmXLNR* gene, a transcription factor, controls xylan degradation and pentose catabolism (D-xylose and L-arabinose) in *A. melanogenum* TN2-1-2. AmXlnR, which has a superfamily-specific motif, Cys-X6-Cys-X5-Cys-X2-Cys-X6-Cys, can function both as an activator of xylanase and β-xylosidase genes by binding to the conserved sequences 5′-GGNT(A/t/g)A-3′ of the promoters of these two genes and a repressor of α-L-arabinofuranosidase genes by binding to the conserved sequences 5′-GGC(T/a)AT-3′ of the promoters of these genes (Zhao et al. 2026).

6.2.4 Inulinase

Inulin is a linear β-(2,1)-linked fructose polymer found as a reserve carbohydrate in Jerusalem artichoke, dahlia tubers, chicory, and yacon root. Inulin sourced from Jerusalem artichoke serves as a renewable, cost-effective, and abundant raw material for producing ethanol, single-cell protein, single-cell oil, ultrahigh fructose syrup, inulo-oligosaccharide, citric acid, 2,3-butanediol, lactic acid, sugar alcohols, pullulan, and various other valuable products (Chi et al. 2011). Moreover, crops containing inulin, such as Jerusalem artichoke—a non-grain crop—possess the

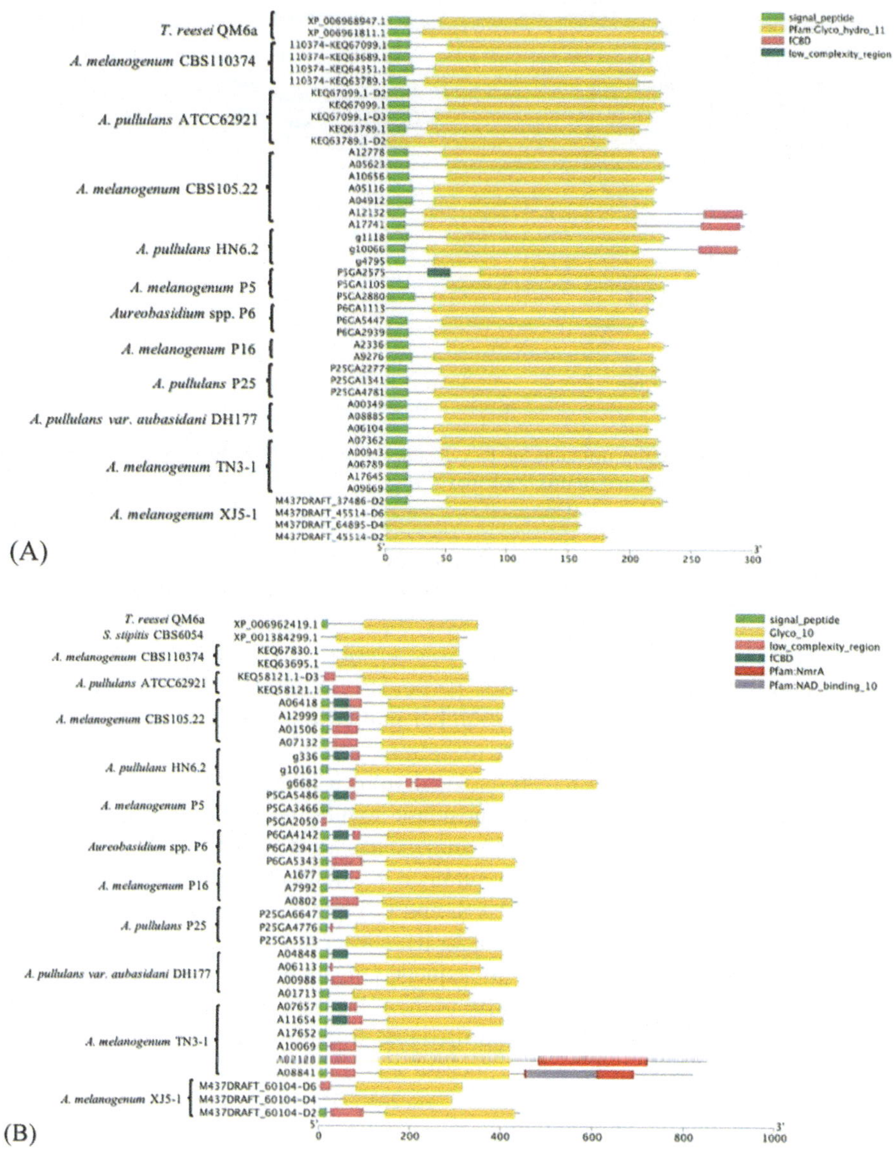

Fig. 6.2 Alignments of the domains of endo-1,4-β-xylanases from GH11 (**A**) and GH10 (**B**) families in all the strains of *Aureobasidium* spp. mentioned in this book, *T. reesei* and *S. stipitis*

ability to withstand insect and fungal infections, and thrive in nutrient-deficient soils. Inulinases are categorized into exo-inulinase and endo-inulinase. Exo-inulinase acts on the terminal fructose residues at the non-reducing end of the inulin molecule, while endo-inulinase targets the internal linkages within the inulin molecule, leading to the production of inulooligosaccharides, such as inulotriose,

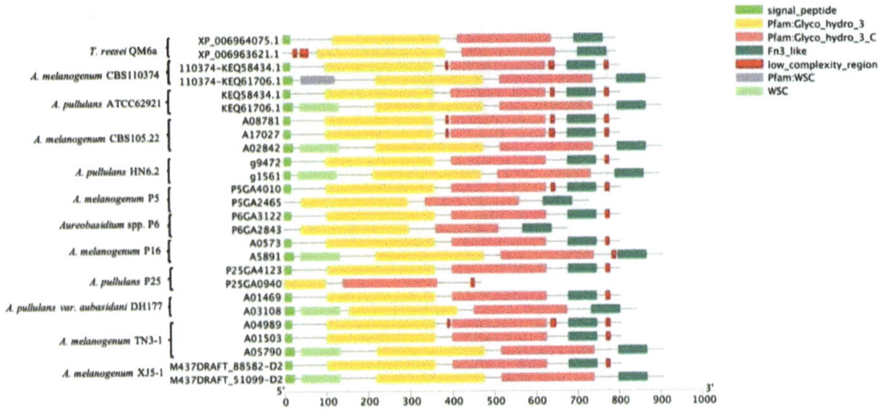

Fig. 6.3 Alignments of the domains of β–xylosidases from GH3 family in all the strains of *Aureobasidium* spp. mentioned in this book and *T. reesei*

inulotetraose, and inulopentaose as primary products. Inulinase, among hydrolytic enzymes, has garnered significant interest due to its versatile applications in the production of fuel ethanol, high fructose syrup, and inulooligosaccharides derived from inulin.

It is well documented that yeasts, particularly *Kluyveromyces marxianus* and *Kluyveromyces fragilis*, demonstrate higher inulinase production than filamentous fungi and bacteria (Chi et al. 2008).

The fungus *A. hainanensis* P6, isolated from a mangrove ecosystem, has the capability to produce β-D-fructofuranosidase, exhibiting inulin-hydrolyzing activity at a level of 31.0 ± 0.8 U/mL. The enzyme has a molecular weight of 47.6 kDa, and its encoding gene (accession number KU308553) harbors the conserved sequences MNDPNGL, RDP, ECP, FS, and Q typical of the glycoside hydrolase GH32 family. However, the enzyme lacks the conserved sequence SVEVF found in inulinases. The complete amino acid sequence of the protein from *A. hainanensis* P6 strain shows a significant similarity to the β-fructofuranosidase sequence found in various other fungal strains. The purified β-D-fructofuranosidase shows the capability to effectively hydrolyze both sucrose and inulin, demonstrating transfructosylating activity when acting on 30.0% sucrose, thereby converting sucrose into fructooligosaccharides (FOS). Following the deletion of the β-D-fructofuranosidase gene, the disruptant retains low inulin hydrolyzing and invertase activities, along with a minimal level of transfructosylating activity. These results suggest the presence of a gene encoding an inulinase in *A. hainanensis* P6 strain (Jiang et al. 2016). The β-D-fructofuranosidase (EC 3.2.1.26) (FFase), invertase, and inulinase enzymes are classified within the glycoside hydrolase GH32 family. FFase has the capacity to hydrolyze sucrose, raffinose, and inulin into glucose and fructose (Rustiguel et al. 2015), whereas invertase specifically catalyzes the hydrolysis of sucrose. Therefore, it has wide applications in the food, beverage, and brewing industries, as well as in the fermentation processes for the production of alcohol, lactic acid, glycerol, and artificial honey. Additionally, the FFase finds utility in cosmetics, pharmaceuticals, and paper production (Jiang et al. 2016).

It has been observed that *K. marxianus* strain KM exhibits significant inulinase activity levels exceeding 125 U/mL, as reported by Liu et al. (2013) and Li et al. (2022). Consequently, the *INU1* gene responsible for inulinase in *K. marxianus* KM is actively expressed in the high pullulan producer *A. melanogenum* strain P16, isolated from the mangrove ecosystem (Ma et al. 2015). The recombinant strain EI36 demonstrates an extracellular inulinase activity of 40.9 U/mL, whereas its wild-type strain P16 produces only 7.6 U/mL of inulinase activity. In a 10-Liter fermentation process, the culture achieves 70.6 ± 1.3 g/L of pullulan from the hydrolysates of inulin (138.0 g/L) within 108 h. Notably, a high inulinase activity of 42.0 U/mL is achieved within the initial 60 h of the process. The majority of reducing sugars derived from inulin, under the action of the recombinant inulinase, are effectively utilized by the recombinant strain EI36. This observation verifies the efficacy of the genetically modified yeast strain of *A. melanogenum* P16 and its associated inulinase for direct pullulan production from inulin, as established by Ma et al. (2015).

The oleaginous yeast *A. melanogenum* strain P10, obtained from the mangrove ecosystem in Hainan Province, China, exhibits a remarkable capability to synthesize substantial amounts of intracellular lipids, making it a promising candidate for biodiesel production, as reported by Wang et al. (2014). To produce high level of intracellular lipids from inulin, the *INU1* gene encoding inulinase from *K. marxianus* strain KM is expressed in the single-cell oil producer *A. melanogenum* strain P10. The resulting transformant, API41, is capable of generating 28.5 U/mL of extracellular inulinase activity, a significant improvement compared to its wild-type P10 strain, which produces only 8.6 U/mL of inulinase activity. In a 10-Liter fermentation process, the transformant API41 accumulates 66.2% (w/w) lipid content in the yeast-like fungal cells and reaches a cell dry weight of 14.4 g/L from the hydrolysates of 80.0 g/L inulin within 120 h. Moreover, the fermentation results in the production of high inulinase activity measuring 23.7 U/mL within the initial 72 h. The lipids stored in the cells of the transformant API41 were extracted and utilized as the raw materials for biodiesel production. Subsequently, the fatty acids present in the extracted lipids from the yeast transformant API41 were subjected to transmethylation process. Analysis reveals that the lipids comprise 30.0% palmitic acid ($C_{16:0}$), 2.1% palmitoleic acid ($C_{16:1}$), 6.1% stearic acid ($C_{18:0}$), 35.6% oleic acid ($C_{18:1}$), and 26.0% linoleic acid ($C_{18:2}$). It was observed that the produced fatty acid methyl esters, intended as biodiesel, exhibit favorable combustion characteristics, as noted by Li et al. (2018). This indicates that inulin can be directly and efficiently utilized for lipid production by the yeast transformant API41, leading to the subsequent preparation of biodiesel from the generated lipids.

All the results mentioned above suggest that the recombinant inulinases produced by *A. melanogenum* effectively hydrolyze inulin into glucose and fructose, which can subsequently be converted into pullulan and intracellular lipids for biodiesel production. Furthermore, our findings indicate that inulin, extracted from the tubers of Jerusalem artichoke, is much more readily decomposed than cellulose and semicellulose from lignocellulosic biomasses. Therefore, inulin extracted from the tubers of Jerusalem artichoke is also an excellent, cheap, and renewable raw material in biotechnology. In the future, it as a none-grain material will replace starch for production of various fermentation products by synthetic biotechnology. This will be of very significant to keep food and grain security in the developing countries.

6.2.5 Pectinase

Pectin is a polysaccharide known for its abundance of essential sugars, mainly galacturonic acid and methanol. It exhibits a high molecular weight, heterogeneous, and acidic structural compositions, making it a significant component found in cereals, vegetables, fruits, and fibers. It forms a matrix within the cell wall, entwining with cellulose and hemicelluloses to create a network structure. Pectic substances are the complex colloidal acidic polysaccharides characterized by extensive galacturonic acid chains interconnected through glycosidic bonds. It can be categorized into various forms, including pectic acid, pectinic acid, pectin, and protopectin. Pectinases are a group of enzymes responsible for the degradation of pectin. They encompass protopectinase, pectin methylesterases, endo-polygalacturonase (EC 3.2.1.15), and exo-polygalacturonase (EC 3.2.1.67), with the latter two encoded by the *PCT1* gene. Protopectinase, a subclass of pectinase, functions to break down insoluble protopectin found in unripe fruits into highly polymerized soluble pectin. Pectin methylesterases are the enzymes responsible for removing methoxyl groups from pectic substances, ultimately leading to the production of pectic acid and methanol, a toxic byproduct. Polygalacturonases (PGs) are the pectinolytic enzymes that catalyze the hydrolytic cleavage of α-1,4-glycosidic linkages between α-D-galacturonic acid units within the pectin scaffold by introducing a water molecule across the oxygen bridge through both endo- and exo-actions. In general, pectin methylesterase is not suitable for clarifying fruit juice and wine, as it can lead to the release of toxic methanol. On the contrary, polygalacturonase, with its non-specific action on pectic substances, serves as a key enzyme enabling the producers to effectively penetrate and break down plant cells, even when plants have developed defenses against such invasion processes. Pectinases are extensively utilized in the food industry for various purposes, including the clarification of fruit juices and wines, enhancement of juice yield, control of haze and cloudiness, oil extraction, and coffee and tea production for the removal of mucilage from coffee beans and the promotion of tea fermentation. Additionally, they are involved in the degumming of plant fibers, have applications in the paper and textile industry, can be integrated into animal feeds, and play a role in wastewater treatment processes. To date, various fungal species, including *Aspergillus* spp. such as *A. niger*, *A. oryzae*, *A. fumigatus*, *A. terreus*, *A. sojoe*, *A. awamori*, *Penicillium* spp., *Fusarium* spp., *Mucor* spp., *Neurospora crassa*, and *Sclerotinia sclerotium*, have been identified as capable producers of pectinase, as reported by Dwivedi et al. (2023). For instance, the highest

pectinase production (15.5 U/mL) by *A. niger* ABT-5, isolated from soil, was attained after 72 h of incubation at a pH of 8.0 and a temperature of 30 °C, as reported by Dwivedi et al. (2023).

Polygalacturonases (EC 3.2.1.15) synthesized by *A. pullulans*, isolated from the Danube River, exhibit the ability to randomly cleave pectate, generating a blend of galactosiduronides with reduced polymerization. In contrast, the exopolygalacturonases (EC 3.2.1.67) from the same strain can specifically cleave pectate at the non-reducing end, resulting in the formation of D-galactopyranuronic acid. The active site configuration of these exopolygalacturonases resembles that of polygalacturonases found in phytopathogenic fungi (Stratilov et al. 2005). Additional strains of *A. pullulans* also exhibit the production of enzymes such as pectin methylesterase, exopolygalacturonases, and pectin lyases. Within the species of *A. pullulans*, D-galacturonic acid acts as the natural inducer for stimulating the formation of pectolytic enzymes, or potentially serves as a precursor to the natural inducer. In contrast, the presence of glucose in the culture functions as a suppressor of polygalacturonase production (Stratilova et al. 1998).

Moreover, *A. pullulans* isolated from the Saharan soil of Algeria demonstrates the capability to produce polygalacturonase with an activity level of 25.8 U/mL utilizing tomato pomace as the primary substrate. The resulting polygalacturonase exhibits stability across a temperature range of 5–90 °C and a wide pH range of 5.0–10, with an optimal temperature of 60 °C and pH of 5.0. Notably, it maintains full residual activity even after 1 h at 60 °C. The polygalacturonase functions as an exo-polygalacturonase, catalyzing the hydrolysis of polygalacturonic acid and releasing galacturonic acid as an end-product. Furthermore, this exo-polygalacturonase has practical applications in improving the clarity of apple and citrus juices within the food industry (Bennamoun et al. 2016).

Polygalacturonase activity produced by *A. pullulans* P56 cultivated on agro-industrial wastes like citrus pectin at 30 °C and pH 5.5 amounts to 8.6 U/mL. The molecular weight of the partially purified enzyme is 54.0 kDa and the peak enzyme activity is observed at 40 °C and pH 4.5 (Oskay 2022).

Even certain strains of *Aureobasidium* spp. are capable of producing cold-active pectinolytic enzymes. For instance, the cold-active pectinolytic enzymes, with activities reaching up to 113.9 U/mg, generated by *A. pullulans* strains are either constitutive or partially constitutive, allowing their production to remain unaffected by high glucose concentrations of 200.0 g/L. Moreover, the presence of SO_2 at 120 mg/L and ethanol at 15% (v/v) does not impact its pectinase production. These distinct cold-active pectinases, generated by *A. pullulans* strains, hold significant potential for serving as food processing aids in low-temperature winemaking. For instance, the presence of *A. pullulans* and its elevated pectinase activity in the GM-R-22 wines lead to higher total anthocyanins, total polyphenolic index, and color intensity than those in the control. This results in a 30% reduction in filtration times, enhanced levels of monomeric anthocyanins during fermentation, and increased polymeric pigments after 9 months of bottle storage (Gabriela Merín and de Ambrosini 2015, 2020). Therefore, *A. pullulans* and its elevated pectinase have many applications in wine industries.

6.2.6 Mannanase

Mannan and heteromannans constitute the primary components of the hemicellulosic fraction in softwoods, and they are commonly found in plant tissues. Mannan is composed of β-1,4-linked D-mannopyranose residues. β-Mannanase (EC 3.2.1.78) serves as the pivotal enzyme responsible for the random cleavage of β-mannosidic bonds within mannan and heteromannan structures. These β-mannanases can be categorized into two types: (1) β-mannanase (1,4-β-D-mannan mannohydrolyase; EC 3.2.1.78) and (2) β-mannosidase (β-D-mannoside mannohydrolyase; EC 3.2.1.25), based on the specific hydrolytic cleavage site along the main-chain polysaccharide. The β-mannanases, specifically endo-β-mannanases, catalyze the hydrolysis of mannan and heteromannans, yielding manno-oligosaccharides. These manno-oligosaccharides can undergo additional cleavage by β-D-mannosidase, β-D-glucosidase, and β-D-galactosidase, leading to the liberation of mannose, glucose, and galactose from both mannan and heteromannans (Lin and Chen 2004). Hence, the comprehensive biodegradation of mannan necessitates a cooperative and synergistic effort among β-mannanase, β-mannosidase, acetyl mannan esterase, β-glucosidase, and α-galactosidase (EC 3.2.1.22). This collaborative action is integral for the breakdown of both the main chains and side chains of mannan.

To date, it has been discovered that *Talaromyces trachyspermus* B168, *Lichtheimia ramose*, *Aspergillus kawachii* IFO 4308, *Trichoderma longibrachiatum* RS1, *Rhizopus microspores*, *A. nidulans*, *Aspergillus terreus* FBCC 1369, *A. niger* CTBS 513.88, and *A. oryzae* RIB4 are capable of producing β-mannanases (Dawood and Ma 2020). For instance, TtMan5A, a thermostable β-mannanase produced by *Trichoderma trachyspermus* isolated from *Picrasma quassioides* roots in Kodaira, Tokyo, Japan, exhibits an optimal temperature of 85 °C and robust catalytic activity across a broad pH range (Ichinose et al. 2017). This enzyme serves as an exceptional candidate for application in drilling operations. In addition, during solid-state fermentation, *A. terreus* SUK-1 isolated from Malaysia exhibits a maximal mannanase activity of 30.2 U/g (Abd Rashid et al. 2012).

Mannanases find many applications in various domains, such as the biobleaching of pulp in the paper industry, bioconversion of biomass wastes into fermentable sugars, enhancement of animal feed quality, viscosity reduction in coffee extracts, drilling activities, detergent production, textile processing, food industry, the development of functional foods, animal feed improvement, and bioethanol production. Mannanase can also be employed to eliminate different food stains that contain guar gum. Moreover, they hold potential for applications in the production of manno-oligosaccharides. These compounds are selectively utilized by intestinal *Bifidobacterium* species and serve as valuable food sweeteners or additives (Lin and

Chen 2004). The generated manno-oligosaccharides also exhibit anti-tumor and immune-modulating activities.

Recent studies have reported that the yeast-like fungi are capable of producing extracellular β-mannanase and cell-associated β-mannosidase. The distinct localizations of these enzymes offer an advantage for manno-oligosaccharide production by preventing the loss of manno-oligosaccharide yield resulting from β-mannosidase activities. Within yeasts and yeast-like fungi, it has been observed that *A. pullulans* are the most effective producers of extracellular β-1,4-mannanase. Notably, they have the capacity to produce both extracellular β-mannanase and cell-associated β-mannosidase. As an illustration, a xylanolytic yeast strain *A. pullulans* NRRL Y 2311-1 has been demonstrated to have the capability to produce all the necessary enzymes for the complete degradation of galactomannan and galactoglucomannan. The endo-β-1,4-mannanase is secreted into the culture broth, while β-mannosidase is strictly intracellular. Additionally, α-galactosidase and β-glucosidase also can be found in both extracellular and intracellular locations. Notably, among these enzymes, only the extracellular β-mannanase and intracellular β-mannosidase are inducible in the presence of mannan and heteromannans (Kremnický and Biely 1997). However, the characterization of the mannanase and the cloning of its gene from *A. pullulans* have not been achieved so far.

When *A. pullulans* NRRL 58524 was cultivated in a medium containing 3.0% (w/v) defatted spent coffee grounds, a mannan-rich waste material, it successfully yields 8.4 ± 0.2 U/mL of mannanase activity. The purified mannanase has an approximate molecular weight of 63.0 kDa, showcasing peak activity at 55 °C and pH 4.0. Furthermore, the mannanase maintains stability even after a 12-h incubation at 55 °C. When the enzyme (84.9 U/g) was utilized to hydrolyze both glucomannan and galactomannan (58.2 mg/100 mg of dry weight) from the defatted spent coffee grounds for 34 h at 55 °C and pH 4.0, the resulting hydrolysate contains mannobiose (16.3 mg/100 mg) and mannotriose (2.9 mg/100 mg) (Malik Ibrahim et al. 2022). Therefore, the produced manno-oligosaccharides can be used to promote the growth of intestinal *Bifidobacterium* species, making them a valuable food sweetener or additive.

6.2.7 Esterases and Lipase

Both esterases (EC 3.1.1.3) and lipases (EC 3.1.1.1) are the enzymes that catalyze the hydrolysis of triglycerides into fatty acids and glycerol. Lipases specifically hydrolyze triglycerides with ten or more carbon atoms at the lipid-water interface. Additionally, under certain conditions, lipases also can catalyze the reverse reaction. Esterases, known as carboxyl ester hydrolases, exhibit activity based on the presence of Ser, His, and Asp residues, with serine typically incorporated into the Gly-X-Ser-X-Gly sequence at the active site. These enzymes catalyze the hydrolysis and formation of short-chain fatty acid esters (<10 carbon atoms). In contrast, lipases can catalyze a diverse array of reactions, encompassing hydrolysis, inter-esterification, alcoholysis, acidolysis, esterification, and aminolysis. Both esterases and lipases find diverse applications across industries such as dairy, bakery, beverages, chemical, pharmaceutical, cosmetic, leather, paper, and cleaning. These enzymes are also utilized in organic syntheses, hydrolysis of fats and oils, modification of fats, flavor enhancement in food processing, resolution of racemic mixtures, and chemical analyses. Esterases are also commonly utilized in the enzymatic depolymerization of polyesters by cleaving ester bonds. They are also applied in soil and wastewater bioremediation, removal of pesticides, synthesis of aroma compounds, and production of optically pure chiral drugs. After proteases and carbohydrases, lipases are recognized as the third-largest group in terms of total sales volume.

Numerous fungi have been well known as excellent lipase producers (Chandra et al. 2020). For instance, studies have reported that *Colletotrichum gloesporioides* isolated from Brazilian savanna soil has the capability to produce 27.7 U/mL of lipase. At present, major commercial lipase producers include *A. niger*, *Candida rugosa*, *Humicola lanuginosa*, *Mucor miehei*, *Rhizopus arrhizus*, *Rhizopus delemar*, *Rhizopus japonicus*, *Rhizopus niveus*, and *Rhizopus oryzae* (Chandra et al. 2020). Moreover, *Penicillium* sp., *Aspergillus* sp., *Humicola* sp., *Sporotrichum* sp., *Saccharomyces* sp., and *Candida* sp. are well known as producers of esterases. For instance, feruloyl esterase from *Aspergillus sydowi* exhibits specific activities of 0.8 U/mg, 0.3 U/mg, and 0.5 U/mg toward methyl ferulate, methyl sinapate, and ρ-nitrophenyl acetate, respectively (Phienluphon et al. 2023).

A. pullulans HN2.3, isolated from a marine environment, demonstrates lipase activity of 8.0 ± 0.2 U/mL. The molecular mass of the lipase produced is 63.5 kDa, with the optimal pH and temperature for enzyme activity reported as 8.5 and 35 °C, respectively. The lipase shows the highest hydrolytic activity toward peanut oil (Liu et al. 2008a). The *LIP1* gene, encoding the lipase, spans 1245 bp and encodes a protein consisting of 414 amino acid residues, including a putative signal peptide of 26 amino acids. The protein sequence deduced from the *LIP1* gene encompasses the lipase consensus sequence (G-X-S-X-G) (Liu et al. 2008a, b). Hence, the lipase, exhibiting a high hydrolytic activity toward peanut oil, holds significant potential for applications in the degradation and recycling of wasted peanut oil for biodiesel production.

The extracellular lipase generated by *A. pullulans* S3, isolated from the stuffed buns steamers, possesses a specific activity of 17.7 U/mg of protein and a molecular weight of 39.5 kDa. The enzyme exhibits its optimal activity at 40 °C and a pH of 7. Furthermore, it demonstrates remarkable stability in certain organic solutions (30%, v/v), such as n-propanol, isopropanol, dimethyl sulfoxide, and hexane. The catalytic activity of the lipase is boosted by Ca^{2+} and activated by the anionic surfactant SDS as well as the non-ionic surfactants Tween 20, Tween 80, and Triton X-100. The produced lipase has the capability to hydrolyze a wide range of edible oils, including peanut oil, corn oil, sunflower seed oil, sesame oil, and olive oil. For instance, if the hydrolytic activity of the lipase towards olive oil is defined as 100.0%, its relative hydrolytic activity against peanut oil is 129.8%, against soybean oil is 79.0%, and against animal fat lard is only 54.8% (Li et al. 2019). This further confirms that the lipase produced by *A. pullulans* exhibits high activity against peanut oil, a widely used cooking oil in China. Consequently, the substantial usage of peanut oil has led to a significant release of unused oil into the environment, resulting in pollution of water discharged from households and restaurants. Therefore, it is crucial to implement effective strategies for collecting and reusing the wasted oils as the raw materials for biodiesel production, thereby mitigating pollution caused by the discharge of the unused oils into the environment.

The purified lipase from *A. pullulans* NAC8 had a subunit molecular weight of 30 kDa, with an optimum pH of 6.5 and an optimum temperature of 50 °C. The enzyme's catalytic efficiency for pNPP and pNPB is 1.53×10^5 and 2.3×10^5 in aqueous media, and 1.74×10^5 and 2.7×10^5 in organic solvents. Its stability against thermal denaturation is between 40 and 70 °C. The immobilized lipase via absorption on glutaraldehyde-activated sodium alginate beads retains 70% of its activity after ten catalytic cycles. In oil-contaminated wastewater, it achieves 83% COD removal, 7.4% lipid hydrolysis, and 15.0% free fatty acid production after five cycles.

Esterase activities in broth cultures of *A. pullulans* isolated from the African tropics range from 0.011 to 0.223 mM/μg protein/min, while activities vary from 1.5 to 12.8 U/mL during the solid-state fermentation. The optimal activities of the esterases are observed at pH 7.6–8.0 and 35 °C (Kudanga et al. 2007).

A. pullulans URM 7059, isolated from the cashew (*Anacardium occidentale* L.) apple peduncle, produces esterase activity of 23.8 ± 0.3 U/mL at 40 °C and pH 7.0 when cultivated using residual glycerol from biodiesel as the sole carbon source. The esterase produced has a molar mass of 50.0 kDa and exhibits stability at neutral pH and temperatures below 30 °C. Its K_m and V_{max} values are 1.4 mM and 218 μM/min for p-nitrophenylcaprylate, and 1.55 mM and 76.7 μM/min for p-nitrophenylbutyrate. The esterase can efficiently degrade MACO-Sty, a novel copolymer composed of castor oil maleate and styrene, within a mere 14 days. This biodegradation process facilitates the recycling of its monomer precursors for industrial purposes. In comparison, the non-enzyme degradation of this copolymer typically takes from 1 to 12 months (Meneses et al. 2021).

The esterase activity produced by *A. pullulans* LABIOTEC 01 in a culture medium containing olive mill wastewater is 14.6 ± 0.5 U/mL. It exhibits an optimal pH of 5.0 and an optimal temperature of 60 °C. According to the results obtained by Cesar Lemes et al. (2019), the esterase enzyme displays stability and high activity when mixed with polar solvents such as dimethyl sulfoxide, dimethylformamide, and methanol. Hence, olive mill wastewater, generated by the olive oil industry with an annual production of around 20 million m^3, can serve as a viable resource for the production of such esterase.

A. melanogenum HN6.2, isolated from the sediment of the sea saltern in Qingdao, China, was discovered to produce a carboxylesterase with an activity of 208.1 ± 2.7 U/mL within 72 h. The purified esterase from *A. melanogenum* HN6.2 has a molecular weight of 60.2 kDa and an isoelectric point (pI) of 4.86. The optimal pH and temperature for the purified esterase are 8.0 and 40 °C, respectively. The purified esterase demonstrates stability at temperatures below 40 °C and within the pH range of 7.5–8.0. The K_m and V_{max} values of the enzyme for ρ-nitrophenyl butyrate were measured as 68.6 μM and 251.4 μM/min, respectively. After disrupting the esterase gene (*Car-Est* gene) in the genomic DNA of the marine-derived yeast strain HN6.2, the esterase activity produced by the disruptant Y44 notably decreases. Furthermore, the transcriptional level of the *Car-Est* gene in the disruptant Y44 is merely 5.2% of that in its wild-type strain HN6.2, affirming that the esterase is indeed encoded by the cloned esterase gene in the yeast strain HN6.2 (Chen et al. 2017). In the putative catalytic site Gly-X-Ser-X-Gly of the esterase from the yeast strain HN6.2, the first "X" is Glu, whereas in the esterases from any other fungal strains, the first "X" in the catalytic site is Phe. The amino acids surrounding the putative catalytic site Gly-X-Ser-X-Gly in the esterase from the yeast strain HN6.2 are Val, Ile, and Ala, whereas in the esterases from any other fungal strains, Ser, Asn, and Gly are observed. These differences may be linked to the high esterase activity and distinctive properties of the esterase produced by the yeast-like fungal strain HN6.2.

Feruloyl esterases (EC 3.1.1.73) target the ester bonds of ferulic and diferulic acids present in plant cell wall polysaccharides. Feruloyl esterases play a crucial role in depolymerizing lignocellulose for the production of fuels and chemicals from plant biomass. These enzymes also have significant applications in the food, pharmaceutical, and paper-pulp industries. Ferulic acid, produced by feruloyl esterases, exhibits antioxidant and photoprotectant properties and is commercially utilized to manufacture feruloylated acylglycerols employed as a sunscreen component. For instance, *A. pullulans* NRRL 62041 RSU is capable of producing 4.2 ± 0.1 U/mg and 0.2 U/mL of feruloyl esterase activity (Rich et al. 2016). This enzyme activity is comparable to the feruloyl esterase activity produced by *A. sydowi*, as mentioned earlier (de Souza et al. 2015). This shows that feruloyl esterase produced by *A. pullulans* NRRL 62041 RSU may also have high potential application in biotechnology.

6.2.8 Protease

Proteases facilitate the hydrolysis of peptide bonds in proteins, breaking them down into peptides and amino acids. Based on their site of action, proteases can be classified into endopeptidases (EC 3.4.21-99) and exopeptidases (EC 3.4.11-19). They are further categorized as acidic proteases (pH 2.0–6.0), neutral proteases (pH 6.0–8.0), and alkaline proteases (pH 8.0–13.0) based on their optimal pH ranges. Fungal proteases offer several advantages over bacterial proteases due to their ability to form mycelium structures and their safety profile. The primary filamentous fungi known for protease production comprise *Aspergillus* species, including *A. oryzae*, *Penicillium* species, such as *P. griseoroseum*, *Rhizopus* species, *Mucor* species like *M. pusillus* and *M. miehei*, *Humicola* species, *Thermoascus* species like *T. aurantiacus*, and *Thermomyces* species such as *T. lanuginosus* (de Souza et al. 2015). Proteases find extensive applications in detergents, leather processing, silver recovery, medical treatments, food processing, animal feeds, chemical industry, and waste treatment (Kurmar and Tagaki 1999). Moreover, proteases play a significant role in advancing high-value products through enzyme-assisted protein digestion from various sources.

The serine alkaline proteases, approximately 32 kDa in size with an optimal pH of 9.0, derived from *A. pullulans* strains 10, HN2-3, and N13d, obtained from diverse marine habitats, exhibit the capacity to enzymatically hydrolyze proteins from various origins. This enzymatic process yields the bioactive peptides displaying angiotensin-converting enzyme (ACE) inhibitory and antioxidant activities. Although the angiotensin-converting enzyme (ACE) inhibitory activity of the crude bioactive peptides generated from shrimp (*Trachypenaeus curvirostris*) muscle using the purified protease from strain HN2-3 reaches the peak at 88.3% compared to that (100%) of the control, the antioxidant activity of the crude bioactive peptides derived from spirulina (*Arthrospira platensis*) powder achieves the highest level at 82.8% under the catalytic action of the purified protease from *A. pullulans* strain N13d compared to that (100%) of the control. Hence, the alkaline proteases produced by the three marine yeasts hold promise for facilitating the production of bioactive peptides from shrimp muscle and spirulina powder (Ni et al. 2008a).

The *ALP2* gene, responsible for encoding the alkaline protease from *A. pullulans* HN2-3, spans 1248 bp and encodes a 415-amino-acid protein (Ni et al. 2008b). As previously noted, the alkaline protease derived from *A. pullulans* HN2-3 demonstrates greater suitability for bioactive peptide production than the proteases from alternative yeasts (Ni et al. 2009). Subsequently, the *cDNAALP2* gene from *A. pullulans* HN2-3 was introduced into *Yarrowia lipolytica* cells, resulting in the display of a highly active protease on the *Y. lipolytica* cells (Ni et al. 2009). Subsequently, the protease exhibited on the cell surfaces was utilized to enzymatically digest the single-cell protein of the marine yeast strain G7a and spirulina. The investigation reveals that the highest ACE inhibitory activity is observed from the single-cell protein of the marine yeast strain G7a (80.8%), whereas the crude bioactive peptides derived from the digestion of spirulina powder exhibit the highest antioxidant activity (74.0%) (Ni et al. 2009). Therefore, the produced bioactive peptides find numerous potential applications in the health-related and medical industries.

It is well established that *Monilinia laxa*, *Botrytis cinerea*, and *Penicillium expansum* are the most significant postharvest pathogens affecting fruits and vegetables. The infection of *M. laxa* can result in significant losses on stone fruits, *B. cinerea* is known to cause gray mold on pome and stone fruits, and the infection of *P. expansum* can induce blue mold decay on pears and apples through the production of patulin, a mycotoxin with immunosuppressive activity against humans. *A. pullulans* PL5 exhibits significant inhibitory activity against *M. laxa* infection on stone fruits, as well as the infection of *B. cinerea* and *P. expansum* on pome fruits. This heightened inhibitory effect is attributed to the extracellular protease synthesized by the antagonist *A. pullulans* PL5, which plays a pivotal role in combating the key postharvest pathogens of apple and peach, including infection of *M. laxa*, *B. cinerea*, and *P. expansum*. Consequently, the alkaline serine protease gene was cloned from the genomic DNA and cDNA of *A. pullulans* PL5. The protease gene was identified to be 1248 bp in size, encoding a 415-amino acid protein with a calculated molecular weight of 42.9 kDa and an isoelectric point (pI) of 4.5. Following the expression of the alkaline serine protease gene in *Escherichia coli* BL21 (DE3), the purified recombinant protease from the recombinant *E. coli* BL21 (DE3) strain exhibits robust casein hydrolysis capability and effectively inhibits the mycelial growth of the previously mentioned pathogens. These activities occur optimally at a pH of 10.0 and a reaction temperature of 50 °C. Moreover, the recombinant protease demonstrates optimal stability at lower temperatures and pH levels ranging between 7.0 and 9.0, as highlighted in a study by Zhang et al. (2012). Therefore, the recombinant protease produced has high potential applications in biocontrol of the postharvest pathogens affecting fruits and vegetables.

6.2.9 Laccases

Laccases are polyphenol oxidases (EC 1.10.3.2) that harbor four copper atoms in their active sites. These active sites feature the conserved amino acid sequences, namely HWH, YHX1H, HPX1HX1HGHX1F, and HCHX1. Laccases can be categorized into yellow, white, and blue variants (lacking an absorption band at 610 nm). Among these, yellow laccases serve as the superior biocatalysts compared to blue laccases, facilitating oxidation reactions without requiring mediators. These enzymes can catalyze the monoelectronic oxidation of substrates by utilizing molecular oxygen, resulting in the formation of water (H_2O). Due to their wide substrate specificities, laccases can efficiently oxidize a diverse range of compounds, encompassing mono-, di-, and polyphenols, aminophenols, methoxyphenols, aromatic amines, ascorbate, various synthetic dyes, and lignin from plant biomass (Ning et al. 2016). This oxidation process involves the concurrent four-electron reduction of oxygen to water. As a result, laccases find many applications in the removal of textile dyes, phenolic compounds, toxic substances, and lignin. Unlike any other oxidases, like peroxidases, laccases do not require H_2O_2 for the oxidation reaction. This characteristic enables them to detoxify phenolic contaminants, such as aromatic amines, into harmless or less harmful byproducts, degrade xenobiotics and phenolic compounds, and decolorize chemically synthesized dyes.

Laccases derived from basidiomycete and ascomycete fungi exhibit significant biotechnological promise due to their exceptional capacity to oxidize a diverse array of aromatic compounds and lignin present in plant lignocellulose. These enzymes can be synthesized constitutively or induced in the presence of lignin, aromatic compounds, and copper. For instance, laccases generated by white-rot fungi and other fungal species can be utilized for lignin biodegradation. In the process of industrial cellulosic pulp preparation, the residual lignin left in the contemporary paper mills is commonly eliminated through the use of ClO_2, which poses environmental hazards due to its toxicity. By employing laccases and xylanases for the removal of residual lignin, the dependency on ClO_2 can be significantly minimized, thereby alleviating environmental pollution. The overexpression of the *LAC1* gene from the basidiomycete *Pycnoporus cinnabarinus* controlled by the glyceraldehyde-3-phosphate dehydrogenase promoter has the potential to produce up to 1.2 g/L of laccase (Aza and Camarero 2023). Nevertheless, the genome editing in this basidiomycete poses significant challenges.

In fact, laccases serve various other functions in fungi, including roles in host-pathogen interactions, sporulation, pigment formation, and morphogenesis. Certain strains of *A. pullulans* are capable of growing on lignin-related aromatic compounds, facilitated by the laccases they produce (Rich et al. 2013).

Laccases derived from strains of *A. pullulans* exhibit an optimal temperature range of 50–60 °C and demonstrate stability for up to an hour at 50 °C. The laccase secreted from *A. pullulans* strain NRRL 50381, yielding 43 ± 1.3 mU/mL of laccase activity, possesses a molecular weight ranging from 60 to 70 kDa. On the other hand, the glycosylated laccase isolated from *A. pullulans* strain NRRL Y-2568, which produces 110 ± 3.2 mU/mL and generates a dark olivaceous pigment, exhibits a molecular weight exceeding 100 kDa following Endo H treatment. The laccases function optimally at temperatures ranging between 35 and 45 °C. Both laccases are glycosylated and exhibit lower specific activities than those from the white rot fungi *Trametes versicolor* and *Pycnoporus cinnabarinus*. Therefore, the laccases from *A. pullulans* differ from those of the lignin-degrading fungi. Nevertheless, the genetic editing of the genomic DNA of white rot fungi to enhance laccase production is still challenging. In contrast, the genomic editing techniques for *Aureobasidium* spp. have been successfully demonstrated (Zhang et al. 2019a, b).

A. melanogenum strain 11-1, which was isolated from a mangrove ecosystem, shows the highest laccase activity, reaching up to 3120.0 ± 170 mU/mL (1.2 U/mg protein) within 5 days under the optimal conditions. The yeast-like fungal strain 11-1's laccase gene (*LAC1*) encodes a protein comprising 570 amino acids. The deduced laccase resulting from the *LAC1* gene harbors four conserved copper-binding domains, which are the characteristics of all the fungal laccases as previously identified. A recombinant yeast strain that overexpresses the *LAC1* gene from the yeast strain 11-1 demonstrates laccase activity at 6005 ± 140 mU/mL. Following Endo H treatment, the molecular weight of the recombinant laccase is approximately 62.5 kDa. The recombinant laccase exhibits optimal activity at a temperature of 40 °C and pH of 3.2. It remains stable at temperatures below 25 °C. The laccase's K_m and V_{max} values for 2,2-azino-bis(3-ethylbenzothiazoline-6-sulfonic acid) (ABTS) were determined to be 6.3×10^2 mM and 177.4 M/min, respectively. The laccase has been demonstrated to have significant decolorization activity on various synthetic dyes. Specifically, the purified laccase exhibits notable decolorizing efficacy toward Bromothymol Blue, Eriochrome Black T, Crystal Violet, Malachite Green, and Methyl Orange, with the corresponding decolorization rates within 24 h of 12.3 ± 1.5%, 14.5 ± 1.4%, 42.0 ± 0.4%, 35.5 ± 1.4%, and 84.1 ± 1.3%, respectively (Aung et al. 2019). The purified laccase generated by the recombinant strain of *A. melanogenum* 11-1 exhibits broad substrate specificity, efficiently catalyzing the decolorization of a diverse range of synthetic dyes. This feature highlights its significant potential for applications in the environmental industry.

Yellow laccases are widely recognized to be superior to blue laccases as stated above. The yellow laccase produced by *A. pullulans* NAC8, derived from soil rich in decaying plant matter, exhibits a molecular weight of 68.4 kDa. The optimal reaction conditions for achieving its maximum activity include a temperature of 45 °C and a pH of 4.5. It remains stable within a temperature range of 45–55 °C after 2 h of incubation and in the presence of up to 10% of all organic solvents. The K_m, V_{max}, K_{cat}, and K_{cat}/K_m values of the purified laccase for guaiacol were determined as 1.1 ± 0.1 mM, 12.7 ± 0.6 μmol/mL/min, 25.3×10^{-1} s^{-1}, and 2.4×10^3/M/s, respectively. The yellow laccase demonstrates efficacy in decolorizing textile dyes and textile wastewater effluents, as well as in oxidizing methoxy, di-, and meta-hydroxy substituted phenol compounds. The rate of oxidation follows the order Guaiacol > Catechol > Tannic acid > Caffeic acid > Gallic acid > Oxalic acid (Ademakinwa and Agboola 2016). As a result, they do not exhibit high specificity for a particular substrate, showcasing the capability to utilize a broad spectrum of environmental pollutants, including textile dyes.

A purified thermostable intracellular laccase, derived from the mycelia of *A. pullulans* NAC8, was employed for the biodegradation of bisphenol A (BPA), which is recognized as an endocrine-disrupting compound. The laccase is composed of a 61.0 kDa monomeric protein and exhibits the optimal pH and temperature at 3.0 and 50 °C, respectively. It remains stable, retaining approximately 79–80% activity, at these optimal pH and temperature conditions for 6 and 8 h. The laccase exhibits the capability to biodegrade Bisphenol A (BPA), with significantly enhanced biodegradation efficiency observed in the presence of 2,2-azino-bis[3-ethylbenzothiazoline-6-sulfonic acid (ABTS) as a mediator. Under such conditions, the enzyme initially oxidizes ABTS to generate a radical, $ABTS^+$, which serves as an electron transfer agent for BPA. Notably, at BPA concentrations of 5.0, 10.0, and 20.0 mg/L, the complete degradation of BPA is achieved, reaching 100%. The BPA treated with laccase exhibits mild to no phytotoxicity towards Sorghum bicolor (Nelson and Anne 2021).

6.2.10 Tannase

Tannins represent the second most prevalent group of phenolic compounds in nature. When derived from plants, tannins typically interact to form stable complexes with polysaccharides and proteins. This can lead to reduced nutritional value in the plant-based foods, decreased food intake, slowed growth rates, reduced efficiency in utilizing fodder, and impaired protein breakdown in animals. Therefore, it is crucial to understand how to efficiently remove tannins during food and feed processing. Tannase (EC 3.1.1.20), featuring a characteristic conserved domain Gly-X-Ser-X-Gly, functions as a tannin acylhydrolase capable of efficiently hydrolyzing ester bonds and carboxyphenol bonds within various tannins, including ellagitannins, gallotannins, and other gallic acid esters. This enzymatic activity leads to the production of small-molecule polyphenol compounds like gallic acid, a significant intermediate in both chemical and pharmaceutical industries. This serine hydrolase can efficiently bind to tannic substrates and establish disulfide bonds, thereby enhancing its catalytic activity. Consequently, tannase can be safely utilized for the enzymatic degradation of tannins, showcasing numerous applications in the feed, food, brewing, and pharmaceutical industries. For instance, it can be employed in the treatment of tannins present in rapeseed meal, green tea, and xylan.

Fungal tannase can be produced by various *Aspergillus* species, including *A. niger*, *A. glaucus*, and *A. ficuum*, as well as by *Rhizopus* spp., *Trichoderma* spp., *Fusarium* spp., *Penicillium* spp., *Candida* spp., and *Saccharomyces* species. For instance, when *A. glaucus* is cultivated in the solid black tea waste, it produces 38.3 mg/mL (100.8 U/mL) of tannase, while growing in liquid culture, the tannase activity reaches 81.76 U/mL. The optimal bioconversion efficiency is achieved at 40 °C with tannic acid concentrations of up to 200 g/L according to the results obtained by Mahmoud Saad et al. (2023).

Recent findings have revealed that several strains of *Aureobasidium* spp. can also produce significant quantities of tannase. For instance, *A. melanogenum* T9, isolated from a red wine starter culture, has the ability to degrade tannins. Its genome harbors *TanA*, *TanB*, and *TanC* genes responsible for the distinct tannases, as well as the *GAD* gene, which encodes gallic acid decarboxylase (Zhang et al. 2019a, b). Consequently, the *TanA* gene, which encodes tannase cloned from *A. melanogenum* T9, was heterologously expressed in *Y. lipolytica*. The purified tannase TanA from the recombinant strain has a molecular weight above 63.0 kDa, displaying a specific activity of 941.4 U/mg of protein. It exhibits the optimal activity at 60 °C and pH 6.0. This thermostable tannase retains 61.3% of its activity after incubation for 12 h at 55 °C. It effectively degrades various gallic acid esters when exposed to 55 °C for 2 h. For instance, compounds such as methyl gallate, propyl gallate, tannic acid, gallocatechin gallate, epicatechin gallate, epigallocatechin gallate, and ester catechins found in green tea can be converted into gallocatechin, epicatechin, and epigallocatechin (Liu et al. 2022).

6.3 Conclusions and Perspectives

Similar to many filamentous fungi, different strains of *A. pullulans* and *A. melanogenum* possess the ability to produce a variety of enzymes such as amylases, cellulases, xylanases, inulinase, pectinase, mannases, proteinase, lipase, esterase, laccase, and tannase (Fig. 6.4). These enzymes actively hydrolyze substrates including amylose, amylopectin, cellulose, xylan, inulin, pectin, mannose, proteins, lipids, lignin, and tannins, as well as synthetic chemicals. This process releases numerous monomers essential for microbial cell growth, produces many useful products in food industries, and treats toxic substances in the polluted environment. Moreover, after the hydrolysis of xylan, proteins, inulin, and mann, the resulting oligomers can serve various functions and exhibit activities beneficial to both the human body and animals. Genomic DNA analysis reveals that *Aureobasidium* species carry all the essential genes encoding these enzymes. This is one of the reasons why *A. pullulans* and *A. melanogenum* are among the most commonly isolated species on earth,

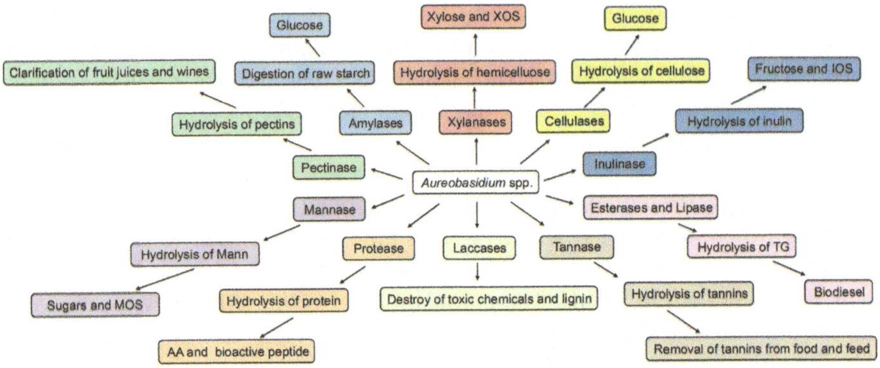

Fig. 6.4 The enzymes produced by *Aureobasidium* spp. and their functions. *XOS* Xylooligosaccharides, *IOS* Inulooligosaccharides, *MOS* Manno-oligosaccharides, *TG* Triglycerides, *AA* Amino acids

especially in the phyllospheres of different plants, with wide distribution across various environments, particularly in extreme habitats, as stated in Chap. 2. Both native strains and their recombinant counterparts of *A. melanogenum* and *A. pullulans* exhibit the capacity to produce significant quantities of pullulan, polymalate, liamocin, lipids, melanin, fumarate, gluconic acid, and siderophores which can be produced from the waste materials of plant biomass. These compounds can serve various purposes in biotechnology. Therefore, *Aureobasidium* species can serve as an ideal fungal chassis cell for fungal biology, fungal biotechnology, metabolic engineering, and synthetic biology. With the discovery and genome sequencing of an increasing number of *Aureobasidium* species, more novel, unique products and enzymes will be uncovered for food industries. Consequently, *Aureobasidium* species and their resulting products, including enzymes, hold significant applications in various sectors of biotechnology.

As stated above, because *Aureobasidium* spp. have pronounced enzyme activity, decomposing cellulose, hemicellulose, xylans, wood lignin, and low-molecular-weight caprolactam in natural environments, multiplying on different plant and animal wastes and producing melanin, they can be used for recycling waste compounds in forestry and agriculture.

References

Abd Rashid JI, Samat N, Yusoff WMW (2012) Screening and optimization of medium composition for mannanase production by *Aspergillus terreus* SUK-1 in solid state fermentation using statistical experimental methods. Res J Microbiol 7:242

Ademakinwa AN, Agboola FK (2016) Biochemical characterization and kinetic studies on a purified yellow laccase from newly isolated *Aureobasidium pullulans* NAC8 obtained from soil containing decayed plant matter. J Genet Eng Biotechnol 14:143–151

Ademakinwa AN, Agunbiade MO, Ayinla ZA, Agboola FK (2019) Optimization of aqueous two-phase partitioning of *Aureobasidium pullulans* α-amylase via response surface methodology and investigation of its thermodynamic and kinetic properties. Int J Biol Macromole 140:833–841

Aung T, Jiang H, Chen CC, Liu GL, Hu Z, Chi ZM, Chi Z (2019) Production, gene cloning, and overexpression of a laccase in the marine-derived yeast *Aureobasidium melanogenum* strain 11-1 and characterization of the recombinant laccase. Mar Biotechnol 21:76–87

Aza P, Camarero S (2023) Fungal laccases: fundamentals, engineering and classification update. Biomole 13:1716

Bennamoun L, Hiligsmann S, Dakhmouche S, Ait-Kaki A, Labbani FZK, Nouadri T, Meraihi Z, Turchetti B, Buzzini P, Thonart P (2016) Production and properties of a thermostable, pH-stable exo-polygalacturonase using *Aureobasidium pullulans* isolatedfrom saharan soil of algeria grown on tomato pomace. Foods 5:1–20

Blumberg Machado T, Corrêa Junior LCS, da Veiga de Mattos MVC, Victoria Gautério G, Julian Kalil S (2021) Sequential alkaline and ultrasound pretreatments of oat hulls improve xylanase production by *Aureobasidium pullulans* in submerged cultivation. Wast Bioma Valoriz 12:5991–6004

Brito Santos G, de Sousa Francisco Filho A, Silva Rodrigues JR, de Souza RR (2022) Cellulase production by *Aspergillus niger* using urban lignocellulosic waste as substrate: Evaluation of different cultivation strategies. J Environ Manage 305:114431

Cesar Lemes A, Silvério SC, Rodrigues S, Rodrigues LR (2019) Integrated strategy for purification of esterase from *Aureobasidium pullulans*. Separat Purifi Technol 209:409–418

Chandra P, Enespa R, Singh P, Arora K (2020) Microbial lipases and their industrial applications: a comprehensive review. Microb Cell Fact 19:169

Chen CC, Chi Z, Liu GL, Jiang H, Hu Z, Chi ZM (2017) Production, purification, characterization and gene cloning of an esterase produced by *Aureobasidium melanogenum* HN6.2. Proc Biochem 53:69–79

Chi ZM, Chi Z, Zhang T, Liu GL, Yue LX (2008) Inulinase-expressing microorganisms and applications of inulinases. Appl Microbiol Biotechnol 82:211–220

Chi ZM, Zhang T, Cao TS, Liu XY, Cui W, Zhao CH (2011) Biotechnological potential of inulin for bioprocesses. Biore Technol 102:4295–4303

Chi Z, Kong CC, Wang ZZ, Wang Z, Liu GL, Hu Z, Chi ZM (2022) The signaling pathways involved in metabolic regulation and stress responses of the yeast-like fungi *Aureobasidium* spp. Biotechnol Adv 55:107898

Christov LP, Prior BA (1996) Repeated treatments with *Aureobasidium pullulans* hemicellulases and alkali enhance biobleaching of sulphite pulps. Enzy Microb Technol 18:244–250

Dawood A, Ma K (2020) Applications of Microbial β-Mannanases. Front Bioeng Biotechnol 8:598630

de Souza PM, de Assis ML, Bittencourt C, Caprara M, de Freitas RPC, de Almeida D, Silveira YF, Fonseca EX, Ferreira Filho AP, Junior PO, Magalhães A (2015) Biotechnology perspective of fungal proteases. Braz J Microbiol 46(2):337–346

Dwivedi S, Yadav K, Gupta S, Tanveer A, Yadav S, Yadav D (2023) Fungal pectinases: an insight into production, innovations and applications. World J Microbiol Biotechnol 39:30

Elyasi Far B, Ahmadi Y, Khosroushahi AY, Dilmaghani A (2020) Microbial alpha-amylase production: progress, challenges and perspectives. Adv Pharm Bull 10(3):350–358

Gabriela Merín M, de Ambrosini VIM (2015) Highly cold-active pectinases under wine-like conditions from non-*Saccharomyces* yeasts for enzymatic production during winemaking. Lett Appl Microbiol 60(5):467–474

Gabriela Merín M, de Ambrosini VIM (2020) Application of a grape surface majority pectinolytic species, *Aureobasidium pullulans*, to low-temperature red winemaking: development and stability of wine colour. J Wine Res 31(3):218–239

Gostinčar C, Ohm RA, Kogej T, Sonjak S, Turk M, Zajc J, Zalar P, Grube M, Sun H, Han J, Sharma A, Chiniquy J, Ngan C, Lipzen A, Barry K, Grigoriev I, Gunde-Cimerman N (2014) Genome sequencing of four *Aureobasidium pullulans* varieties: biotechnological potential, stress tolerance, and description of new species. BMC Genom 15:549

Gostinčar C, Curk M, Zajc J, Gunde-Cimerman N (2019) Fifty *Aureobasidium pullulans* genomes reveal a recombining polyextremotolerant generalist. Environ Microbiol 21(10):3638–3652

Hirasawa H, Shioya K, Mori K, Tashiro K, Aburatani S, Shida Y, Kuhara S, Ogasawara W (2019) Cellulase productivity of *Trichoderma reesei* mutants developed in Japan varies with varying pH conditions. J Biosci Bioeng 128(3):264–273

Ichinose H, Suzuki K, Michikawa M, Sato H, Yuki M, Kamino K (2017) Purification, cloning, functional expression, structure, and characterization of a thermostable β-mannanase from *Talaromyces trachyspermus* B168 and its efficiency in production of mannooligosaccharides from coffee wastes. J Appl Glycosci 65:13–21

Jia SL, Zhang M, Liu GL, Chi ZM, Chi Z (2023) Novel chromosomes and genomes provide new insights into evolution and adaptation of the whole genome duplicated yeast-like fungus TN3-1 isolated from natural honey. Funct Integrat Genom 23:206

Jiang H, Ma Y, Chi Z, Liu GL, Chi ZM (2016) Production, purification, and gene cloning of a β-fructofuranosidase with a high inulin-hydrolyzing activity produced by a novel yeast *Aureobasidium* sp. P6 isolated from a mangrove ecosystem. Mar Biotechnol 18:500–510

Khonzue P, Laothanachareon T, Rattanaphan N, Tinnasulanon P, Apawasin S, Paemannee A, Ruanglek V, Tanapongpipat S, Champreda V, Eurwilaichitr L (2011) Optimization of xylanase production from *Aspergillus niger* for biobleaching of eucalyptus pulp. Biosci Biotechnol Biochem 75(6):1129–1134

Kremnický L, Biely P (1997) β-Mannanolytic system of *Aureobasidium pullulans*. Arch Microbiol 167:350–355

Kudanga T, Mwenje E (2005) Extracellular cellulase production by tropical isolates of *Aureobasidium pullulans*. Can J Microbiol 51:773–776

Kudanga T, Mwenje E, Mandivenga F, Read JS (2007) Esterases and putative lipases from tropical isolates of *Aureobasidium pullulans*. J Bas Microbiol 47:138–147

Kurmar CG, Tagaki H (1999) Microbial alkaline protease: from bioindustrial viewpoint. Biotechnol Adv 17:561–594

Li XL, Ljungdahl LG (1994) Cloning, sequencing, and regulation of a xylanase gene from the fungus *Aureobasidium pullulans* Y-2311–1. Appl Environ Micrpbiol 60:3160–3166

Li XL, Zhang ZQ, Dean JFD, Eriksson KEL, Ljungdahl LG (1993) Purification and characterization of a new xylanase (APX-II) from the fungus *Aureobasidium pullulans* Y-2311–1. Appl Environ Microbiol 59(10):3212–3218

Li HF, Chi ZM, Wang XH, Ma CL (2007a) Amylase production by the marine yeast *Aureobasidium pullulans* N13d. J Ocean Univ Chin 6:61–66

Li HF, Chi ZM, Duan XH, Wang L, Sheng J, Wu LF (2007b) Glucoamylase production by the marine yeast *Aureobasidium pullulans* N13d and hydrolysis of potato starch granules by the enzyme. Proc Biochem 42:462–465

Li HF, Chi ZM, Wang XH, Duan XH, Ma LY, Gao LM (2007c) Purification and characterization of extracellular amylase from the marine yeast *Aureobasidium pullulans* N13d and its raw potato starch digestion. Enzy Microb Technol 40:1006–1012

Li YF, Jiang H, Hu Z, Liu GL, Chi ZM, Chi Z (2018) Overexpression of an inulinase gene in an oleaginous yeast, *Aureobasidium melanogenum* P10, for efficient lipid production from inulin. J Mol Microbiol Biotechnol 28:190–200

Li Y, Liu TJ, Zhao MJ, Zhang H, Feng FQ (2019) Screening, purification, and characterization of an extracellular lipase from *Aureobasidium pullulans* isolated from stuffed buns steamers. J Zhejiang Univ-Sci B (Biom Biotechnol) 20(4):332–342

Li F, Wang M, Chi Z, Zhang Z, Wang X, Xing M, Chi ZM, Liu GL (2022) A novel transcriptional activation mechanism of inulinase gene in *Kluyveromyces marxianus* involving a glycolysis regulator KmGcr1p with unique and functional Q-rich repeats. Mol Microbiol 117(5):1063–1079

Lin TC, Chen C (2004) Enhanced mannanase production by submerged culture of *Aspergillus niger* NCH-189 using defatted copra based media. Proc Biochem 39:1103–1109

Liu ZQ, Li XY, Chi ZM, Wang L, Li J, Wang XH (2008a) Cloning, characterization and expression of the extracellular lipase gene from *Aureobasidium pullulans* HN2-3 isolated from sea saltern. Antonie van Leeuwenhoek 94:1245–1255

Liu ZQ, Chi ZM, Wang L, Li J (2008b) Production, purification and characterization of an extracellular lipase from *Aureobasidium pullulans* HN2.3 with potential application for the hydrolysis of edible oils. Biochem Eng. J 40:445–451

Liu GL, Chi Z, Chi ZM (2013) Molecular characterization and expression of microbial inulinase genes. Crit Rev Microbiol 39:152–165

Liu L, Guo J, Zhou XF, Li Z, Zhou HX, Song WQ (2022) Characterization and secretory expression of a thermostable tannase from *Aureobasidium melanogenum* T9: potential candidate for food and agricultural industries. Front Bioeng Biotechnol 9:769816

Ma ZC, Liu NN, Chi Z, Liu GL, Chi ZM (2015) Genetic modification of the marine-isolated yeast *Aureobasidium melanogenum* P16 for efficient pullulan production from inulin. Mar Biotechnol 17:511–522

Mahmoud Saad M, Mahmoud Saad A, Mohamed Hassan H, Ibrahim EI, Abdelraof M, Ali BA (2023) Optimization of tannase production by *Aspergillus glaucus* in solid-state fermentation of black tea waste. Biores Bioproc 10:73

Maitan Vieira M, Kadoguchi E, Segato F, da Silva SS, Chandel AK (2020) Production of cellulases by *Aureobasidium pullulans* LB83: optimization, characterization, and hydrolytic potential for the production of cellulosic sugars. Preparat Biochem Biotechnol 17:99393

Malik Ibrahim SNM, Bankeeree W, Prasongsuk S, Punnapayak H, Lotrakul P (2022) Production and characterization of thermostable acidophilic β-mannanase from *Aureobasidium pullulans* NRRL 58524 and its potential in mannooligosaccharide production from spent coffee ground galactomannan. 3 Biotech 12:237

Manitchotpisit P, Skory CD, Leathers TD, Lotrakul P, Eveleigh DE, Prasongsuk S, Punnapayak H (2011) α-Amylase activity during pullulan production and α-amylase gene analyses of *Aureobasidium pullulans*. J Ind Microbiol Biotechnol 38:1211–1218

Meneses DP, Paixao LMN, Fonteles TV, Gudina EJ, Rodrigues LR, Fernandes FAN, Rodrigues S (2021) Esterase production by *Aureobasidium pullulans* URM 7059 in stirred tank and airlift bioreactors using residual biodiesel glycerol as substrate. Biochem. Eng J 168:10795

Mulay YR, Deopurkar RL (2018) Purification, characterization of amylase from indigenously isolated *Aureobasidium pullulans* Cau 19 and its bioconjugates with gold nanoparticles. Appl Biochem Biotechnol 184:644–658

Na Nongkhai S, Piemthongkham P, Bankeeree W, Punnapayak H, Lotrakul P, Prasongsuk S (2023) Xylooligosaccharides produced from sugarcane leaf arabinoxylan using xylanase from *Aureobasidium pullulans* NRRL 58523 and its prebiotic activity toward *Lactobacillus* spp. Heliyon 9:e22107

Nelson AA, Anne FT (2021) A novel acid-stable intracellular laccase from *Aureobasidium pullulans*: Purification, characterization and application in the removal of Bisphenol A from solutions. Biocataly Agricul Biotechnol 33:10196

Ni XM, Chi ZM, Liu ZQ, Yue LX (2008a) Screening of protease producing marine yeasts for production of the bioactive peptides. Acta Oceanol Sin 27:1–10

Ni XM, Chi ZM, Ma CL, Madzak C (2008b) Cloning, characterization, and expression of the gene encoding alkaline protease in the marine yeast *Aureobasidium pullulans* 10. Mar Biotechnol 10:319–327

Ni XM, Yue LX, Chi ZM, Li J, Wang XH, Madzak C (2009) Alkaline protease gene cloning from the marine yeast *Aureobasidium pullulans* HN2-3 and the protease surface display on *Yarrowia lipolytica* for bioactive peptide production. Mar Biotechnol 11:81–89

Ning YJ, Wang DD, Chen QJ, Ling ZR, Wang SN, Wang WP, Zhang GQ, Zhu ML (2016) An extracellular yellow laccase with potent dye decolorizing ability from the fungus *Leucoagaricus naucinus* LAC-04. Int J Biol Macromol 93:837–842

Ohta K, Moriyama S, Tanaka H, Shige T, Akomoto H (2001) Purification and characterization of an acidophilic xylanase from *Aureobasidium pullulans* var. *melanigenum* and sequence analysis of the encoding gene. J Biosci Bioeng 92(3):262–270

Ohta K, Fujii S, Higashida C (2013) Characterization of a glycoside hydrolase family-51 α-L-arabinofuranosidase gene from *Aureobasidium pullulans* ATCC 20524 and its encoded product. J Biosci Bioeng 116(3):287–292

Oskay M (2022) Production, partial purification, and characterization of polygalacturonase from *Aureobasidium pullulans* P56 under submerged fermentation using agro-industrial wastes. Curr Microbiol 79:296

Pant S, Ritika, Nag P, Ghati A, Chakraborty D, Rocha Maximiano M, Luiz Franco O, Kumar Mandal A, Kuila A (2022) Employment of the CRISPR/Cas9 system to improve cellulase production in *Trichoderma reesei*. Biotechnol Adv 60:108022

Patipong T, Lotrakul P, Padungros P, Punnapayak H, Bankeeree W, Prasongsuk S (2019) Enzymatic hydrolysis of tropical weed xylans using xylanase from *Aureobasidium melanogenum* PBUAP46 for xylooligosaccharide production. 3 Biotech 9:56

Pedersen H, Beyer M, Nielsen J (2000) Glucoamylase production in batch, chemostat and fed-batch cultivations by an industrial strain of *Aspergillus niger*. Appl Microbiol Biotechnol 53:272–277

Phienluphon A, Kondo K, Mikami B, Nagata T, Katahira M (2023) Structural insights into the molecular mechanisms of substrate recognition and hydrolysis by feruloyl esterase from *Aspergillus sydowii*. Int J Biol Macromole 253:12718

Rensink S, van Nieuwenhuijzen EJ, Sailer MF, Struck C, Han AB, Wosten HAB (2024) Use of *Aureobasidium* in a sustainable economy. Appl Microbiol Biotechnol 108:202

Ribeiro Leite RS, Ferreira Alves-Prado H, Cabral H, Carlos Pagnocca F, Gomes E, Da-Silva R (2008) Production and characteristics comparison of crude β-glucosidases produced by microorganisms *Thermoascus aurantiacus* and *Aureobasidium pullulans* in agricultural wastes. Enzy Microb Technol 43:391–395

Rich JO, Leathers TD, Anderson AM, Bischoff KM, Manitchotpisit P (2013) Laccases from *Aureobasidium pullulans*. Enzy Microb Technol 53(1):33–37

Rich JO, Manitchotpisit P, Peterson SW, Liu S, Leathers TD, Anderson AM (2016) Phylogenetic classification of *Aureobasidium pullulans* strains for production of feruloyl esterase. Biotechnol Lett 38:863–870

Roberto Ribeiro E, Amadeu Barreto R, Ambrósio de Menezes CL, Anas Zaiter M, Boscolo M, da Silva R, Gomes E, da Silva RR (2026) A plant biomass–degrading enzyme consortium from Aureobasidium leucospermi LB86 with biotechnological potential. Biol 81:126

Rong YJ, Zhang L, Chi ZM, Wang XH (2015) A carboxymethyl cellulase from a marine yeast (*Aureobasidium pullulans* 98): Its purification, characterization, gene cloning and carboxymethyl cellulose digestion. J Ocean Univ China 14:913–921

Rustiguel CB, AtílioJorge J, Guimarães LHS (2015) Characterization of a thermo-tolerant mycelial β-fructofuranosidase from *Aspergillus phoenicis* under submerged fermentation using wheat bran as carbon source. Biocatal Agric Biotechnol 4:362–369

Shen XJ, Sun R (2021) Recent advances in lignocellulose prior-fractionation for biomaterials, biochemicals, and bioenergy. Carboh Polym 261:117884

Stratilov E, Dzúrov M, Breierov E, Omelkov J (2005) Production and biochemical characterization of polygalacturonases produced by *Aureobasidium pullulans* from forest soil. Ann Microbiol 56:35–40

Stratilova E, Breierova E, Vadkertiova R, Machova E, Malovıkova A, Slavikova E (1998) The adapt ability of the methylotrophic yeast *Candida boidinii* on media containing pectic substances. Can J Microbiol 44:116–120

Victoria Gauterio G, Cardoso Vieira M, da Silva LGG, Hübner T, Virginia A, Sanzo L, Juliano Kalil S (2018) Production of xylanolitic enzymes and xylooligosaccharides by *Aureobasidium pullulans* CCT 1261 in submerged cultivation. Indu Crop Prod 125:335–345

Victoria Gauterio G, da Silva LGG, Hübner T, da Rosa Ribeiro T, Juliano Kalil SJ (2020) Maximization of xylanase production by *Aureobasidium pullulans* using a by-product of rice grain milling as xylan source. Biocatal Agric Biotechnol 23:101511

Victoria Gauterio G, da Silva LGG, Hübner T, da Rosa Ribeiro T, Juliano Kalil S (2021) Xylooligosaccharides production by crude and partially purified xylanase from *Aureobasidium pullulans*: Biochemical and thermodynamic properties of the enzymes and their application in xylan hydrolysis. Proc Biochem 104:161–170

Vignolle GA, Mach RL, Mach-Aigner AR, Derntl C (2021) Genome sequence of the black yeast-like strain *Aureobasidium pullulans* var. *aubasidani* CBS 100524. Microbiol Res Announ 10:e01293–20

Wang D, Chi Z, Zhao SF, Chi ZM (2011) Disruption of the acid protease gene in *Saccharomycopsis fibuligera* A11 enhances amylolytic activity and stability as well as trehalose accumulation. Enzy Microb Technol 49:88–93

Wang CL, Li Y, Xin FH, Liu YY, Chi ZM (2014) Single cell oil production by *Aureobasidium pullulans* var. *melanogenum* P10 isolated from mangrove systems for biodiesel making. Proc Biochem 49:725–731

Wang P, Jia SL, Liu GL, Chi Z, Chi ZM (2022) *Aureobasidium* spp. and their applications in biotechnology. Proc Biochem 116:72–83

Wei L, Liu J, Wen HQJ (2015) Engineering *Scheffersomyces stipitis* for fumaric acid production from xylose. Biores Technol 187:246–254

Wei X, Zhang M, Wang GY, Liu GL, Chi ZM, Chi Z (2023) The ornithine-urea cycle involves fumaric acid biosynthesis in *Aureobasidium pullulans* var. *aubasidani*, a green and eco-friendly process for fumaric acid production. Synth Syst Biotechnol 8:33–45

Yegin S (2016) Xylanase production by *Aureobasidium pullulans* on globe artichoke stem: process optimization, enzyme characterization, and application in saccharification of lignocellulosic biomass. Preparat Biochem Biotechnol 12:24245

Yegin S, Altinel BV, Tuluk K (2018) A novel extremophilic xylanase produced on wheat bran from *Aureobasidium pullulans* NRRL Y-2311-1: effects on dough rheology and bread quality. Food Hydrocoll 81:389–397

Yegin S, Altinel B, Tuluk K (2024) Exploitation of *Aureobasidium pullulans* NRRL Y-2311-1 xylanase in mulberry and rice flours–based gluten-free cookie formulation: Effects on dough properties and cookie characteristics. J Food Sci 89:2645–2658

Yu XH, Gu ZX (2014) Direct production of feruloyl oligosaccharides and hemicellulase inducement and distribution in a newly isolated *Aureobasidium pullulans* strain. World J Microbiol Biotechnol 30:747–755

Zhang L, Chi ZM (2007) Screening and identification of a cellulase producing marine yeast and medium and fermentation condition optimization for cellulase production. J Ocean Univ Chin Sup II 37:101–108

Zhang D, Spadaro D, Valente S, Garibaldi A, Maria Lodovica Gullino M (2012) Cloning, characterization, expression and antifungal activity of an alkaline serine protease of *Aureobasidium pullulans* PL5 involved in the biological control of postharvest pathogens. Int J Food Microbiol 153:453–464

Zhang LL, Li J, Wang YL, Liu S, Wang ZP, Yu XJ (2019a) Integrated approaches to reveal genes crucial for tannin degradation in *Aureobasidium melanogenum* T9. Biomole 9:439–445

Zhang Z, Lu Y, Chi Z, Liu GL, Jiang H, Hu Z, Chi ZM (2019b) Genome editing of different strains of *Aureobasidium melanogenum* using an efficient Cre/loxp site-specific recombination system. Fung Biol 123:723–731

Zhao WH, Liu YJ, Zhang LL, Uke A, Kosugi A, Chi Z, Chi ZM, Liu GL (2026) AmXlnR, a transcription factor involved in xylan degradation and pentose catabolism, enhances pullulan production from xylose in *Aureobasidium melanogenum*. Synthet Syst Biotechnol 12:471–480

Chapter 7
Production, Biosynthesis, Regulation, and Applications of Exopolysaccharides by *Aureobasidium* spp.

Abstract Exopolysaccharides produced by different strains of *Aureobasidium* spp. include pullulan, β-glucan, and fructooligosaccharides which have been commercially produced and widely used. It has been clear how to synthesize and overproduce fructooligosaccharides from sucrose under the catalysis of β-fructofuranosidase produced by *Aureobasidium* spp., and the expression of the β-fructofuranosidase gene is activated by the transcriptional activator encoded by the gene *FTR1*. However, only recently, the pullulan synthesis pathway and regulation have been elucidated, and it has been found that the multidomain α-glucan synthetase encoded by the *Amags2* gene is responsible for pullulan biosynthesis, and the key α-glucan synthetase contains Amy_D, Gys_D, Big_5 with the conserved sequence LQS and EPST_D domains. The pullulan biosynthesis is regulated by glucose catabolite repression in the SNF1 signaling pathway, nitrogen catabolite repression in the TOR signaling pathway, Msn2 in the cAMP-PKA signaling pathway, TOR signaling pathway, SNF1 signaling pathway and MAPK-HOG signaling pathway, pH signaling pathway, pullulan synthase, transcriptional activator NsdD and fructosyltransferases. The molecular weight of pullulan can be controlled by amylase, glucoamylase, and isopullulanase, so that the molecular weight of pullulan can be greatly increased by deletion of the *AMY* gene encoding α-amylase, the *GLY* gene encoding glucoamylase, and the *IPU* gene encoding isopullulanase. The produced high molecular weight of pullulan has high values in biotechnology. But it is still not very clear how β-glucan is synthesized in *Aureobasidium* spp.

Keywords Pullulan · β-Glucan · Fructooligosaccharides · Biosynthesis · Regulation

Z. Chi et al., *Biology and Synthetic Biotechnology of Aureobasidium spp.*,
https://doi.org/10.1007/978-981-92-2607-8_7

7.1 Introduction

It has been well known that different strains of *Aureobasidium* spp. can produce pullulan, β-glucan, and fructooligosaccharides. In fact, since the 1970s, pullulan has been produced on a large scale in Japan and is being used worldwide, and the chemical structure of pullulan was identified in the 1960s. Since 2020, the pullulan synthesis pathway and regulation have been basically elucidated, and the molecular weight of pullulan has been edited by using synthetic biology, metabolic engineering, and genome editing as stated in Chap. 5 in this laboratory. The high molecular weight of pullulan obtained has many unique properties and applications in nanobiotechnology and medical industries. Although β-glucans produced by *A. pullulans* have been used in healthcare and medical industries, its synthetic pathway, regulation and overproduction must be further clarified by synthetic biology, metabolic engineering, and genome editing. Fructooligosaccharides can be easily synthesized from sucrose under catalysis of β-fructofuranosidase or fructosyltransferase.

7.2 Production, Biosynthesis, Regulation, and Applications of Pullulan

Pullulan is a linear α-D-glucan built of maltotriose subunits and linked by (1 → 6)-α-D-glucosidic linkages, while internal glucoses in the maltotriose subunits are linked by α-(1 → 4) glycosidic bonds (Fig. 7.1). The regular alternation of α-(1, 4) and α-(1, 6) bonds results in its enhanced solubility, film-forming property, structural flexibility, and oxygen impermeability. The side-chain radicals of pullulan are stable, and the cross-linking reaction with other substances hardly occurs.

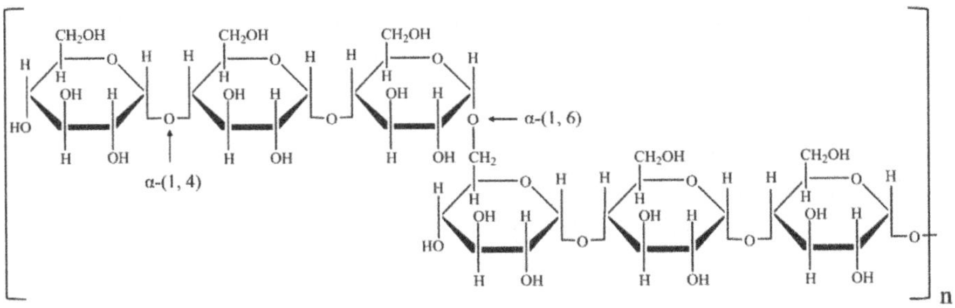

Fig. 7.1 Molecular structure of pullulan

7.2.1 Pullulan Overproduction by Metabolic Engineering

After screening over 100 strains of *Aureobasidium* spp. isolated from a mangrove ecosystem for exopolysaccharide production, *A. melanogenum* P16 was found to produce a high level of pullulan (67.4 g/L) from 120.0 g/L sucrose. Notably, no melanin was observed during its fermentation and growth (Ma et al. 2014). To enable direct and efficient conversion of inulin into pullulan, the *INU1* gene encoding exo-inulinase from the genomic DNA of *K. maximus* was integrated into the genomic DNAs and actively expressed in this high-pullulan-producing yeast-like fungal strain. One transformant, called the recombinant strain EI366, shows an exo-inulinase activity of 40.9 U/mL, compared to only 7.6 U/mL of exo-inulinase activity produced by its parent strain P16. Approximately 99.3% of the exo-inulinase produced by the engineered EI366 strain is secreted into the culture broth. In a 10-liter fermentation experiment, the engineered strain EI366 achieves a pullulan titer of 70.6 g/L from 140.0 g/L inulin within 108 h, while high exo-inulinase activity (42.03 U/mL) is reached within 60 h. The added inulin can be actively hydrolyzed by the extracellular inulinase, and most of the resulting reducing sugars are utilized for pullulan biosynthesis and cell growth (Ma et al. 2015). These results demonstrate that the engineered EI366 is suitable for direct high pullulan production from inulin which is also a green and renewable raw material in the fermentation industry like starch, cellulose, and xylan.

The primary components of natural honey are fructose and glucose. As noted in Chap. 2, only the osmophilic yeasts can thrive in such an environment. Among the *Aureobasidium* spp. strains isolated from the natural honey, only *A. melanogenum* TN3-1 produces a high level of extracellular pullulan (66.8 g/L) at a flask level within 120 h when grown in a medium containing 120.0 g/L glucose, not sucrose, as the only carbon source. In a 10-Liter batch fermentation with a medium containing 140.0 g/L glucose, this native strain yields 110.3 g/L of pullulan and 14.3 g/L of cell dry weight within 132 h (Xue et al. 2019a). This means that the obtained *A. melanogenum* TN3-1 can produce a large amount of pullulan from glucose, not sucrose which is more expensive than glucose. For example, another strain, *A. melanogenum* TN1-2, isolated from the same honey produces 114.0 g/L pullulan from sucrose, but not from glucose (Jiang et al. 2018).

Another yeast-like fungal strain, *A. melanogenum* 13-2, isolated from the Taklimakan Desert, was found to produce a high level of extracellular pullulan from sucrose. Under the optimal conditions, 73.3 g/L of pullulan is produced at a flask level within 5 days. In a 10-Liter fermenter, the titer in the culture reaches 78.1 g/L pullulan within 120 h (Jiang et al. 2019). The purified pullulan has a molecular weight of 7.7×10^5 g/mol. All the pullulan production by various *Aureobasidium* spp. strains is summarized in Table 7.1, which shows that only *A. melanogenum* TN3-1 can produce a very high concentration of pullulan from glucose.

In order to produce high levels of pullulan by *Aureobasidium* spp., a high glucose concentration in the culture is required. As shown in Table 7.1, so far, most researchers have used sucrose rather than glucose as the raw material for pullulan

Table 7.1 Summary of pullulan production by different strains of *Aureobasidium* spp.

Strains	Carbon sources	Titers of pullulan	References
EI36	Inulin	70.6 g/L	Ma et al. (2015)
A. melanogenum P16	Sucrose	67.4 g/L	Ma et al. (2014)
A. pullulans P56	Sucrose	51.5 g/L	Youssef et al. (1998)
A. pullulans P56	Sucrose	54.2 g/L	Goksungur et al. (2005)
A. pullulans CGMCC 1234	Glucose	55.5 g/L	Sheng et al. (2013)
A. pullulans RBF 4A3	Sucrose	88.6 g/l	Sharma et al. (2013)
A. pullulans RBF 4A3	Glucose	66.8 g/L	Sharma et al. (2013)
A. pullulans CJ001	Glucose	31.2 g/L	Wu et al. (2013)
A. pullulans Y68	Sucrose	52.5 g/L	Duan et al. (2008)
A. pullulans ATCC 201253	Sucrose	60.7 g/L	Cheng et al. (2011)
Transformant G63	Sucrose	80.2 g/L	Chen et al. (2017)
A. melanogenum TN3-1	Glucose	110.3 g/L	Xue et al. (2019a)
A. melanogenum V6 strain	Sucrose	102.3 g/L	Xue et al. (2019b)
A. melanogenum F44 strain	Sucrose	101.4 g/L	Xue et al. (2019b)
A. melanogenum AMY-PKS-11	Glucose	103.5 g/L	Chen et al. (2020)
A. melanogenum 13-2	Sucrose	78.1 g/L	Jiang et al. (2019)
A. melanogenum TN2-1-2	Xylose	56.8 g/L	Zhao et al. (2026)
A. melanogenum TN1-2	Sucrose	114.0 g/L	Jiang et al. (2018)

production. This is due to the fact that high glucose concentrations can inhibit expression of the genes related to pullulan biosynthesis (such as those encoding phosphoglucomutase (PGM), UDPG-pyrophosphorylase (UGP), and glycosyltransferases) by triggering carbon catabolite repression (CCR). However, sucrose from sugarcane and sugar beet is relatively limited in supply and costly. In contrast, glucose is easily derived from the enzymatic hydrolysis of widely available starch in this world and is less expensive than sucrose. Therefore, *A. melanogenum* TN3-1 has high potential application in fermentation industries as discussed below.

Furthermore, using sucrose for pullulan production by *Aureobasidium* spp. can lead to the accumulation of fructooligosaccharides (FOSs) in the culture, which negatively affects pullulan biosynthesis (as discussed below). Therefore, obtaining the osmophilic and osmotolerant *Aureobasidium* spp. strains—such as *A. melanogenum* TN3-1 listed in Table 7.1—that can both tolerate high glucose concentrations and efficiently synthesize pullulan from glucose is biotechnologically significant.

These traits are linked to the specific physiological features of *A. melanogenum* TN3-1: the genomes of *A. melanogenum* TN3-1 are highly duplicated; its cell walls are very thick, which alleviates glucose repression on pullulan biosynthesis; and its cells contain numerous small vacuoles, an important characteristic of the osmotolerant yeasts and fungi (Xue et al. 2019a; Chen et al. 2020). Additionally, *A. melanogenum* TN3-1 accumulates more glycerol, trehalose, and glycogen, which are also the characteristics of the osmotolerant fungi and yeasts, and produces significantly more pullulan than any other *Aureobasidium* spp. strains. It can even grow on the YPD plates containing up to 70.0% (w/v) glucose. In contrast, *A. melanogenum*

P16, isolated from the mangrove ecosystem as stated above, grows only weakly on the plates with 20.0% (w/v) glucose, and the higher concentrations of glucose completely inhibit the growth of the P16 strain. This clearly demonstrates that *A. melanogenum* TN3-1 is far more tolerant of high glucose concentrations than *A. melanogenum* P16 and any other strains of *Aureobasidium* spp. Collectively, these results strongly indicate that *A. melanogenum* TN3-1 is indeed a hyper-osmotolerant yeast-like fungus and is the most suitable candidate for industrial pullulan production using a high-concentration glucose medium.

7.2.2 Elucidation of Pullulan Biosynthesis Pathway by Genome Editing and Synthetic Biology

It has been a big question how pullulan is biosynthesized in *Aureobasidium* spp. since its chemical structure was identified in 1964. From Fig. 7.1, it can be observed very clearly that glucose residue is a basic unit of pullulan molecule while maltotriose is a subunit of pullulan and maltotrioses are linked by (1 → 6)-α-D-glucosidic linkages whereas glucoses residues in the maltotriose subunit are linked by α-(1 → 4) glycosidic bonds. So, in fact, like starch and glycogen, pullulan is also an α-glucan occurring in fungal and yeast cells. It has been well demonstrated that if glucose takes part in glucan biosynthesis, glucose must be activated to be UDP-glucose in order to make glucose be implicated in pullulan and any other glucan biosynthesis in microorganisms. First, for its activation, 6-phosphate-glucose from EMP pathway is transformed into 1-phosphate-glucose under the catalysis of α-phosphoglucose mutase and the formed 1-phosphate-glucose and UTP are reacted to be converted into UDP-glucose and phosphate under the catalysis of UDPG-pyrophosphorylase. Then, UDP-glucose formed is used as a general glycosyl donor in pullulan and any other glucan biosynthesis in the fungal and yeast cells (Duan et al. 2008).

It has been well known that fungal glycogen is mainly composed of α-D-glucose residues that are linearly linked via α-1,4-glycosidic bonds and its branches are created via α-1,6-glycosidic bonds while pullulan is α-glucan. Therefore, the molecular structures of both glycogen and pullulan are very similar. So, it is speculated that there should be similarities in the biosynthetic mechanisms of these two α-glucans. It has been evidenced that the biosynthesis of glycogen occurs in the cytoplasm, including initiation, elongation, and branching processes. Therefore, the glycogenins required for the initiation step of glycogen biogenesis may be most likely to be involved in pullulan biosynthesis. In the genomes of *Aureobasidium* spp., glycogenin is encoded by two genes, *GLG1* (GenBank accession number: MG825857.1) and *GLG2* (GenBank accession number: MG825858.1), in *A. melanonum* P16, a high pullulan-producing yeast as mentioned above. Disruption of the *GLG1* gene alone, the *GLG2* gene alone, and both the two genes greatly reduces the amount of pullulan produced, and complementation of the genes in the disruptants restores pullulan production, confirming that the *GLG1* and *GLG2* genes are indeed involved in the biosynthesis of pullulan. This means that like glycogen biosynthesis, the formed short α-(1, 4)-glucosyl chain as a primer autoglucosylated by Glg1 and Glg2 proteins participates not only in the biosynthesis of glycogen but also in pullulan biosynthesis in *Aureobasidium* spp. (Chen et al. 2020a). However, loss of function of these two genes does not result in a complete pullulan deficiency, which indicates that the pullulan primers may come from alternative sources.

It has been reported that sterol glycosides (glycolipids) can be used as the primers for plant cellulose biosynthesis and may also be implicated in the pullulan biosynthesis. Two glycolipid glucosyltransferase genes, ceramide glucosyltransferase gene (*GCS1*) (GenBank accession number:MG208875.1) and sterol glucosyltransferase gene (*SGT1*) (GenBank accession number: MG208876.1), involved in glycolipid biosynthesis, were identified in the genome of *A. melanogenum* P16. Finally, the *GCS1* and *SGT1* genes are confirmed to be involved in the biosynthesis of pullulan by gene knockout, complementary expression, and transcription analysis. Therefore, the formed long α-(1, 4)-glucosyl chain as a pullulan primer is biosynthesized on the basis of the short α-(1, 4)-glucosyl chain as the primer under the catalysis of Gcs1 and Sgt1 proteins. As mentioned above, as an extracellular α-glucan, pullulan biosynthesis may be similar to that of cell wall α-glucan. It was identified that the putative α-glucan synthase is encoded by three genes, *AmAGS1*, *AmAGS2*, and *AmAGS3*, in the P16 genome. Deletion of all three genes found that only the *AmAGS2* gene is involved in the biosynthesis of pullulan. After bioinformatic analysis, the AmAgs2 protein encoded by the *AmAGS2* gene (GenBank accession number:MH917125.1) contains various domains such as Amy_D, Gys_D, Big_5, and EPST_D containing multiple transmembrane regions at the C-terminus of the AmAgs2 protein. Deletion of the *AmAGS2* gene totally blocks the biosynthesis of pullulan. So, the *AmAGS2* gene is the key gene and plays a crucial role in the biosynthesis of pullulan in the P16 strain. The intracellular Gys_D domain of the AmAgs2 protein is thought to catalyze the formation of the long α-(1, 4) glucosyl chain (pullulan precursor) on the basis of glucosylceramide and sterol glycosides catalyzed by the Gcs1 and Sgt1 as shown above, or extend the short α-(1,4) glucosyl chain catalyzed by glycogenins. The EPST_D domain located in the multi-transmembrane region embedded in the cell membrane is thought to be responsible for transferring the intracellular long α-(1, 4) glucosyl chain (pullulan precursor) to the outside cell membrane. The extracellular Amy_D domain is thought to be responsible for the hydrolysis of the long α-(1, 4) glucosyl chain to release maltotriose and transferring it to the lipid carrier, Lph-glucose, to form α-(1, 6) glycoside bonds of the pullulan molecule. The Big_5 domain containing the only conserved amino acids LQS is thought to be required for pullulan biosynthesis, probably because it enhances the AmAgs2 specifically to be bound to the pullulan precursor and promotes the Amy_D domain of the AmAgs2 to cut the maltotriose from the pullulan precursor and transfer the released maltotriose to be connected together to form pullulan molecules on Lph-glucose (Fig. 7.2). Finally, the whole process is

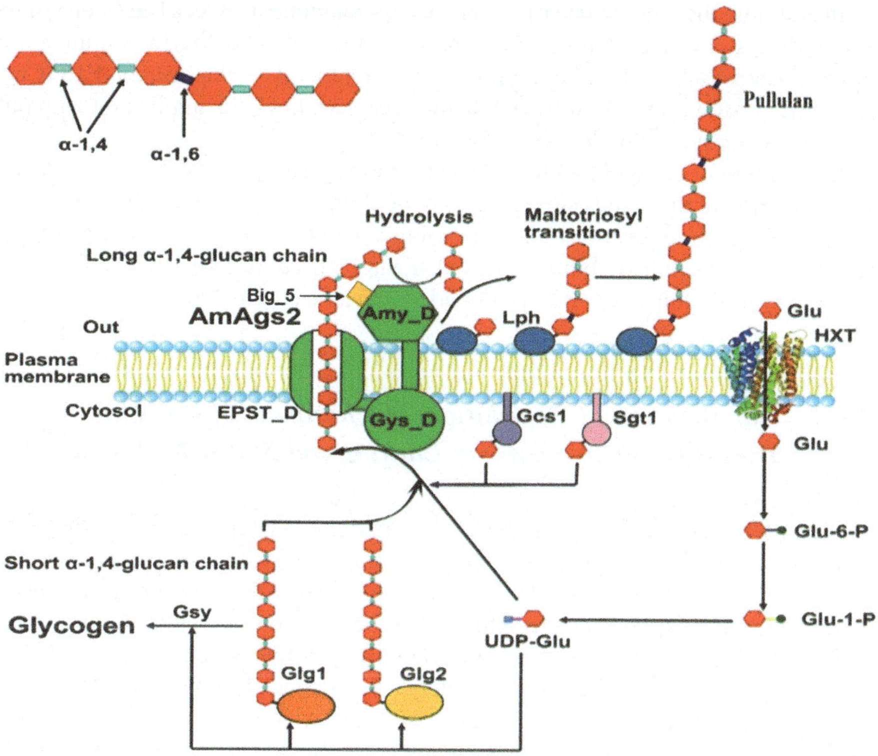

Fig. 7.2 The proposed pullulan biosynthesis pathway (Chen et al. 2020b; Wei et al. 2021). (Note: The Big_5 domain should contain the conserved amino acids LQS)

repeated many times to complete the biosynthesis of pullulan molecules. However, it is completely unknown what determines the repeat times and molecular weight of the synthesized pullulan molecule. Furthermore, it should be mentioned that it is still challenging to identify the function of each domain in pullulan biosynthesis because the AmAgs2 is a multidomain α-glucan synthetase with high molecular weight and its EPST_D domain is localized in the plasma membrane (Chen et al. 2020a, b).

In order to elucidate the function of the *AmAGS2* gene from the P16 strain, the big gene was heterologously overexpressed in *A. melanogenum* CBS102.55, a glycogen producer, but not a pullulan producer. The obtained transformant CBS2 can produce 13.5 g/L pullulan, which demonstrates that the AmAgs2 protein indeed plays a crucial role in pullulan biosynthesis. The hypothesis that glycogenins are involved in the biosynthesis of pullulan needs to be further confirmed. Based on all the mentioned-above results, the biosynthesis pathway of pullulan is proposed and shown in Fig. 7.2 (Chen et al. 2020b; Wei et al. 2021).

In addition, the *UGT1* gene (4774 bp, accession number: KX421262), one of the glucosyltransferase genes for pullulan production, is also involved in pullulan biosynthesis and control of pullulan molecular weight by *A. melanogenum* P16 (Chen et al. 2017). However, it is still completely unknown how it is implicated with pullulan biosynthesis and molecular weight control.

Elucidation of the pullulan biosynthesis pathway can pave the way for a deeper understanding of the metabolic regulation of pullulan biosynthesis, which has many biotechnological applications and physiological functions, for enhanced pullulan biosynthesis by synthetic biology, and also can be used as a reference for elucidation of any other fungal exopolysaccharide biosynthesis pathways.

7.2.3 *Clarification of Regulation Mechanisms of* Pullulan Biosynthesis *by Genome Editing and Synthetic Biology*

Numerous studies have reported that the pullulan synthetase gene (*PUL1*) is involved in pullulan biosynthesis in *Aureobasidium* spp. (Ma et al. 2015). However, for a long time, the role of the pullulan synthetase in pullulan biosynthesis and its regulation remained unclear. Recently, it has been found that in the high-yielding strain *A. melanogenum* P16, as mentioned above, the open reading frame (ORF) of the *PUL1* gene is 592 bp long and encodes 178 amino acids, interrupted by a 55-bp intron. Its promoter contains one CAAT box, one TATA box, and a 5′-HGATAR-3′ motif, meaning expression of this gene would be repressed by the preferred nitrogen source. The deduced protein includes an 18-amino-acid signal peptide and five putative N-glycosylation sites. Genetic removal of the *PUL1* gene from the genomic DNAs in the strain P16 yields a recombinant DP1088, which produces only 34.7 g/L of pullulan from sucrose—significantly less than that produced by its parent strain P16. This confirms that this gene indeed has an essential role in pullulan biosynthesis and its regulation. Furthermore, a fused Gfp-Pul1 protein is found to be localized to the cell membrane and vacuolar surface, suggesting that pullulan synthesis may occur in the cytosol. Overexpression of the *PUL1* gene in the transformant G144 substantially enhances pullulan production, yielding over 72.0 g/L of pullulan from sucrose compared to 65.5 g/L pullulan produced by its parent strain P16 under the identical conditions. The purified pullulan displays a narrow molecular mass distribution with an average molecular weight of 4.4×10^5 g/mol. Finally, Pul1 is predicted to act as an auxiliary protein that binds to AmAgs2, the key enzyme for pullulan biosynthesis as mentioned above, thereby facilitating increased pullulan production (Hansali et al. 2024).

It is well evidenced that pullulan biosynthesis in all high pullulan-producing yeast-like fungal strains is subject to glucose catabolite repression (CCR) via the SNF1 signaling pathway and nitrogen catabolite repression (NCR) via the TOR signaling pathway. To relieve CCR, sucrose, rather than glucose, is typically used as the carbon source for pullulan production (Table 7.1). Furthermore, most complex natural and cheap media extracted from plants, animals, and yeast cells contain high concentrations of nitrogen sources, making them unsuitable for pullulan synthesis due to NCR. Consequently, obtaining nitrogen-derepressed mutants capable of high pullulan production in such natural media is important (Kang et al. 2021). During pullulan production from sucrose, a significant amount of fructooligosaccharides (FOSs) accumulates concurrently under the catalysis of fructosyltransferases (FTases) as mentioned above, which negatively impacts pullulan biosynthesis (Chi et al. 2019). Regarding CCR, the key proteins mediating the transcriptional repression are Mig1, encoded by the *MIG1* gene in yeasts, and CreA, encoded by the *CREA* gene in fungi.

Our study found that disrupting the *CREA* gene in the genomic DNA of the strain DG411 of *A. melanogenum* P16 relieves CCR on pullulan biosynthesis. In this disruptant, the transcriptional levels of several key genes are enhanced. These include five genes (*GluTs*) encoding glucose transporters, the *UGT1* and *PUL1* genes responsible for pullulan biosynthesis as mentioned above, and the genes encoding α-amylase, glucoamylase, and isopullulanase which determine the molecular weight of the produced pullulan. During a 10-Liter fermentation, the DG411 strain produces 64.9 g/L of pullulan from 120 g/L glucose, significantly more than 50.4 g/L pullulan produced by its parent strain P16 under the identical conditions. These results confirm that relieving glucose repression through the *CREA* gene removal indeed can enhance pullulan production from glucose by the derepression mutant. This finding opens possibilities for improving the industrial production of this exopolysaccharide through genetic engineering and synthetic biology (Wang et al. 2017).

It has been well known that in filamentous fungi, AreB functions as a transcriptional repressor, while AreA acts as a transcriptional activator during nitrogen catabolite repression (NCR). In the pullulan-producing P16 strain, we achieved full relief of NCR by simultaneously deleting the native *AreB* and *AreA* genes, mutating the DNA sequence encoding the AreA phosphorylation sites, and overexpressing the genetically modified *AreA* gene. In the *A. melanogenum* P16 strain as mentioned above, AreA and AreB are the GATA-type zinc finger proteins, with AreA containing two phosphorylation sites. The disruptant DA66, lacking the *AreB* gene in its genomic DNA, produces 34.8 g/L of pullulan in a high-nitrogen medium containing $(NH4)_2SO_4$—an increase of 20.75% compared to the parent P16 strain. Separately, when the two phosphorylation sites in the *AreA* gene were individually mutated and the mutated gene was expressed in the P16 strain, the resulting transformants EPA211 and EPR88 also show enhanced pullulan production, indicating that both dephosphorylated forms of AreA improve pullulan synthesis in these mutants. Consequently, the novel transformant DEA122 cultivated in the high-nitrogen medium, featuring overexpression of the mutated *AreA* gene and complete deletion of the *AreB* gene, produces substantially more pullulan (37.6 g/L) than either its parent P16 strain or the DA66 mutant without the *AreB* gene in its genomic DNA. It is also found that the transcriptional levels of the *GluK*, *PGM*, *UGT1*, *PUL1*, *AreA*, and four *GluT* genes are markedly higher in the DA66 mutant than in its parent P16 strain. Similarly, their expression levels in the novel transformant DEA122 are significantly elevated compared to those in the DA66 mutant. Furthermore, it is interesting to note that the mutated AreA protein, the activator, is localized to the nucleus in the novel transformant DEA122, greatly stimulating the pullulan biosynthesis-related gene expression and pullulan biosynthesis, whereas the native AreA is distributed throughout the cytoplasm in its parent P16 strain, producing less pullulan than its the novel transformant DEA122. These findings strongly demonstrate that once NCR on pullulan biosynthesis is alleviated, overexpression of the mutated *AreA* gene promotes the pullulan biosynthesis-related gene expression and pullulan production in *A. melanogenum* (Kang et al. 2021). This is one efficient method to greatly enhance pullulan biosynthesis by relieving NCR in the yeast-like fungal strain grown in the high-nitrogen medium.

Genomic and transcriptomic analysis (under high- and low-nitrogen conditions) of the P16 strain suggests that the global transcriptional activator Msn2 in the nuclei of the yeast-like fungal cells may also control pullulan biosynthesis. A search for orthologs of *MSN2/4* genes in *A. melanogenum* P16 genomic DNA reveals only a single *MSN2* gene in its genome. Deletion of the single *MSN2* gene yields the mutant *Δmsn255*, which exhibits greatly reduced pullulan production (2.1 g/L), and its morphological features shift from the yeast-like fungal cells to filamentous growth. In contrast, its parent *A. melanogenum* P16 strain produces 57.6 g/L of pullulan and maintains a normal yeast-like cellular morphology. The produced cell mass in the *Δmsn255* mutant, however, is not significantly impaired. The *Δmsn255* mutant also displays a rough, dry colony surface with invasive filamentous growth into the medium, whereas the parent P16 strain colony is smooth and moist due to the abundant pullulan secretion. These results indicate that in addition to pullulan biosynthesis, cell and colony morphologies are also controlled by the global transcriptional activator Msn2. To investigate Msn2's role during pullulan biosynthesis, a C-terminal GFP fusion under control of the native *MSN2* promoter was constructed. The resulting MO177 mutant, carrying the Msn2-GFP fusion, shows enhanced pullulan production (62.2 g/L pullulan) and exhibits a green fluorescent signal (Msn2-GFP) co-localized with the nuclear staining (blue) in the yeast-like fungal cells. These findings strongly confirm that the functional Msn2-GFP fusion protein localizes to the nucleus and can promote pullulan biosynthesis (Yang et al. 2020; Chi et al. 2022).

To identify the upstream regulators of the *MSN2* gene in the P16 strain, all the key genes within the cAMP-PKA, TOR, SNF1, and MAPK-HOG signaling pathways available in the yeast-like fungal cells were knocked out. It is very strange to confirm that only the cAMP-PKA signaling pathway plays a crucial role in regulating pullulan biosynthesis, the Msn2 activity, and its subcellular localization. It has been well evidenced that the major genes involved in the cAMP-PKA pathway include *AC*, *BCY1*, *PDE1*, *PDE2*, *TPK1*, and *TPK2* genes. Based on this signaling pathway principle, the disruptants Δ*bcy1*1, Δ*pde1*/Δ*pde2–9*, Δ*tpk1*/Δ*bcy1–3*, and Δ*tpk2*/Δ*bcy1–7* are expected to exhibit high cAMP-PKA activity, indicating they have high phosphorylating activity against Msn2 and make the phosphorylated Msn2 localized in the cytoplasm during the signal transduction. However, these mutants produce only 1.6, 4.3, 5.2, and 2.6 g/L of pullulan, respectively, whereas their parent P16 strain produces over 52.2 g/L. In contrast, the mutants Δ*ac-2*,

Δ*tpk1–7*, Δ*tpk2–8*, and Δ*tpk1*/Δ*tpk2–16*—predicted to have low cAMP-PKA activity against Msn2 based on the signaling pathway principle—produce more pullulan (60.6, 62.1, 64.7, and 62.9 g/L, respectively) than their parent P16 strain, meaning that the dephosphorylated Msn2 is localized in the nuclei and can promote expression of pullulan biosynthesis-related genes and pullulan biosynthesis. Simultaneously, it was found that in the mutants (Δ*bcy1*1-*MG13*, Δ*tpk1*Δ*bcy1*-*MG7*, Δ*tpk2*Δ*bcy1*-*MG1*, and Δ*pde1*Δ*pde2*-*MG3*) that produces low pullulan levels, the expressed GFP-tagged Msn2 protein is localized to the cytoplasm of these yeast-like fungal cells. In these same mutants, the transcript levels of the key pullulan biosynthesis genes (*AmAGS2, GCS1, GLG1, GLG2, PGM, UGP1, UGT1, SGT1,* and *PUL1*) as mentioned above are significantly downregulated. While in the mutants Δ*ac-2*, Δ*tpk1–7*, Δ*tpk2–8*, and Δ*tpk1*/Δ*tpk2–16* that produce higher pullulan, the expressed GFP-tagged Msn2 protein is localized to the nuclei of these yeast-like fungal cells and can enhance expression of the pullulan biosynthesis-related genes and pullulan biosynthesis. These findings strongly demonstrate that pullulan biosynthesis is indeed governed by the cAMP-PKA signaling pathway via the transcriptional activator Msn2 (Yang et al. 2020; Chi et al. 2022).

Previous research also has indicated that the global transcriptional Msn2 can specifically bind to the conserved DNA sequences C4T and AG4 in the promoters of its target genes in the genomic DNAs. Analysis of the promoters of pullulan biosynthesis-related genes in the genomic DNAs of *A. melanogenum* P16 reveals that they indeed contain the conserved AG4 or C4T motifs. This suggests that Msn2 may regulate the expression of all these genes. Finally, the *UGP1* gene, encoding UDP-glucose pyrophosphorylase as mentioned above, is identified as a downstream target of Msn2. This enzyme produces UDP-glucose from glucose and UTP, the general glucosyl donor for glucan, including pullulan biosynthesis in the fungal cells. So, during active and high pullulan biosynthesis in *A. melanogenum* P16, the low cAMP-PKA activity triggers Msn2 activation via dephosphorylation, leading to accumulation of the dephosphorylated Msn in the nucleus. The nuclear Msn2 then promotes *UGP1* gene expression and production of UDP-glucose. So, the high Ugp1 enzyme activity subsequently directs UDP-glucose flux toward pullulan biosynthesis. Conversely, when pullulan biosynthesis is halted due to high cAMP-PKA activity, the phosphorylated Msn2 remains in the cytoplasm of the yeast-like fungal cells. This results in low Ugp1 activity, redirecting glucose flux into the EMP pathway and TCA cycle to support cell growth, ATP production, and lipid accumulation, not pullulan biosynthesis (Yang et al. 2020). This is the major role of the global transcriptional activator Msn2 in pullulan biosynthesis and regulation of the yeast-like fungal cells (Chi et al. 2022).

It is well known that the MAPK-HOG signaling pathway regulates glycerol and trehalose synthesis and vacuole formation, which are closely linked to both pullulan biosynthesis and high osmotic tolerance in *A. melanogenum*. Consequently, deletion of the key HOG pathway genes in the high pullulan-producing yeast *A. melanogenum* TN3-1 results in mutants with reduced pullulan biosynthesis and increased sensitivity to high glucose concentrations. In particular, inactivation of both the

VSP11 and *VSP12* genes—which regulate vacuolar fission/fusion balance—in the double mutant *ΔDV-55* leads to a significant reduction in pullulan production (to less than 7.4 g/L) and heightened sensitivity to high glucose. Reintroducing the *VSP11* gene into the mutant *ΔDV-55* restores pullulan production and glucose tolerance by the transformant EV-22, as described in Chap. 4, while its cell growth remains unaffected. The *ΔDV-55* mutant exhibits fewer but larger vacuoles compared to the transformant EV-22 and its parent TN3-1. Additionally, the mutant *ΔDV-55* shows impaired growth on the YPD plates containing 40% (w/v) glucose, whereas the transformant EV-22 and their parent strain TN3-1 grow normally even at 60% (w/v) glucose. The double mutant *ΔDV-55* also displays elevated melanin levels and a swollen cell morphology, meaning that its cell wall and cell integrity are also negatively affected. Thus, the intracellular glycerol, trehalose, and vacuoles—all of which are regulated by the MAPK-HOG pathway—are closely associated with high pullulan biosynthesis and high osmotic tolerance in *A. melanogenum* (Chen et al. 2020). All these also determine the unique polyextremotolerance of *A. melanogenum* TN3-1 isolated from the natural honey sample (Chi et al. 2022).

Recent studies have revealed that the *NSDD1* (Never in sexual development D) gene, which encodes a global GATA-type transcription factor, is also closely involved in pullulan biosynthesis. In *A. melanogenum* 13-2, as mentioned above, the *NSDD1* gene (GenBank accession number: MH708228.1) is 1539 bp in length and encodes a 475-amino-acid protein containing a typical GATA-type zinc finger DNA-binding domain (C-X(2)-C-X(18)-C-X(2)-C). This domain reportedly binds specifically to the [A/T/C]GATA[A/G] DNA sequence in the target gene promoters. Removal of the *NSDD1* gene from the genomic DNA in strain 13-2 results in the mutant PKN88, which produces only 26.8 g/L of pullulan, compared to 64.7 g/L produced by the parental strain. Concurrently, expression levels of the pullulan biosynthesis-related genes were also downregulated. Overexpression of the *NSDD1* gene in the mutant PKN88 restores pullulan production to a level 15.98%, higher than that of the parent strain 13-2 and normalizes the transcriptional levels of the relevant pullulan biosynthesis-related genes. This means that the GATA-type global transcription factor positively regulates the expression of the pullulan biosynthesis-related genes. Potential NsdD1-binding sites (A/T/C]GATA[A/G] are indeed identified in the promoter regions of the key pullulan biosynthesis-related genes, including *AmAGS2*, *GluK*, *PUL1*, and *UGT1*, as mentioned above. These findings indicate that the GATA-type transcription factor NsdD1 acts as a key positive regulator for pullulan biosynthesis in *A. melanogenum* 13-2 (Chi et al. 2024). As shown below, the GATA-type transcription factor NsdD1 also acts as a key negative regulator for melanin biosynthesis in *A. melanogenum.* So, it has a dual role in regulation of metabolite biosynthesis in *Aureobasidium* spp. However, the dual regulatory mechanism underlying how the single GATA-type transcriptional factor NsdD1 governs the positive role in pullulan biosynthesis and the negative role in melanin biosynthesis remains to be fully identified (Chi et al. 2024; Moon et al. 2023).

The PacC1-mediated pH signaling pathway is the most extensively studied pH regulatory system in fungi, wherein the C_2H_2-type zinc finger transcription factor PacC1 serves as the key pH-responsive transcriptional activator. This signaling pathway comprises six components—PalA, PalB, PalC, PalF, PalH, and PalI—which facilitate the processing and activation of PacC1. Active PacC1 binds to the specific DNA sequence (5′-GCCARG-3′; R = A or G) in the promoters of its target genes. It has been reported that pullulan biosynthesis is also regulated by this pH signaling pathway. For instance, in the Δ*pacc* disruptant of *A. pullulans* HIT-LCY3T, both pullulan production and the expression levels of pullulan biosynthesis-related genes are reduced. In contrast, overexpression of the native *PACC1* gene in *A. pullulans* HIT-LCY3T enhances expression of these pullulan biosynthesis-related genes, resulting in the production of 17.8 g/L pullulan by the overexpressing strain. These findings demonstrate that the *PACC1* gene also plays a definite role in regulating pullulan biosynthesis (Zhang et al. 2023a, 2023b). However, the regulatory mechanism underlying how the PacC1 governs the positive role in pullulan biosynthesis also remains to be fully clarified. But *A. pullulans* HIT-LCY3T produces much less pullulan than any other yeast-like fungal strains (Table 7.1).

It has been reported that two cell wall α1,2-mannosyltransferases encoded by the *GT66* and *GT77* genes in *A. melanogenum* P16 are also essential for its normal colony morphology, cellular morphology, cell integrity, glycogen accumulation, and pullulan biosynthesis. For example, removal of the two genes from the genomic DNAs of *A. melanogenum* P16 causes its abnormal colony morphology, damaged cell integrity, reduced glycogen accumulation, and less pullulan biosynthesis by the disruptants while overexpression of the two genes in the disruptants of *A. melanogenum* P16 restores its normal colony morphology, cellular morphology, glycogen accumulation, and pullulan biosynthesis by the overexpressing strain. These findings demonstrate that cell wall integrity is a prerequisite for pullulan biosynthesis and glycogen accumulation in this strain (Chen et al. 2018). This functional link is plausible because pullulan biosynthesis is closely associated with glycogen synthesis, and the key multidomain enzyme for pullulan production, AmAgs2 mentioned above, localizes to the cell membrane (Fig. 7.2).

It is well established that α-amylase hydrolyzes the α-(1 → 4) glucosidic bonds in pullulan and starch molecules, while glucoamylase cleaves both α-(1 → 4) and α-(1 → 6) bonds in these two molecules. In contrast, isopullulanase specifically hydrolyzes α-(1 → 6) bonds in pullulan molecules. After pullulan is hydrolyzed by these three enzymes, low-molecular-weight oligosaccharides will be released, causing a reduced molecular weight of pullulan. The bioinformatic analysis confirm that the genomic DNAs of all the strains of *Aureobasidium* spp. indeed contains the *AMY1* gene encoding α-amylase, the *GLY1* gene encoding glucoamylase, and the *IPU1* gene encoding isopullulanase for hydrolysis of pullulan.

Under the optimal conditions, the molecular weight (Mw) of pullulan produced by *A. melanogenum* P16 in a 10-Liter fermentation reaches up to (4.4×10^5) Da, with a concentration of 67.4 g/L pullulan. To further increase the Mw of the pullulan synthesized by the P16 strain, the effects of α-amylase, glucoamylase, and isopullulanase on Mw of the produced pullulan were investigated by disrupting the corresponding *AMY1*, *GLY1*, and *IPU1* genes. It was found that the Mw of pullulan produced by the resulting triple-gene knockout mutant increases to (3.0×10^6) Da, with an released oligosaccharide concentration of 38.4 g/L. These results demonstrate that simultaneous knockout of α-amylase, glucoamylase, and isopullulanase genes significantly enhances the molecular weight of pullulan yielded by the P16-derived mutants. This work establishes a foundation for developing high-molecular-weight, high-concentration pullulan production via synthetic biology to meet market demands in pharmaceutical capsules, food, and cosmetics industries as discussed below (Liu et al. 2018). In fact, like any other strains of *Aureobasidium* spp. *A. melanogenum* P16 produces high molecular weight of pullulan in the presence of high concentration of sugars to reduce osmotic pressure in the medium, and after depletion of the sugars in the medium, *A. melanogenum* P16 starts to secrete α-amylase, glucoamylase, and isopullulanase which degrade the high-molecular-weight pullulan to release oligosaccharides and monosaccharides for continual cell growth and metabolism. This phenomenon can be easily observed during pullulan fermentation by all the strains of *Aureobasidium* spp. and is one way to resist high osmotic pressure in the medium.

As mentioned above, *A. melanogenum* TN3-1, which produces a very high concentration of pullulan from glucose, exhibits higher α-amylase, glucoamylase, and isopullulanase activities than *A. melanonum* P16. Most notably, the α-amylase activity in strain TN3-1 is 510.4% higher than that in the P16 strain. Since α-amylase can actively hydrolyze the α-(1 → 4) linkages in pullulan molecules, the resulting polymer produced by the TN3-1 strain has a lower molecular weight (1.6×10^5 g/mol) compared to that from the P16 strain (4.4×10^5 g/mol). Additionally, strain TN3-1 produces a considerable amount of melanin as a byproduct, which negatively impacts the appearance of the pullulan and complicates downstream processing, thereby increasing pullulan production costs and negatively affecting the quality of the produced pullulan (Chen et al. 2020).

To increase the molecular weight of the produced pullulan and prevent melanin production by strain TN3-1, the duplicated α-amylase genes (*AMY1* gene) for pullulan degradation and polyketide synthase genes (*PKS1* genes) for melanin biosynthesis, as stated in Chap. 10, are simultaneously abolished in the genomic DNAs of *A. melanogenum* TN3-1, resulting in the double mutant Amy-Pks-111. This double mutant Amy-Pks-111 converts 140.0 g/L of glucose into 103.5 g/L of melanin-free pullulan with a molecular weight of (3.3×10^5) g/mol. In this double mutant Amy-Pks-111, α-amylase activity and expression of both the *AMY1* and *PKS1* genes are significantly reduced, while expression of various pullulan biosynthesis-related genes is upregulated (Chen et al. 2020). As mentioned below, this produced high-molecular-weight pullulan also can be used for preparing the medical capsules for storing medicine and drug (Chen et al. 2019).

As described in Chap. 6, *Aureobasidium* spp. produce various enzymes for hemicellulose xylan degradation, including xylanase, β-xylosidase, α-L-arabinofuranosidase, α-glucuronidase, lytic polysaccharide monooxygenases, and ferulic acid esterase. In *A. melanogenum*, it has been reported that xylan degradation and pentose (D-xylose and L-arabinose) utilization are regulated by the transcription factor AmXlnR, encoded by the *AmXLNR1* gene. The AmXlnR1, which contains a superfamily-specific motif (Cys-X6-Cys-X5-Cys-X2-Cys-X6-Cys), serves as both an activator and a repressor: it activates xylanase and β-xylosidase gene expression by binding to the conserved promoter sequence 5′-GGNT(A/t/g)A-3′, whereas it represses expression of α-L-arabinofuranosidase gene by binding to the conserved promoter sequence 5′-GGC(T/a)AT-3′. This means that like NsdD1 mentioned above, AmXlnR1 is also a dual transcriptional actor in *A. melanogenum.* It has been reported that the marine-derived yeast-like fungus *A. melanogenum* TN2-1-2 efficiently co-utilizes glucose and xylose for pullulan production. Overexpression of the *AmXLNR1* gene alone in the *A. melanogenum* TN2–1-2 EX55 strain enhances pullulan synthesis from xylose, yielding 56.8 g/L of pullulan, and upregulates the genes involved in xylose metabolism, including L-arabinose reductase gene (*LAR1*), D-xylose reductase gene (*XYR1*), xylitol dehydrogenase gene (*XDH1*), and D-xylulokinase gene (*XKI1*) (Zhao et al. 2026). So, the regulatory mechanism underlying how the AmXlnR1 governs the dual role in xylan degradation and pentose utilization has been partially clarified. In addition, xylose must be transformed into fructose-6-phosphate and glyceraldehyde-3-phospahte via pentose phosphate pathway. The obtained fructose-6-phosphate and glyceraldehyde-3-phosphate will enter the central glycolysis pathway (EMP) to form glucose-6-phosphate via pentose phosphate pathway (PPP) (Fig. 8.9) as shown in the next chapter. Finally, glucose-6-phosphate is implicated with pullulan biosynthesis as mentioned above.

7.2.4 Pullulan Applications

Disruption of the pullulan synthase gene (*PUL1*) in *A. melanogenum* 13-2, as mentioned in Table 7.1, makes a mutant DAG277 lose the ability to synthesize any pullulan. The mutant DAG277 is found to be more sensitive to radiation of UV light, high NaCl concentration, heat treatment, strong oxidation of H_2O_2, and desiccation than its parent strain 13-2, suggesting that the produced pullulan can play an important role in resistance of the yeast-like fungal cells to various stresses. This means that the produced pullulan has many functions and an obviously protective effect on its producer.

The regular alternation of (1,4)-α and (1, 6)-α bonds makes pullulan molecules have two distinctive properties: structural flexibility and enhanced solubility. Because of such properties, pullulan has many highly potential applications in different technological processes, such as food production and packaging and pharmaceutical industries.

In biomedical and pharmaceutical industry, pullulan can be used as drug and gene delivery, 3D bioprinting and tissue engineering, and drug loading using pullulan hydrogel for controlled drug release. Due to its non-irritating properties, pullulan has been used in hair styling, lotions, powders, shampoos, and facial masks in the cosmetics industry.

Pullulan also can be used as a thickening, gelling, and stabilizing agent in food industry because pullulan films have high translucency and low oxygen porosity and can replace petroleum-based packaging of food. Pullulan-based hydrogels have been used in the removal of heavy metals, dyes, and pesticides from wastewater. Pullulan also can be used in cancer and tumor targeting, vaccination, stem cell therapy, medical imaging, nanoparticle synthesis, and chromatography.

So far, gelatin obtained from animals has been a common material for preparing drug capsules, but gelatin capsules have the disadvantages of poor stability and potential contamination by animal diseases and viruses. Some religious believers, vegetarians, diabetic persons, and patients with restricted diet refuse to use the gelatin capsules. The capsule shell made of pullulan and other EPSs can replace the gelatin capsules with many shortcomings, and the polysaccharides produced by microorganisms are green, environmentally friendly, and safe. However, the pullulan used to prepare such capsules should have good stability and structural strength, which are determined by the molecular weight (Mw) of the pullulan as mentioned above. Especially pullulan with high molecular weight (Mw) is highly valuable because of its high viscosity and malleability, which will undoubtedly expand the scope of its commercial applications because the nanomaterials, oxygen impermeable film, and drug capsules made of the high Mw pullulan have strong strength and are very stable. Furthermore, pullulan can replace the traditional gelatin in the capsule preparation, with several advantages such as avoiding many potential security risks and overcoming the cross-linking reaction between gelatin capsules and drugs. For example, the high Mw pullulan produced by the mutant Amy-Pks-111, as mentioned above, is used to make the capsule shells successfully with the concentration of 180.0 g/L, with the aid of 0.8 g/L gellan gum, 18.0 g/L glycerol, and 1.0 g/L potassium citrate. The prepared capsule shells have various advantages such as high strength, good oxygen barrier properties, raw materials availability, tightness, lightness, and high water resistance and can be suitable for all consumers. Therefore, such prepared pullulan capsule shells will have highly potential applications in food and pharmaceutical industries (Chen et al. 2019).

Especially, the prepared high Mw-pullulan-nanocellulose, high Mw-pullulan food inner packaging film (Fig. 7.3), and high Mw-pullulan-nanocellulose-lignin food outer packaging (Fig. 7.4) in this laboratory can be used to replace the widely used chemical plastic bag to avoid plastic waste accumulation in the environments and relieve the environmental pressure (Ding et al. 2020).

The mechanical properties, barrier properties, humidity stability, and heat-sealing strength of the high Mw-pullulan-nanocellulose composite films can be enhanced by nanocellulose. The food packaging prepared by the high Mw-pullulan-nanocellulose composite films can be used to store coffee, milk powder, biscuits, olive oil, etc. for a long time after adding a paper outer packaging. Furthermore, it can be used as either instant edible packaging or ordinary inner packaging and will be quickly degraded by microorganisms in an open environment, while the high Mw pullulan-nanocellulose-lignin food outer packaging has the ability of UV shielding and oxidation resistance and can withstand environmental humidity and temperature changes (Figs. 7.3 and 7.4).

Furthermore, the prepared high Mw-pullulan-nanocellulose hard capsules (Fig. 7.5) in this laboratory can be used to replace the gelatin hard capsules which have many drawbacks, opening a new way for the development of plant hard

(A) (B)

Fig. 7.3 High Mw pullulan-nanocellulose composite film (**A**) and high Mw pullulan film (**B**) prepared in this laboratory

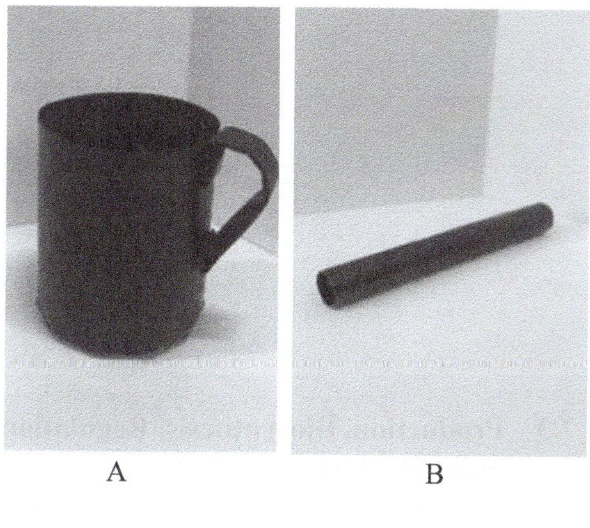

A B

Fig. 7.4 The high Mw pullulan-nanocelluloselignin composite film straw (**B**) and high Mw-pullulan-nanocellulose-lignin composite film disposable cup (**A**) prepared in this laboratory

capsules. Moreover, the disintegration of the high Mw pullulan-nanocellulose hard capsules in the simulated gastric fluid can be promoted. In addition, the humidity stability and hardness of the hard capsules are improved, and the mechanical properties and barrier properties of the hard capsule films are enhanced due to the hydrogen bond interaction between pullulan, nanocellulose, and high acyl gellan gum.

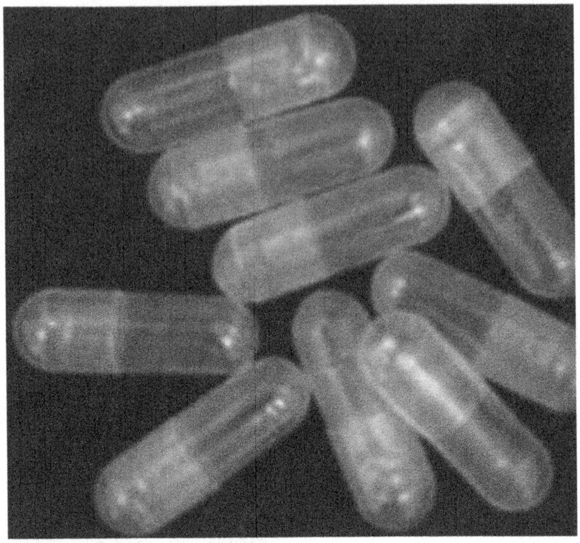

Fig. 7.5 The high-Mw pullulan-high gellan gum nanocellulose hard capsule shells prepared in this laboratory

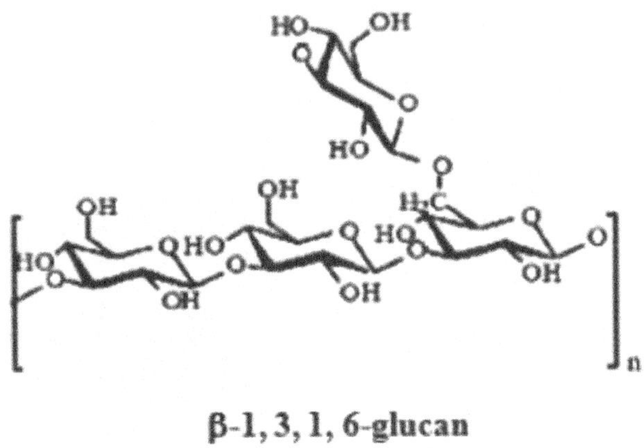

Fig. 7.6 Chemical structure of β-1,3 β-1,6-glucan

7.3 Production, Biosynthesis, Regulation, and Applications of β-Glucans

β-glucans, the acidic glucan, are a bioactive polymer in which glucose is linked by β-glycosidic bonds. They can be divided into β-1,3 and β-1,6-glucan based on their β-glycosidic bonds. However, β-glucans produced by *A. pullulans* contain a 1,3-β-D-glucan backbone with a single 1,6-β-D-glucopyranosyl branching unit at every third residue (Fig. 7.6).

7.3.1 Production of β-Glucans

It has been reported that *A. pullulans* can produce water-soluble β-glucan and water-insoluble β-glucan and has been used to produce soluble extracellular β-glucan for industrial purposes because of its high productivity compared to that produced by any other organisms, such as mushrooms. However, some strains of *A. pullulans* only produce low yield of β-glucans compared to pullulan productivity as stated above. For example, the genetically modified strain of *A. pullulans* with overexpressing *AREA* gene only produces 5.2 g/ L of β-glucans (Zhang et al. 2023a, b). A M2 strain mutated by UV radiation can produce 9.2 g/L β-glucan (Moriya et al. 2013). The *PUL1* gene disruption makes *A. pullulans* NP 1221 stop pullulan production, but produce 2.0 g/L pure β-glucan (Kang et al. 2010). *A. pullulans* D7 obtained following multiple rounds of UV mutagenesis and domestication can produce 4.84 g/L of β-glucan (Liu et al. 2015).

7.3.2 Biosynthesis Pathway of β-Glucans

It has been reported that UDP (uridine-diphospho)-glucose is also the major precursor for β-glucan biosynthesis in *A. pullulans.* Killer toxin resistant 6 (Kre6) and its paralog are thought to be involved in the biosynthesis of cell wall β-(1 → 6)-D-glucan. In addition, in *S. cerevisiae* enzymes, both K6LP1 and K6LP2 belong to the glycoside hydrolase family, which are responsible for β-glucan synthesis. It has been evidenced that both K6LP1 and K6LP2 in *A. pullulans* are partially involved in the side branch formation and secretion of extracellular β-glucan and K6LP1 might act solely as a major molecule related to the cell wall assembly by synthesizing β-(1 → 6)-D-glucan because the disruptants: *Δk6lp1*, *Δk6lp2*, and *Δk6lp1Δk6lp2* mutants produce less β-glucan than their wild type strain and the degree of branching in the extracellular β-glucan produced by the mutants is also decreased (Uchiyama et al. 2018). It has been reported that in *S. cerevisiae*, a β-1,3-glucan synthase is composed of both a catalytic subunit, Fks1 or Fks2, or Fks3, and a regulatory subunit, Rho1p, the GTP-binding protein. The *FKS1* and *FKS2* genes encode the integral membrane proteins, which are the subunits of 1,3-β-D-glucan synthase with 16 transmembrane helices, and β-1,3-glucan chains are synthesized by the Fks1 together with Fks2 using UDP-glucose as a substrate, transferring glucose to the growing chain of β-1,3-glucan via β-1,3-glycosidic linkages and translocating the synthesized β-1,3-glucan into the extracellular space (Hu et al. 2023). It should be noted that the catalytic subunit, Fks1p or Fks2p, or Fks3p, does not have significant sequence homology to any other functionally glycosyltransferases and has

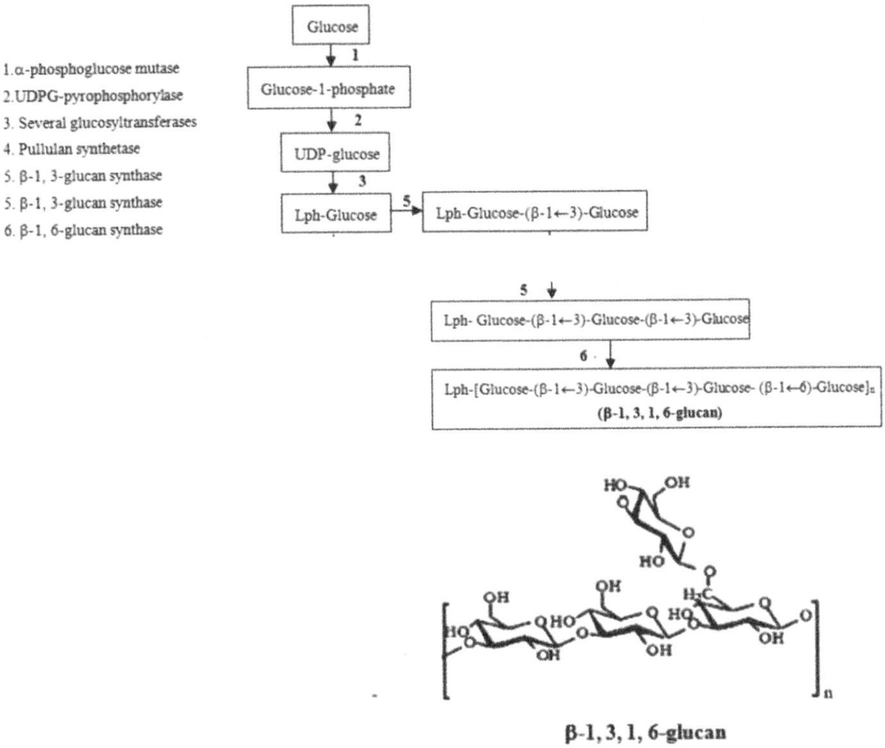

Fig. 7.7 The possible synthetic pathway of β-1,3, 1,6-glucan in *A. pullulans*

been classified to form a unique family of GT48, although they also catalyze glycosyltransfering reaction. Now it has been known that in *S. cerevisiae*, the β-1,3-glucan synthase catalyzes self-priming in the absence of an exogenously added priming substrate. Subsequent elongation proceeds by transferring glucosyl from UDP-glucose to the β-1,3-glucan primer until the length of β-1,3-glucan chain reaches 6550 ± 760 mer, followed by the release of (1,3)-β-D-glucan from the cell membrane (Fig. 7.7). This is called the GS catalytic cycle. This 1,3-β-D-glucan synthase (UDP-glucose: 1,3-β-D-glucan 3-β-D-glucosyltransferase; EC 2.4.1.34), which is a membrane enzyme activated by GTP, may also be responsible for biosynthesis of the backbone 1,3-β-D-glucan in *A. pullulans* (Chhetri et al. 2020) (Fig. 7.7). However, the relationship of β-glucan biosynthesis-related genes with the production of these extracellular β-glucans in *A. pullulans* is still unclear and is being awaited to be elucidated by deletion and complementation of the related genes, such as the key *PKS* genes encoding 1,3-β-D-glucan synthases and the gene encoding 1,6-β-D-glucan synthase as shown in Fig. 7.7. This work is being done in this laboratory.

7.3.3 Regulation of β-Glucan Biosynthesis

It has been reported that nitrogen catabolite repression (NCR) on β-glucan biosynthesis occurs in *A. pullulans*, and the presence of sodium nitrate, ammonia, and glutamate as the nitrogen sources has an obvious influence on β-glucan biosynthesis. Similarly, overexpression of the GATA transcriptional activator AreA in the nitrogen signaling pathway can enhance β-glucan biosynthesis (Zhang et al. 2023a, b).

The *MAL31* gene encodes a sugar transporter, Mal31, which is responsible for maltose uptake as well as maltotriose export in yeast cells. The disruption of the *MAL31* gene in the β-glucan-producing *A. pullulans* can significantly decrease pullulan production by reduced *AMAgs2* gene expression and increase biomass of the mutant and β-glucan production (5.0 g/L) by enhanced UDP-glucose production and phosphoglucomutase, UDPG-pyrophosphorylase, and β-glucan synthetase activities (Chen et al. 2022).

7.3.4 Applications of β-Glucan

Due to their excellent bioactivity, such as antioxidant, anti-aging, skin health promotion, anti-inflammatory, immunostimulatory, anti-tumor, anti-metastatic, anti-virus, anti-bacterial, anti-diabetic, and anti-food allergy activities, β-glucans have been widely used in many areas, such as medicine, food, health-promoting food additive, cosmetics, feed supplement, and anticancer agent. β-Glucans are also used as important dietary substances because they can lower the plasma cholesterol level and enhance the hematopoietic response. They also can be used in both modern medicine and traditional oriental therapeutic strategies. β-glucan can become an immune stimulator because it can be recognized by the innate immune cells in mammals to protect against fungal infections, so that its anti-tumor, anti-infectious disease, and anti-inflammatory activities mentioned above are closely related to the modulation of the immune system.

7.4 Production, Biosynthesis, Regulation, and Application of Fructooligosaccharides by *Aureobasidium* spp.

Fructooligosaccharides (FOSs) can be divided into two types according to their chemical structures: glucose-fructose (GFn) type and fructose-fructose (Fructan) type (Fig. 7.8). In general, the fructan type FOSs can be obtained by inulin hydrolysis using the bacterial endo-inulinase, whereas GFn can be gotten through the transfructosylation reaction under the catalysis of either β-fructofuranosidase or fructosyltransferase with sucrose as a substrate (Jiang et al. 2016).

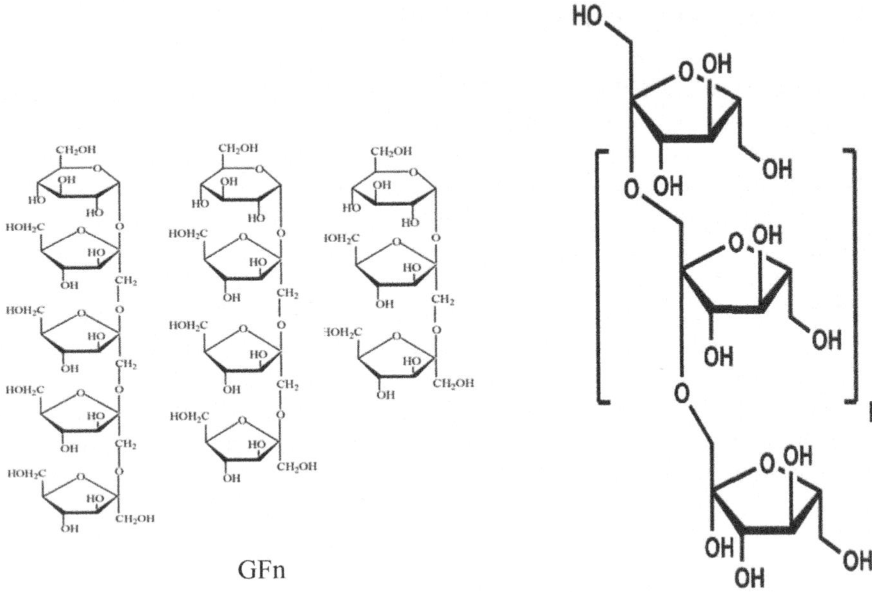

GFn

Fructan

Fig. 7.8 The chemical structures of FOSs

7.4.1 Overproduction of Fructooligosaccharides by Metabolic Engineering

At present, commercial FOSs are mainly prepared by fructosylation using β-fructofuranosidase or fructosyltransferase. For example, when *A. pullulans* NCYC 4012 is cultivated in the airlift bioreactor, the highest concentration of total FOS of 65.0 g/L and yield of FOS of 0.58 g/g of sugar-cane molasses are obtained (Fernandes et al. 2024).

A. melanogenum 11-1 strain with high β-fructofuranosidase activity (281.7 U/mL) was isolated from a mangrove ecosystem. Its genome contains β-fructofuranosidase1 gene, and its encoded β-fructofuranosidase1 has the conserved motif A (IGDP), motif D (RDP), and motif E (ET), as well as 11 N-glycosylation sites. After the β-fructofuranosidase1 gene is overexpressed in the yeast strain 11–1, the obtained transformant 333 produces 557.7 U/mL of β-fructofuranosidase activity. The molecular weight of the β-fructofuranosidase1 in which all the carbohydrates were removed by the Endo-H is 82.4 kDa. Within 7 h of the transfructosylation reaction using sucrose as the substrate, the yield of FOS is

0.66 g of FOS/g of sucrose, and percentages of GF2, GF3, and GF4 are 79.5%, 18.9%, and 1.6%, respectively (Aung et al. 2019).

Then, the whole cells of the engineered yeast strain C1–77 of *A. melanogenum*, in which glucose repression is relieved and the transcriptional activator *FTR1* gene is overexpressed and which can produce high β-fructofuranosidase activity, are used to overproduce FOSs from sugar-cane molasses, not sucrose, due to the high price of the substrate sucrose. It has been well known that sugar-cane molasses, as the main by-product of the sugar industry, contain a large amount of reducing sugar and sucrose. Compared with sucrose and other carbohydrates, it has received increasing attention due to its low cost and the by-product of the sugar industry. When the over-expressing strain C1-77 is cultured in a 10-liter fermenter using sucrose or sugar-cane molasses, as the carbon sources, respectively, the crude β-fructofuranosidase activity of the C1-77 strain reaches 2414 U/mL and 2100 U/mL within 72 h of the cultivation. Then, cane molasses is used instead of sucrose substrate to synthesize FOSs. In the molasses solution at pH 4.5 and 350 g/L of sugar-cane molasses, the transfructosylation reaction under the catalysis of the crude β-fructofuranosidase (375 U/g of molasses) produced by the over-expressing strain C1-77 grown in molasses was performed at 50 °C and 100 rpm for 4 h. Finally, 0.58 g FOSs/g of sugar-cane molasses is obtained within 4 h of the reaction, and the contents of GF_2, GF_3, and GF_4 are 38.7%, 49.3%, and 12.0%, respectively. This means that 178.6 μL of the C1-77 cell suspensions is enough to convert 1.0 g of cane molasses sugar into FOS, whereas at least 670 μL of the parent strain cell suspensions is needed to convert 1.0 g of cane molasses sugar into FOS (Zhang et al. 2019). This means that the overexpressing strain C1-77 can transform sugar-cane molasses into FOS much more efficiently than its wild-type strain.

This also strongly proves that the yeast whole cells with high β-fructofuranosidase activity can be used to efficiently convert cane molasses into FOSs, avoiding the use of the purified and expensive enzyme, which can greatly reduce production costs.

By the way, endo-inulinase from *A. niger*, *Aspergillus arachidicola*, the recombinant *E. coli*, and the recombinant *Yarrowia lipolytica* can directly hydrolyze inulin into fructan on a large scale (Song et al. 2026).

7.4.2 Biosynthesis of Fructooligosaccharides

Generally, the GFn are made up of GF2 (kestose), GF3 (nystose), and GF4 (1-β-fructofuranosyl nystose). During biosynthesis, the fructosyl residues from sucrose are transferred to sucrose and are bound at the β (2 → 1) position of sucrose under the catalysis of β-fructofuranosidase (FFase) or fructosyltransferase (FTase) (Fig. 7.9).

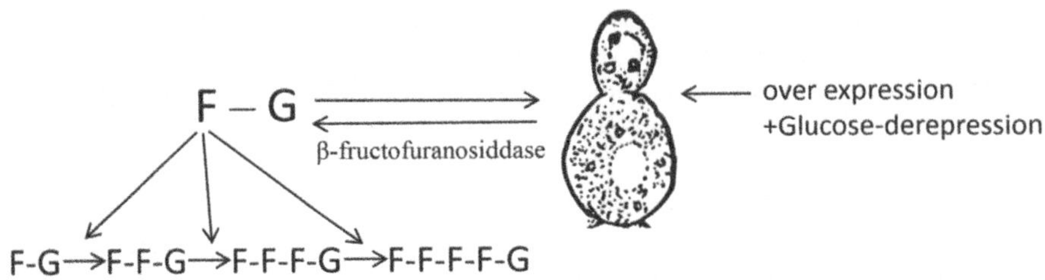

Fig. 7.9 Biosynthesis of FOSs under catalysis of β-fructofuranosidase from *Aureobasidium* spp. *F* fructose, *G*: glucose

7.4.3 Clarification of Regulation Mechanisms of Fructooligosaccharide Biosynthesis by Genome Editing and Synthetic Biology

It has been evidenced that transcription of certain genes such as *β-fructofuranosidase 1* gene and biosynthesis of β-fructofuranosidase in *A. melanogenum* 333 are easily inhibited by high concentrations of glucose released from sucrose or molasses. For example, *A. melanogenum* 333 grown at glucose concentrations higher than 60.0 g/L can significantly decrease β-fructofuranosidase activity. This is due to the fact that the 5'-SYGGRG-3′ site in the promoter of the glucose-repressed genes such as *β-fructofuranosidase 1* gene used in this study can be bound by the C_2H_2-type transcriptional repressor CreA, the key protein in GCR, which can decrease the transcriptional expression of the related genes. In addition, as the sucrose concentration used in the fermentation medium is 180.0 g/L, under such conditions, expression of the related genes is easily repressed by glucose released from sucrose hydrolysis under catalysis of β-fructofuranosidase. So, it is necessary to disrupt the *CREA* gene to promote transcription of the glucose-repressed genes, including the *β-fructofuranosidase1* gene. After deletion of the *CREA* gene in *A. melanogenum* 333, the obtained mutant D288 enhances β-fructofuranosidase activity. Moreover, it was also found that the enzyme activity in the disruptant strain D288 is not inhibited by 100 g/L glucose. This means that glucose de-repression significantly enhances the production of β-fructofuranosidase by the engineered *A. melanogenum* D288. It was also found that no expression of the *CREA* gene in the disruptant D288 is observed, while the relative transcriptional level of that in the parent strain 333 is 100%, and the transcriptional level (463%) of the *β-fructofuranosidase1* gene in the disruptant D288 achieves much higher level than that of the *β-fructofuranosidase1* gene in the parent strain 333 (100%). In addition, when the disruptant D288 is cultivated in the 10-liter fermenter, the β-fructofuranosidase activity produced by the engineered yeast cells reaches 2100 U/mL within 48 h, while the activity of β-fructofuranosidase produced by the starting strain in which the glucose repression is not relieved is only 557.7 U/mL within 72 h of the cultivation.

In order to further enhance the β-fructofuranosidase activity, the transcriptional activator *FTR1* gene was overexpressed in the disruptant strain D288 as mentioned above. The gene *FTR1* encoding the transcriptional activator specific for expression of the *β-fructofuranosidase1* gene was cloned and characterized. It is found that the encoded protein is a $Zn(II)_2Cys_6$ type zinc finger protein, and there is also a conserved region (Pfam00172) bound to the specific DNA sequence $TCGGN_{9\text{-}12}GG$. Finally, it was evidenced that the β-fructofuranosidase activity in the over-expressed strain C1–77 reaches 1750 U/mL. At the same time, after comparison of the transcriptional levels of those genes, it was also found that transcriptional levels (3103% and 342%) of the *FTR1* gene and *β-fructofuranosidase1* gene in the over-expressed strain C1–77 are greatly enhanced compared to those of the *FTR1* gene and *β-fructofuranosidase1* gene in the D288 strain (100% and 100%, respectively).

Meanwhile, FOSs are also accumulated by the high pullulan-producing *A. melanogenum* P16 strain cells when they are grown in the medium with high concentration of sucrose. When transcriptional levels of the genes encoding FFase in the glucose-derepressed mutant DG411 obtained above and its wild-type strain P16 were determined, the results show that CreA has a negative effect on the expression of the FFase gene and FOSs accumulation. It was found that the inducer FTR1 factor encoded by the *FTR1* gene that activates expression of the FFase gene is a $Zn(II)_2Cys_6$ finger protein, and the promoter of the *FTR1* gene contains one 5′-HGATAR-3′ sequence and one 5′-SYGGRG-3′ sequence, suggesting that its expression is regulated by CCR and NCR. After disruption of the *FTR1* gene, FOSs level (12.1 g/L) produced by the disruptant DF33 is reduced by 21.27%, but pullulan titer (77.0 g/L) is improved by 20.8%, and pullulan molecular weight (3.8×10^5)is decreased by 19.1% compared to those produced by the P16 strain. All the results indicate that the transcriptional activator FTR1 could negatively regulate the pullulan biosynthesis and positively increase the molecular weight of pullulan, so that it is important to reduce FOSs accumulation in order to increase pullulan production by the P16 strain grown in the medium with high sucrose (Chi et al. 2019).

A. pullulans FRR 5284 grown in molasses-based medium produces cells and broth with a total transfructosylating activity of 123.6 U/mL. The sucrose medium containing glucose reduces both transfructosylating and hydrolytic activities of the produced cells and broth. When *A. pullulans* FRR 5284 is grown in pure glucose medium, the produced cells and broth have very low levels of transfructosylating activities and have hydrolytic activities. These results indicate that *A. pullulans* FRR 5284 produces both constitutive and inducible enzymes in the sucrose-rich media, such as molasses, while it only yields constitutive enzymes in the glucose media (Sheauly Khatun et al. 2021). This means that the transfructosylating activity in *A. pullulans* is also repressed by glucose and induced by sucrose.

7.4.4 Application of Fructooligosaccharides

FOS are carbohydrates resistant to the hard conditions of gastrointestinal digestion, reaching the colon almost intact for subsequent selective metabolization by beneficial bacteria in the human intestine. As the nondigestible food ingredients, FOSs can promote the growth of beneficial gastrointestinal microbes, such as bifidobacterium and lactic acid bacteria, provide protection against cancer, cardiovascular disease, diabetes and metabolic problems. In addition, they also have important applications in feed and food additives. Because different strains of *A. pullulans* can secrete high activity of FFase and FTase, they can be used to synthesize a large amount FOSs from sucrose and the sucrose-rich raw materials, such as molasses.

7.5 Conclusions and Perspectives

Although the key *AmAGS2* gene encoding multidomain AmAgs2 for pullulan biosynthesis has been cloned and characterized, the functions of each domain need to be further clarified and confirmed. Furthermore, all the genes for β-glucan biosynthesis and regulation must be deeply confirmed by the relevant gene deletion and complementation. In this case, we believe that β-glucan overproduction can be reached in the future. The physiological functions of FOSs need to be compared with those of fructan from inulin and xylooligosaccharides from xylan to determine which one is best for human and animal health and commercial applications.

References

Aung T, Jiang H, Liu GL, Chi Z, Hu Z, Chi ZM (2019) Overproduction of a β-fructofuranosidase1 with a high FOS synthesis activity for efficient biosynthesis of fructooligosaccharides. Int J Biol Macromol 130:988–996

Chen X, Wang QQ, Liu NN, Liu GL, Chi Z, Chi ZM (2017) A glycosyltransferase gene responsible for pullulan biosynthesis in *Aureobasidium melanogenum* P16. Int J Biol Macromol 95:539–549

Chen TJ, Chi Z, Jiang H, Liu GJ, Hu Z, Chi ZM (2018) Cell wall integrity is required for pullulan biosynthesis and glycogen accumulation in *Aureobasidium melanogenum* P16. BBA-General Subjects 1862:1516–1526

Chen L, Chi Z, Liu GL, Xue SJ, Wang ZP, Hu Z, Chi ZM (2019) Improved pullulan production by a mutant of *Aureobasidium melanogenum* TN3-1 from a natural honey and capsule shell preparation. Int J Biol Macromol 141:268–277

Chen L, Wei X, Liu GL, Hu Z, Chi ZM, Chi Z (2020) Glycerol, trehalose and vacuoles had relations to pullulan synthesis and osmotic tolerance by the whole genome duplicated strain *Aureobasidium melanogenum* TN3-1 isolated from natural honey. Int J Biol Macromol 164:131–140

Chen TJ, Liu GL, Chen L, Yang G, Hu Z, Chi ZM, Chi Z (2020a) Alternative primers are required for pullulan biosynthesis in *Aureobasidium melanogenum* P16. Int J Biol Macromol 147:10–17

Chen TJ, Liu GL, Wei X, Wang K, Hu Z, Chi Z, Chi ZM (2020b) A multidomain α-glucan synthetase 2 (AmAgs2) is the key enzyme for pullulan biosynthesis in *Aureobasidium melanogenum* P16. Int J Biol Macromol 150:1037–1045

Chen X, Wang Y, Zhang X, Wu Y, Zhang G, Wang C, Zou X, Wang D, Wei G (2022) MAL31, a sugar transporter involved in pullulan biosynthesis in *Aureobasidium pullulans*. J Biotechnol 359:176–184

Cheng KC, Demirci A, Catchmark J (2011) Pullulan: biosynthesis, production, and applications. Appl Microbiol Biotechnol 92(1):29–44

Chhetri A, Loksztejn A, Nguyen H, Pianalto KM, Kim MJ, Hong J, Alspaugh A, Yokoyama K (2020) Length specificity and polymerization mechanism of (1,3)-β-D-glucan synthase in fungal cell wall biosynthesis. Biochemist 59:682–693

Chi Z, Liu NN, Jiang H, Wang QQ, Chen TJ, Liu GL, Hu Z, Chi ZM (2019) Relationship between β-D-fructofuranosidase activity, fructooligosaccharides and pullulan biosynthesis in *Aureobasidium melanogenum* P16. Int J Biol Macromol 125:1103–1111

Chi Z, Kong CC, Wang ZZ,Wang Z,Liu GL, Hu Z, Chi ZM (2022)The signaling pathways involved in metabolic regulation and stress responses of the yeast-like fungi *Aureobasidium* spp. Biotechnol Adv 55:107–898

Chi Z, Wei X, Ge N, Jiang H, Liu GL, Chi ZM (2024) NsdD, a GATA-type transcription factor is involved in regulation and biosynthesis of macromolecules melanin, pullulan, and polymalate in *Aureobasidium melanogenum*. Intern J Biol Macromole 268:131820

Ding YY, Jiang F, Chen L, Lyu WC, Chi ZM, Liu CG, Chi Z (2020) An alternative hard capsule prepared with the high molecular weight pullulan and gellan: processing, characterization, and in vitro drug release. Carbohydr Polym 237:116–172

Duan X, Chi ZM, Wang L, Wang XH (2008) Influence of different sugars on pullulan production and activities of α-phosphoglucose mutase, UDPG-pyrophosphorylase and glucosyltransferase involved in pullulan synthesis in *Aureobasidium pullulans* Y68. Carbohydr Polym 73(4):587–593

Fernandes S, Dias B, Gonçalves DA, Nobre C, Belo I, Lopes M (2024) Co-production of gluconic acid and Fructo-oligosaccharides by *Aureobasidium pullulans* from sugarcane, molasses: Effect of oxygen transfer rate in stirred tank and airlift bioreactors. Food Bioprocess Technol 17:1321–1334

Goksungur Y, Dagbaglı S, Uçan A (2005) Optimization of pullulan production from synthetic medium in a stirred tank reactor by response surface methodology. J Chem Technol Biotechnol 80:819–827

Hansali K, Wang P, Zhao SF, Wang P, Ma ZC, Chi Z, Chi ZM (2024) Overexpression of the pullulan synthetase gene enhanced pullulan production and its molecular weight by a mutant of *Aureobasidium melanogenum* P16. Int J Biol Macromol 282:137013

Hu X, Yang P, Chai C, Liu J, Sun H, Wu Y, Zhang M, Zhang M, Liu X, Yu H (2023) Structural and mechanistic insights into fungal beta-1,3-glucan synthase FKS1. Underw Nat 616(7955):190–198

Jiang H, Ma Y, Chi Z, Liu GL, Chi ZM (2016) Production, Purification, and gene cloning of a β-fructofuranosidase with a high inulinhydrolyzing activity produced by a novel yeast *Aureobasidium* sp. P6 isolated from a mangrove ecosystem. Mar. Australas Biotechnol 18(4):500–510

Jiang H, Xue SJ, Li YF, Liu GL, Chi ZM, Hu Z, Chi Z (2018) Efficient transformation of sucrose into high pullulan concentrations by *Aureobasidium melanogenum* TN1-2 isolated from a natural honey. Food Chem 15:29–35

Jiang H, Chen TJ, Chi Z, Hu Z, Liu GL, Sun Y, Zhang SH, Chi ZM (2019) Macromolecular pullulan produced by *Aureobasidium melanogenum* 13-2 isolated from the Taklimakan desert and its crucial roles in resistance to the stress treatments. Int J Biol Macromol 135:429–436

Kang B, Yang H, Choi N, Ahn K, Park C, Yoon B, Kim M (2010) Production of pure β-glucan by *Aureobasidium pullulans* after pullulan synthetase gene disruption. Biotechnol Lett 32(1):137–142

Kang XX, Wang QQ, Chi Z, Liu GL, Hu Z, Chi ZM (2021) The GATA type transcriptional factors regulate pullulan biosynthesis in *Aureobasidium melanogenum* P16. Int J Biol Macromol 192:161–168

Liu H, Li S, Zeng Q, Zeng J, Gao S, Wang M, Du C, Wang T, Li P, Wang R, Wang J (2015) High-throughput screening of *Aureobasidium pullulans* for high β-glucan production using UV mutagenesis combined with artificial fluorescent labeling. Int J Biol Macromol 319:145421

Liu NN, Chi Z, Liu GL, Chen TJ, Jiang H, Hu Z, Chi ZM (2018) α-Amylase, glucoamylase and isopullulanase determine molecular weight of pullulan produced by *Aureobasidium melanogenum* P16. Int J Biol Macromol 117:727–734

Ma ZC, Fu WJ, Liu GL, Wang ZP, Chi ZM (2014) High-level pullulan production by *Aureobasidium pullulans* var. *melanogenium* P16 isolated from mangrove system. Appl Microbiol Biotechnol 98(11):4865–4873

Ma ZC, Liu NN, Chi Z, Liu GL, Chi ZM (2015) Genetic modification of the marine-isolated yeast *Aureobasidium melanogenum* P16 for efficient pullulan production from inulin. Mar Biotechnol 17:511–522

Moon H, Khan KH, Yu KJ (2023) Upstream regulation of development and secondary metabolism in *Aspergillus* species. Cell 12:2

Moriya N, Moriya Y, Nomura H, Kusano K, Asada Y, Uchiyama H, Park E, Okabe M (2013) Improved beta-glucan yield using an *Aureobasidium pullulans* M-2 mutant strain in a 200-L pilot scale fermentor. Targeting industrial mass production. Biotechnol Bioproc E 18:1083–1089

Sharma N, Prasad GS, Choudhury AR (2013) Utilization of corn steep liquor for biosynthesis of pullulan, an important exopolysaccharide. Carbohydr Polym 93(1):95–101

Sheauly Khatun M, Hassanpour M, Mussatto SI, Harrison MD, Robert E, Speight RE, Hara IMO, Zhang Z (2021) Transformation of sugarcane molasses into fructooligosaccharides with enhanced prebiotic activity using whole-cell biocatalysts from *Aureobasidium pullulans* FRR 5284 and an invertase-defcient *Saccharomyces cerevisiae* 1403-7A. Bioresour Bioprocess 8:85

Sheng L, Zhu G, Tong Q (2013) Mechanism study of tween 80 enhancing the pullulan production by *Aureobasidium pullulans*. Carbohydr Polym 97(1):121–123

Song JZ, Gui ZT, Fang YM, Qin YX, Liu YJ, Chi ZM, Liu GL (2026) Distribution and transcriptional regulation of fungal inulinases and their biotechnological applications for inulin biorefinery. Bioresour Technol 439:133381

Uchiyama H, Iwai A, Dohra H, Ohnishi T, Kato T, Park EY (2018) The effects of gene disruption of Kre6-like proteins on the phenotype of β-glucan-producing *Aureobasidium pullulans*. Appl Microbiol Biotechnol 102:4467–4475

Wang QQ, Lu Y, Ren ZY, Chi Z, Liu GL, Chi ZM (2017) CreA is directly involved in pullulan biosynthesis and regulation of *Aureobasidium melanogenum* P16. Curr Genet 63:471–485

Wei X, LiuGL JSL, Chi Z, Hu Z, Chi ZM (2021) Pullulan biosynthesis and its regulation in *Aureobasidium* spp. Carbohydr Polym 251:117076

Wu J, Zhong F, Li Y (2013) Preparation and characterization of pullulan–chitosan and pullulan–carboxymethyl chitosan blended films. Food Hydrocoll 30(1):82–91

Xue SJ, Chen L, Jiang H, Liu GL, Chi ZM, Hu Z, Chi Z (2019a) High pullulan biosynthesis from high concentration of glucose by a hyperosmotic resistant, yeast-like fungal strain isolated from a natural comb-honey. Food Chem 286:123–128

Xue SJ, Jiang H, Chen L, Ge N, Liu GL, Hu Z, Chi ZM, Chi Z (2019b) Over-expression of Vitreoscilla hemoglobin (VHb) and flavohemoglobin (FHb) genes greatly enhances pullulan production. Int J Biol Macromol 132:701–709

Yang G, Liu GL, Wang SJ, Chi ZM, Chi Z (2020). Pullulan biosynthesis in yeast-like fungal cells is regulated by the transcriptional activator Msn2 and cAMP-PKA signaling pathway. Intern J Biol Macromole 157: 591–603

Youssef F, Biliaderis CG, Roukas T (1998) Enhancement of pullulan production by *Aureobasidium pullulans* in batch culture using olive oil and sucrose as carbon sources. Appl Biochem Biotechnol 74(1):14–30

Zhang S, Jiang H, Xue S, Ge N, Sun Y, Chi Z, Liu GL, Chi ZM (2019) Efficient conversion of cane molasses into fructooligosaccharides by a glucose derepression mutant of *Aureobasidium melanogenum* with high β-fructofuranosidase activity. J Agric Food Chem 67:13665–13672

Zhang K, Lin C, Zhao S, Wang W, Zhou W, Ru X, Cong H, Yang Q (2023a) The role of pH transcription factor Appacc in upregulation of pullulan biosynthesis in *Aureobasidium pullulans* using potato waste as a substrate. Int J Biol Macromol 242:124797

Zhang K, Zhou W, Wang W, Zhao S, Lin C, Ru X, Guan J, Cong H, Yang Q (2023b) Area gene regulates the synthesis of β-Glucan with antioxidant activity in the *Aureobasidium pullulans*. Foods 12:660

Zhao WH, Liu YJ, Zhang LL, Uke A, Kosugi A, Chi Z, Chi ZM, Liu GL., AmXlnR, a transcription factor involved in xylan degradation and pentose catabolism, enhances pullulan production from xylose in *Aureobasidium melanogenum*. Synthet Syst Biotechnol 2026; 12: 471–480.

Chapter 8
Production, Biosynthesis, Regulation, and Applications of Organic Acids by *Aureobasidium* spp.

Abstract The main organic acids produced by *Aureobasidium* spp. include polymalate, malic acid, fumarate, and gluconate. Although polymalate, malic acid, and gluconate have been commercially produced and applied in various sectors of Biotechnology, only recently, biosynthesis and regulation of polymalate, malic acid, and fumarate in *Aureobasidium* spp. have been elucidated at molecular levels by synthetic biotechnology. It has been confirmed that the PMA synthetase (PMAs) encoded by the *PMAs* gene is the key enzyme for polymerization of malate into polymalate, and the PMA synthetase carries Activation (A) domain, Transfer (T) domain, and Concentration (C) domain. Polymalate biosynthesis is controlled by the Ca^{2+} signaling pathway, the TORC1 (nitrogen) signaling pathway, the global regulatory protein LaeA, and the transcriptional activator NsdD. The ornithine-urea cycle (OUC) and acetyl-glutamate cycle (AGC) are involved in fumarate biosynthesis in *A. aubasidani*, and fumarate biosynthesis in *A. aubasidani* is also regulated by the Ca^{2+} signaling pathway. These can make it possible to overproduce polymalate, malic acid, and fumarate. Gluconate is synthesized directly from glucose only under the catalysis of extracellular glucose oxidase. So, it is very easy to overproduce gluconate in industry.

Keywords Polymalate · Fumarate · Gluconic acid · Biosynthesis · Regulation

8.1 Introduction

The main organic acids produced by *Aureobasidium* spp. include polymalate, malic acid, fumarate, and gluconate. Only recently, biosynthesis and regulation of polymalate, malic acid, and fumarate in *Aureobasidium* spp. have been elucidated at molecular levels by metabolic engineering, genome editing, and synthetic biotechnology. They also have many applications in various sectors of biotechnology.

Z. Chi et al., *Biology and Synthetic Biotechnology of Aureobasidium spp.*,
https://doi.org/10.1007/978-981-92-2607-8_8

8.2 Production, Biosynthesis, Regulation, and Applications of Polymalate and Malic Acid

Poly(β-L-malic acid) (PMA), synthesized and secreted by *Aureobasidium* spp., is a linear anionic biopolymer which is formed by the interlinkage of L-malic acid via ester bonds. In nature, this anionic biopolymer has three different isomers (Fig. 8.1A). Because PMA has so many excellent biochemical properties, such as biocompatibility, biodegradability, water solubility, non-immunogenicity, and chemical modifiability, it and its derivatives are being used as a novel drug delivery system, surgical suture, advanced biomaterials (nanoparticles and nanoconjugates), and biodegradable plastics. Besides, its hydrolysis product, L-malic acid (Fig. 8.1B), also has wide applications in food, pharmaceuticals, agriculture, and chemical industries.

8.2.1 *Overproduction of Polymalate and Malic Acid by Metabolic Engineering*

It has been reported that *A. melanogenum*, ATCC62921 isolated from the offshore wetlands in Fort Myers, Florida, USA, produces 51.4 g/L and 4.3 g/L PMA in the cultures with and without $CaCO_3$, respectively, meaning that the presence of $CaCO_3$ greatly enhances PMA biosynthesis. Only the hydrolysates of the extracellular products contain a large amount of released malic acids, while the hydrolysates of the intracellular products contain no malic acid, which indicate that PMA only exists in extracellular fermentation culture (Wang et al. 2020). Mw of the purified PMA produced by *A. melanogenum* ATCC62921 is 3.8×10^4 Da. The genome

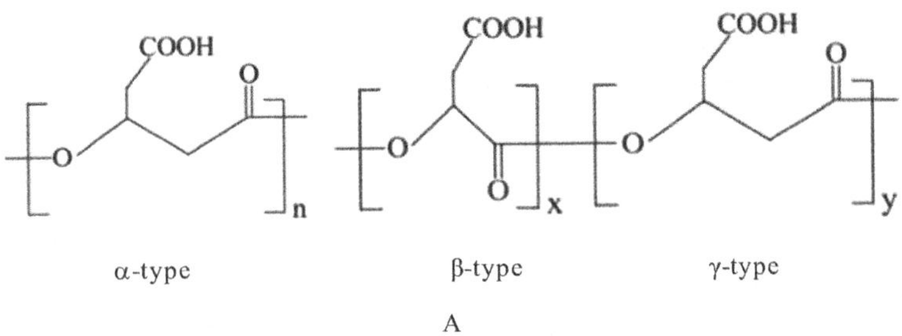

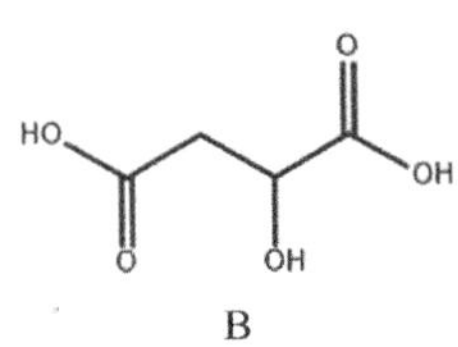

Fig. 8.1 The three different isomers of PMA (**A**) and malic acid (**B**)

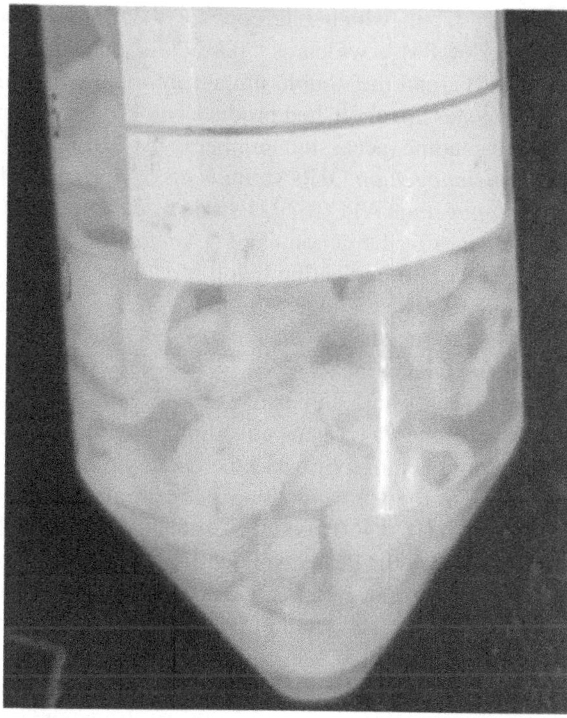

Fig. 8.2 The noodle-like PMA sample produced by *A. melanogenum* OUC

(accession number is GCA_016807825.1) size of *A. melanogenum* ATCC62921 strain is 25.3 Mpb. So, it is a haploid strain.

The concentration of extracellular products produced by the whole genome duplicated yeast-like fungal *A. melanogenum* OUC strain isolated from the mangrove ecosystem in Hainan is 17.2 g/L. The purified products were obtained by repeated precipitation with methanol/water, anion-exchange resin adsorption, NaCl elution, dialysis, and lyophilization. Finally, the purified extracellular product was identified as a polymalic acid by HPLC, HPLC-MASS, and NMR. It is very strange to note that the purified PMA produced by *A. melanogenum* OUC can form hydrogel and film, and the precipitated PMA looks like noodle (Fig. 8.2), whereas the purified PMA produced by *A. melanogenum* ATCC62921 cannot form hydrogel and film, and the precipitated PMA is powder-like, suggesting PMA produced by the OUC strain has a high molecular weight. Determination of the viscosity of the solution by using the Ubbelohde capillary viscometer clearly indicates that the viscosity (74.7 cP) of the solution of the purified PMA produced by *A. melanogenum* OUC is much higher than that (4.1 cP) of the solution of the purified PMA produced by *A. melanogenum* ATCC62921. Meanwhile, measurements of their Mws indeed shows that Mw of the purified PMA produced by *A. melanogenum* OUC is 3.9×10^5 Da, indicating that Mw (3.8×10^4 Da) of PMA produced by *A. melanogenum* ATCC62921 was much lower than that of the purified PMA produced by *A. melanogenum* OUC. It was found that the single mutants *Δpmas1–10* and

Δpmas2–47, in which one copy of the *PMAs* genes was deleted, only produce 8.5 g/L of PMA which is 50.8% less than that of their wild-type OUC strain ($P < 0.01$), and the double mutant *Δpmas-12* in which both the two copies of the *PMAs* genes are abolished produces no PMA and non-malic acids (Qi et al. 2021).

The genome (accession number is JACXJB000000000-JACXJC000000000) size of *A. melanogenum* OUC strain is 48.6 Mpb, much higher than that (25.3 Mpb) of *A. melanogenum* ATCC62921 strain as described in Chap. 5. It was found that most of the genes in the genomes of *A. melanogenum* OUC strain are duplicated genes (15231, 85.5%), indicating that it is a whole genome duplication and diploid strain. This suggests that a lower titer (17.2 g/L) of PMA and a higher Mw (3.9×10^5 Da) of PMA produced by *A. melanogenum* OUC may be related to its highly duplicated genomes (Qi et al. 2021).

Removal of both copies of the *PKS* genes for melanin biosynthesis and overexpression of the *PYC1* gene encoding pyruvate carboxylase, the *VGB* gene encoding Vitreoscilla hemoglobin, and the *CRZ2* gene coding for the specific transcriptional activator in the Ca^{2+} signaling pathway in *A. melanogenum* OUC make the new transformant Crz466 produce 34.6 g/L of extracellular Ca^{2+}-PMA with Mw of 4.9×10^5 Da at the flask level. During a 10-Liter fermentation, the new transformant Crz466 can produce 35.6 ± 1.2 g/L of Ca^{2+}-PMA and 13.9 g/L of cell mass within 168 h, leading to the yield of 0.36 g/g of glucose and the productivity of 0.21 g/L/h (Qi et al. 2021).

The *cnb* gene encoding the calcineurin regulatory subunit B from *Beauveria bassiana* was overexpressed in the multi-adapted strain of *A. pullulans* obtained under high concentrations of waste xylose-containing mother liquor and at low pH, and the constructed AE599 strain was used to convert the waste xylose-containing mother liquor for efficient PMA production (49.47 g/L) under low pH control. Finally, the produced PMA was hydrolyzed using sulfuric acid to produce pure monomer L-malic acid (Li et al. 2023). The recombinant strain overexpressing cytosolic malate dehydrogenase (Mdh) from *Aspergillus oryzae* and pyruvate carboxylase (Pyc) from *A. oryzae* and endogenous PMA synthetase (PMAs) can produce 96.7 g/L PMA from glucose in the fed-batch fermentation (Qin et al. 2024). The rubberwood enzymatic hydrolysates + glucose and rubberwood enzymatic hydrolysates + xylose can be transformed by *A. pullulans* NRRL Y-2311-1 to produce 59.9 g/L and 53.7 g/L of PMA, respectively (Zeng et al. 2024).

A mutant of *A. pullulans* CCTCC M2012223, in which the endogenous malate dehydrogenase gene was overexpressed and the key genes for the pullulan, melanin, and liamocin biosynthetic pathways were deleted to enhance conversion of NADPH to NADH, can produce the highest PMA titer (194.3 g/L) from glucose during the fed-batch fermentation. Furthermore, a total of 86.19% L-malic acid, with a purity of 99.7%, was obtained from hydrolysis of the produced PMA (Li et al. 2024). A thermotolerant yeast -like fungal *A. melanigenum* DA22 islated from the mangrove tree leaves in Hainan, China, was found to have ability to keep sustained cell growth and high PMA production (56.1 g/L) from glucose at 35 °C, higher temperaure than any other *A. melanogenum* during 5-Liter bioreactor. The thermoprotective heat shock proteins, antixoidant enzymes and the memebrane associated transcriptional factor SPT23 may be related to the thermotplerance in this strain. This high

Table 8.1 The PMA titers and the apparent molecular weights (Mws) by different strains of *Aureobasidium* spp.

Strains	Carbon sources	PMA titers (g/L)	Mw (Da)	References
A. pulluans FJ-D2 mutant	Waste xylose	57.1	/	Feng et al. (2019)
A. pulluans MCW	Glucose	152.15	2.1×10^5	Wang et al. (2015)
A. hainanensis sp. P6	Glucose	118.3	8.1×10^4	Ma et al. (2013)
A. pulluans ZD-3d	Glucose	62.3	4.6×10^3	Zhang et al. (2011)
A. pulluans ZX-10	Glucose	87.6	/	Zou et al. (2019)
A. pulluans ipe-1	Glucose	57.0	/	Cao et al. (2016)
A. melanogenum ATCC62921	Glucose	51.4	3.8×10^4	Wang et al. (2020)
Aureobasidium sp. A-91	Glucose	80.0	$6.0–11.0 \times 10^3$	Nakajima-Kambe et al. (1996)
A. melanogenum OUC	Glucose	17.2 ± 0.3	3.9×10^5	Qi et al. (2021)
A. melanogenum Crz46	Glucose	34.6 ± 0.3	4.9×10^5	Qi et al. (2021)
A. pulluans	Glucose	9.7	$5.1–7.9 \times 10^3$	Manitchotpisit et al. (2012)
A. pulluans YJ6–11	Corncob hydrolysate	80.4	8.7×10^3	Zou et al. (2016a, b)
A. pulluans YJ6–11	Xylose	80.4	8.4×10^3	Zou et al. (2016a, b)
A. pulluans ipe-1	JA tuber	24.9	/	Cao et al. (2019)
A. pulluans ipe-1	Hydrolysates of bagasse	23.2	/	Cao et al. (2020)
A. melanogenum GXZ-6	Malt syrup	124.1	/	Zeng et al. (2019)
A. pulluans CCTCC M2012223	Hydrolysate of raw sweet potato	57.5	/	Zan and Zhou (2013)
A. pulluans GXL-1	Liquefied corn starch	49.0	/	Zeng et al. (2020)
A. pulluans ZX-10	Soy molasses	71.9	/	Cheng et al. (2017)
A. pulluans ZX 10	Sugarcane juice	116.3	/	Wei et al. (2017)
A. pulluans NRRL 50383	Pretreated wheat straw	50.0		Leathers and Manitchotpisit (2013)
A. pullulans ZX-10	Waste xylose	49.47		Li et al. (2023)
A. pulluans NRRL Y2311–1	Barley straw hydrolysate	43.5	/	Yegina et al. (2019)

JA: Jerusalem artichoke

temperature PMA fermentation technique can offer the reduced cooling energy demands and improved inherent resistance to bacterrial contamination (Yang et al. 2026).

PMA production from glucose and any other materials by any other strains of *Aureobasidium* spp. is shown in Table 8.1. It can be seen from Table 8.1 that the engineered *A. melanogenum* Crz466 can produce the highest Mw of PMA, and the mutant of *A. pullulans* CCTCC M2012223 can produce the highest amount of PMA. It also can be seen from Table 8.1 that in addition to glucose, malt syrup, waste xylose mother liquor, sugarcane molasses, sugarcane juice, soy molasses, Jerusalem artichoke hydrolysate, bagasse hydrolysates, potato starch hydrolysates, corncob hydrolysates, and corn fiber, wheat straw hydrolysates, and cassava hydrolysate have also been used for PMA production by different strains of *Aureobasidium* spp. in order to greatly reduce the cost of substrates for PMA production.

8.2.2 Elucidation of Biosynthesis Pathway of Polymalate and Malic Acid by Genome Editing and Synthetic Biology

Although PMA and its derivatives have a great potential application as a novel drug delivery system and advanced biomaterials, it is interesting and urgent to know how PMA is synthesized in *Aureobasidium* spp., which kinds of enzymes and their encoding genes are involved in PMA biosynthesis, which signaling pathways regulate PMA biosynthesis, and from where its monomer (L-malic acid) comes. As the unique bio-extracellular and biodegradable polyester material known at present, elucidation of the PMA biosynthesis pathway will have an important impact on the development of biopolyester materials, PMA overproduction by metabolic engineering and synthetic biology, and a deeper understanding of PMA molecular regulation.

As shown above, because the haploid *A. melanogenum* ATCC62921 is a high PMA producing yeast-like fungal strain, it is used to elucidate PMA biosynthesis pathway. As the hydrolysates of the extracellular PMA produced by *A. melanogenum* ATCC62921 contain only a large amount of released L-malic acids, it is speculated that the site of PMA biosynthesis is on the cell membrane or in the periplasmic space. So, the transmembrane domain should be contained in the protein sequence of the key enzyme involved in the PMA polymerization pathway. In addition, as PMA is one kind of biopolyester, an NRPS may be implicated in its biosynthesis. Moreover, according to the known metabolic pathways in eucaryotic cells, L-malic acid, the monomer of PMA, may originate from the TCA cycle in mitochondria, as well as from the glyoxylate pathway and the reductive pathway in the cytoplasm.

For searching for the main sources of monomer (L-malic acid) for PMA biosynthesis, deletion of the key genes involved in synthesis and degradation of L-malic acid in TCA cycle confirms that L-malic acid in PMA molecule mainly comes from TCA cycle in mitochondria (Fig. 8.3). Additionally, it was found that the *PYC1* gene encoding pyruvate carboxylase and cyMDH gene for cytoplasmic malate dehydrogenase in the reductive TCA cycle at cytoplasm and the genes for the glyoxylate pathway also play an important role in the synthesis of PMA (Fig. 8.3). First, the genes of *ICL* (encoded isocitrate lyases) and *MSE* (encoded malate synthase) in the

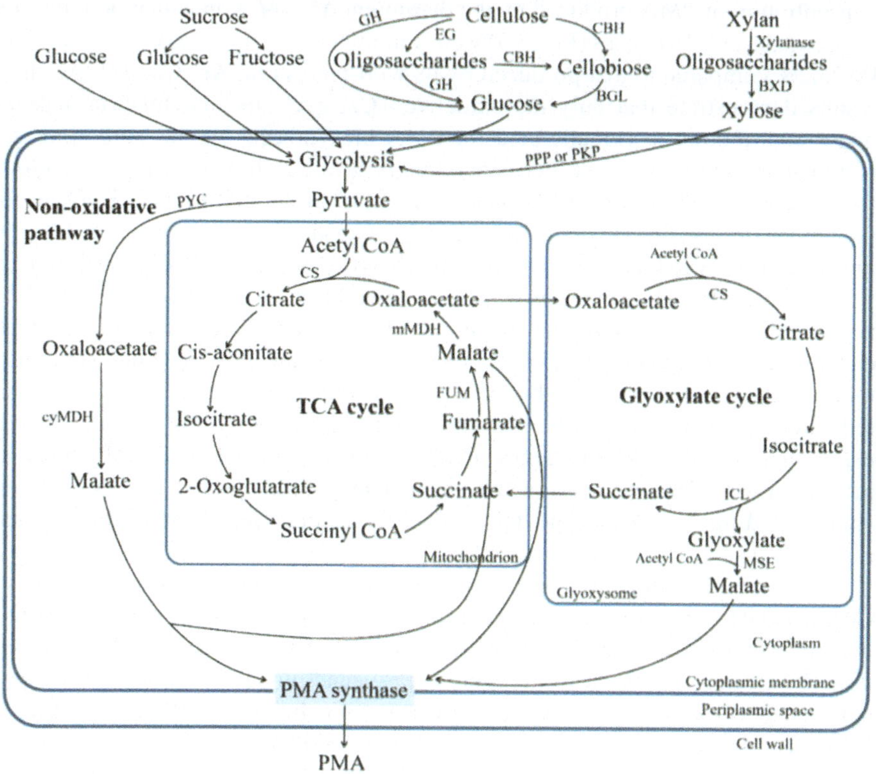

Fig. 8.3 Malate biosynthesis pathways from different sugars in eucaryotic cells. *EG* endo-β-1,4-glucanase, *GH* glucohydrolase, *CBH* cellobiohydrolase, *BGL* β-1,4-glucosidase, *BXD*: β-D-xylanohydrolase, *PPP* pentose phosphate pathway, *PKP* phosphoketolase reaction pathway, *CS* citrate synthase, *cyMDH* malate dehydrogenase in cytoplasm, *mMDH* malate dehydrogenase in mitochondrion, *PYC* pyruvate carboxylase, *FUM* fumarase, *ICL* isocitrate lyase, *MSE* malate synthase

glyoxylate pathway were deleted. Compared with the concentration of PMA produced by the wild type strain ATCC62921, the PMA concentration produced by the *ICL* and *MSE* gene (Fig. 8.3) knockout mutants decreased by 5.4% and 2.7%, respectively. Besides, the *cyMDH* gene is the key gene in the reductive TCA cycle at cytoplasm (Figure 3). After deletion of the *cyMDH* gene, the PMA titer was decreased by 9.97% in the mutant *ΔcyMDH-6*. These results show that the glyoxylate pathway and the reductive TCA cycle at cytoplasm are indeed involved in the synthesis of PMA, but deletion of such genes did not greatly influence PMA biosynthesis. Therefore, these two pathways are not the main sources of L-malic acid for PMA biosynthesis. However, deletion of the *mMDH* gene (Fig. 8.3) encoding mitochondrial malate dehydrogenase makes the mutant not grow in the fermentation medium with $CaCO_3$, but grow slowly in the YPD medium. Meanwhile, the concentration of PMA produced by the disruptant *ΔAcyM*-3, in which both the *ICL* gene and the *cyMDH* genes (Fig. 8.3) were simultaneously knocked out, is decreased by 23.2% compared to that produced by its wild-type strain ATCC62921. All these results demonstrate that only the oxidative TCA cycle in mitochondria makes a major contribution to L-malic acid for PMA biosynthesis. At the same time, the complete removal of the *PYC* gene encoding pyruvate carboxylase (Fig. 8.3) allows the mutant to grow on the YPD plate, but not to grow and produce PMA in the medium with $CaCO_3$. However, the partial removal of the *PYC* gene enables the mutant to greatly reduce PMA production. This reveals that the *PYC* gene also plays some roles in PMA biosynthesis (Fig. 8.3). In addition, it was also found that deletion of the genes for NAD-dependent malate enzyme, NADP-dependent malate enzyme, and phosphoenolpyruvate kinase (Fig. 8.3) makes all the mutants produce more PMA than that produced by the wild-type strain ATCC62921, suggesting that reduction in consumed L-malic acid can increase PMA biosynthesis. Indeed, overexpression of malate dehydrogenase (Mdh) and pyruvate carboxylase (Pyc) and an increase in NADH and acetyl-CoA can enhance PMA production by different strains of *Aureobasidium* spp. (Qi et al. 2021; Qin et al. 2024) (Table 8.1 and Fig. 8.3).

It has been regarded that PMA may be synthesized via two routes: (1) L-Malyl-CoA may be formed from malate under the catalysis of a L-malyl-CoA ligase (Mcl), then the L-malyl-CoA may be polymerized into PMA by an unknown PMA synthetase. However, so far, the key PMA polymerase genes involved in this pathway have not been detected, so that it is still unknown about its function in PMA biosynthesis. It has been reported that the gene encoding malyl-CoA ligase (Mcl) was thought to be involved in PMA biosynthesis, but it has not been confirmed whether it is involved in PMA synthesis or not. (2) A malyl-AMP ligase and a malyl-transferase (polymerase) in analogy to an NRPS may be involved in PMA biosynthesis (Chi et al. 2016). For the first route, deletion of the gene encoding Mcl made the mutant *ΔMCL-6* still produce the same amount of PMA as its wild-type strain. Furthermore, the hydrolysates of the extracellular products produced by this mutant still contain high amount of L-malic acid and D-malic acid. This demonstrates that the Mcl is not implicated in PMA biosynthesis in the wild-type strain ATCC62921. In order to confirm that PMA is synthesized by the second route, the *PPTase* gene, which is involved in the posttranslational modification of most non-ribosomal peptide synthase (NRPS) and polyketide synthase (PKS), was disrupted in the wild-type strain ATCC62921. It was found that the mutant *ΔPPTase-44* obtained could grow on the YPD plate with lysine, but could not grow in the PMA production medium without lysine and $CaCO_3$. This indicates that the PMA synthetase has an ACP domain, which is indeed activated through a phosphopantetheine (PP) addition by the PPTase. Therefore, the key polymerase involved in PMA synthesis should have the following characteristics: First, it should be a transmembrane protein; secondly, it should have an ACP domain which should be activated by PPTase; finally, there should be a significantly difference in expression levels of the related genes in the ATCC62921 strain grown in the medium with and without $CaCO_3$. Based on the

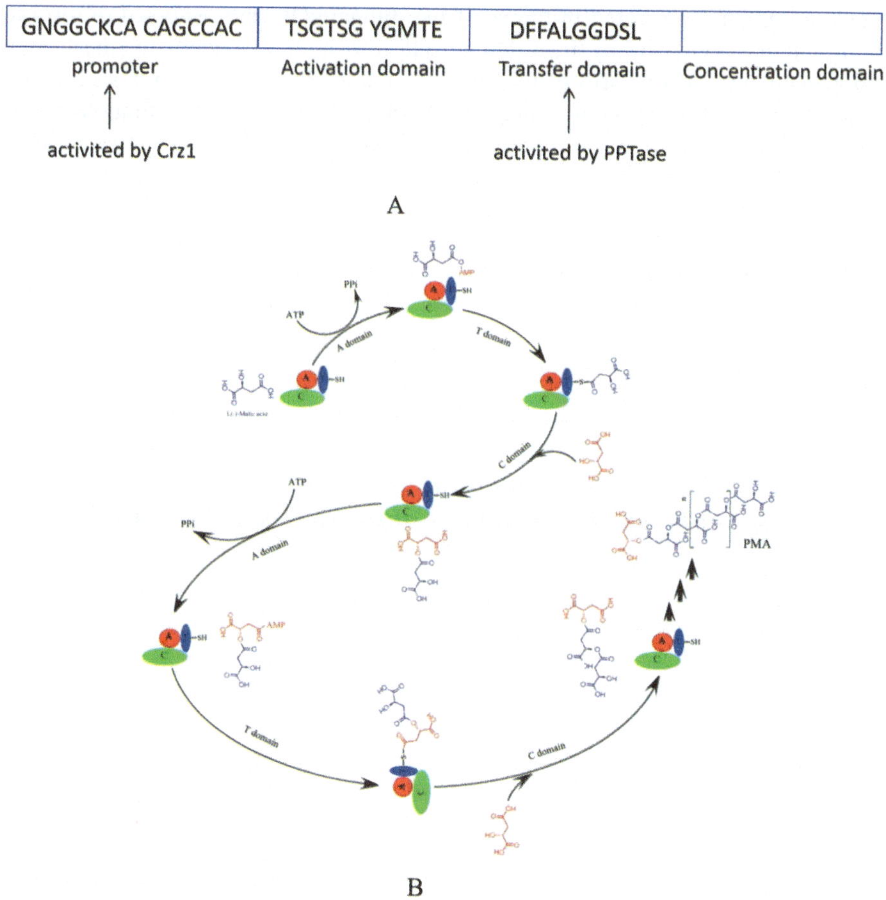

Fig. 8.4 The main domains of the PMAs (**A**) and synthetic pathway of polymalate (PMA) (**B**) in *A. melanogenum*

characteristics mentioned above and searching of the NRPSs by using the database of anti-SMASH, finally, it was found that the gene (*PMAs* gene; GenBank accession number: MN551082.1) may be responsible for PMA biosynthesis in the wild-type strain ATCC62921. The size of the *PMAS* gene is 5052 bp without intron and the gene codes for a protein with 1684 amino acids (Fig. 8.4A). In order to confirm this hypothesis, this gene was totally removed from the genome of the wild-type strain ATCC62921. It was found that there is no D- and L-malic acid released from the extracellular products produced by the mutant *ΔPMAs-33*. These results strongly demonstrate that the PMA synthetase (PMAs) encoded by the *PMAs* gene is the key enzyme for polymerization of malate into PMA. Like the polylysine synthetase for polymerization of lysine into polylysine, the PMA synthetase also has the conserved domains, such as Activation (A) domain, Transfer (T) domain, and Concentration (C) domain (Fig. 8.4A). Based on the function of the conserved domains, the malic acid and ATP accumulated in the yeast-like fungal cells are catalyzed to form the activated malyl-O-AMP by the A domain and subsequently loaded onto the 4′-phosphopantetheine (4′-PP) arm of the adjacent T domain with AMP release, thereby resulting in the formation of a malyl-S-enzyme. The C domain continuously catalyzes an ester bond formation between two malic acid substrates activated as the maly-S-enzyme in PMA molecules (Fig. 8.4B). Thus, the A and C domains are crucial to the NRPS mechanism underlying malate activation and ester bond formation for PMA biosynthesis (Wang et al. 2020; Qi et al. 2022). So, we predict that the polymerization mechanism for PMA biosynthesis by the PMA synthetase is similar to that for polylysine biosynthesis by a polylysine synthetase (Geng et al. 2014). Indeed, it has been confirmed that overexpression of the *PMAs* gene in *A. pullulans* enhances PMA biosynthesis (Qin et al. 2024).

8.2.3 Clarification of Regulation Mechanisms of Polymarate Biosynthesis by Genomic Editing and Synthetic Biology

It has been well evidenced that adding $CaCO_3$ to the culture medium is helpful to the biosynthesis of PMA (Table 8.1), suggesting that Ca^{2+} can enhance the synthesis of PMA, and the Ca^{2+} signaling pathway (Fig. 8.5) may control PMA biosynthesis. In order to examine the effect of the Ca^{2+} signal pathway on PMA synthesis, disrupting the key genes of *CnaA*, *CnaB*, and *CRZ1* involved in the Ca^{2+} signal pathway (Fig. 8.5) found that deletion of the *CnaA* gene and *CnaB* gene in the wild-type strain ATCC62921 results in the mutants *ΔCna A-33* and *ΔCna B-22* that can grow on the YPD plate normally, but cannot grow and produce any PMA in the PMA production medium with $CaCO_3$. These results indicate that calcineurin encoded by the *CnaA* gene and the *CnaB* gene plays an important role in maintaining the growth of the strain ATCC62921 during the active PMA synthesis. Inactivation of the *CRZ1* gene encoding the key transcriptional activator (Crz1) in the Ca^{2+} signaling pathway (Fig. 8.5) allows the mutant *ΔCrz-177* to yield much less PMA than its wild-type strain ATCC62921. Analysis of the upstream promoter of the PMA synthetase gene (Fig. 8.4A) shows that in its −89 bp and − 98 bp regions of the promoter, there are two Crz1 binding sequences. Furthermore, the expression level of the PMA synthetase gene in the mutant *ΔCrz-177* is significantly reduced compared to that of the PMA synthetase gene in its wild-type strain. This demonstrates that the Ca^{2+} signaling pathway indeed regulates PMA biosynthesis. Determination of the subcellular localization of the Crz1 in the transformant CRZ1-GFP-55 expressing the fused *CRZ1-GFP* gene grown in the PMA production media with $CaCO_3$ and without $CaCO_3$ clearly shows that when the transformant expressing the *CRZ1-GFP* gene is grown in the PMA production medium with $CaCO_3$, the expressed Crz1 is localized in the nuclei, inducing expression of the *PMAs* gene, in turn promoting PMA biosynthesis. When the transformant expressing the *Crz1-GFP* gene is grown in the PMA production medium without $CaCO_3$, the expressed Crz1 is localized in the

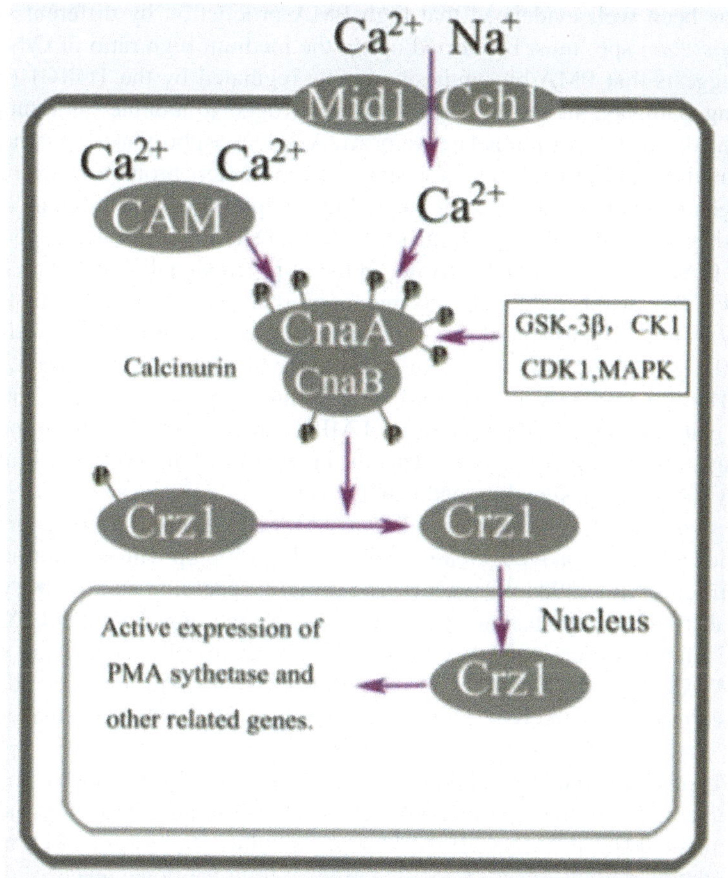

Fig. 8.5 The catalytic functions of all the enzymes involved in the Ca^{2+} signaling pathway. *CAM* calmodulin, *Cch1* Ca^{2+} transporter in the Ca^{2+} channel, *Mid1* another Ca^{2+} transporter in the Ca^{2+} channel, *CnaA* Ca^{2+}/calcineurin A subunit catalyzes dephosphorylation of Crz1 in the presence of Ca^{2+}, and the activated Crz1 is localized in the nucleus, *CnaB* Ca^{2+}/calcineurin B subunit also catalyzes dephosphorylation of Crz1 in the presence of Ca^{2+}, *Crz1* crazy protein, the key transcriptional activator in the Ca^{2+} signaling pathway

cytoplasm, showing that the Ca^{2+} signal pathway indeed has an important influence on subcellular localization of Crz1 and PMA synthesis (Wang et al. 2020). Complementation of the *PMAs* gene from *A. melanogenum* OUC in the mutant *Δpmas*, in which all the *PMAs* gene has been removed from *A. melanogenum* ATCC 62921, shows PMA biosynthesis in the transformant ATCC62921EPMAs1–77 is partially restored while the mutant *Δpmas* cannot synthesize any PMA as mentioned above. This means that the high-Mw polymalic acid biosynthesis in the whole genome duplicated *A. melanogenum* OUC is also controlled by the PMA synthase, and both copies of the *PMAs* genes play the same important roles in PMA biosynthesis (Qi et al. 2021, 2022).

It has been well evidenced that high PMA production by different strains of *Aureobasidium* spp. must be carried out in the medium high ratio of C/N sources. This suggests that PMA biosynthesis may be regulated by the TORC1 (nitrogen) signaling pathway. In the absence of rich nitrogen glutamine, ammonium, and amino acids, the transcriptional activator AreA is dephosphorylated, and the dephosphorylated transcriptional activator stays in the nuclei, promoting expression of many genes involved in nitrogen catabolite derepression and pullulan and PMA biosynthesis (Kang et al. 2022, Song et al. 2020). Deletion of the *AREA* gene encoding the GATA-transcriptional activator in the nitrogen signaling pathway in *A. pullulans* CCTCC M2012223 makes its mutant produce 9.5 g/L of PMA and 8.8 g/L of cell dry weight; expression of the *AREA* gene allows the recombinant strain to produce 20.7 g/L PMA and 19.2 g/L of cell mass, while its wild-type strain only produces 18.6 g/L of PMA and 13.9 g/L of cell mass. At the same time, the glucose kinase gene in the glycolytic pathway (EMP) is also overexpressed in the overexpressing strain (Song et al. 2020). This means that PMA biosynthesis is also regulated by the nitrogen signaling pathway.

The size of the *NSDD* gene encoding a GATA-type transcription factor from the genomic DNA of *A. melanogenum* ATCC62921 is 1398 bp with two introns, encoding a proteins with 1398 amino acids. The protein (accession number: KEQ5978.1) deduced from the gene has one conserved GATA zinc finger domain C-X2-C-X_{18}-C-X_2-C motif which can specifically be bound to the conserved nucleates 5′-HGATAR-3′ in the promoter of the regulated genes and a conserved QSLPSI sequence motif near the N-terminus. Disruption of the *NSDD* gene makes the disruptant *Δnsd62K355*, in which the *NSDD* gene was totally removed, yield only 27.3 g/L of PMA, while its parent strain can produce 61.8 g/L PMA. In contrast, complementation of the native *NSDD* gene in the disruptant *Δnsd62K355* allows the complemented strain CMNSDPEN211 to accumulate 69.4 g/L of PMA in the culture. In the disruptant *Δnsd62K355,* the relative transcriptional levels of the *NSDD* gene, *PYC1* gene, *mMDH* gene, *cyMDH* gene, and *PMAs* gene are decreased to 0.0%, 90.3%, 98%, 94.6%, and 37.2%, respectively, while in the complemented strain CMNSDPEN21, the relative transcriptional levels of the *NSDD* gene, *PYC1* gene, *mMDH* gene, *cyMDH* gene, and *PMAs* gene are increased to 135.1%, 118.7%, 199.5%, 109.1%, and 336.4%, and the relative transcriptional levels of all the genes in their wild-type strain keep 100% (Chi et al. 2024). These data demonstrate that PMA biosynthesis is also controlled by the GATA-type transcription factor NsdD.

LaeA [loss of aflR (Aflatoxin regulatory gene) expression] is a DNA methyltransferase and one of the most conserved global regulatory proteins. It is localized in the nucleus of the fungal cells and regulates a variety of biological processes, including the regulation of redox state, cytochrome P450 pathways, secondary metabolism such as malic acid and citric acid production, and morphologies in various fungi. Furthermore, the gene contains a conserved sadenosylmethionine (SAM)-binding structural domain that can methylate the histone in the chromosome. It was found that overexpression of the *AplaeA* gene in *A. pullulans* HIT-LCY3T yields a mutant that increases PMA production and the yeast-like fungal cell number. At the same time, the expression of the core genes, including the *PYC1* gene encoding pyruvate carboxylase, the *MS* gene encoding malate synthase, and the *MDH* gene encoding malate dehydrogenase for PMA biosynthesis in the mutant, is also enhanced (Guan et al. 2025). This means that PMA biosynthesis is also regulated by the global regulatory protein LaeA.

As mentioned above, PMA biosynthesis is also influenced by the expression of the *PYC1* gene, *MDH* genes, and *PMAs* gene, concentration of NAD and L-malic acid, and oxygen level of the yeast-like fungal cells.

8.2.4 Applications of Polymalate and Malic Acid

Polymalic acid (PMA) is an anionic polyester polymerized from malic acid monomers. It has many applications in food, pharmaceutical, cosmetics, chemical, and other fields because of its characteristics of safety, non-toxicity, biocompatibility, water solubility, and chemical modification. At present, there are chemical synthesis and microbial synthesis to obtain polymalic acid. As the chemical synthesis processes are complex and energy consuming, the microbial synthesis has become the focus of attention.

The unique PMA film can be prepared from the high-molecular-weight PMA produced by the Crz466 constructed in this laboratory. Based on the tensile strength (TS) (5.94 Mpa), elongation at beak (EB) (35.8%), oxygen barrier property (PV), and water vapor permeability (WVP) of the high-Mw PMA film, the mixture containing 10 g/L of glycerol (plasticizer) and 90.0 g/L of the high-Mw PMA produced by the Crz466 strain was used to prepare the high-Mw PMA film. Under these conditions, PV and WVP values of the film are 2.2 meq/kg and 1.6 $ng{\cdot}m/m^2{\cdot}s{\cdot}Pa$, respectively, which indicates that the PMA film has excellent (high) mechanical properties and barrier properties. In addition, the prepared high-Mw PMA films are apparent, colorless, and transparent (Fig. 8.6) (Qi et al. 2023).

Then, the grapes were coated with the high-Mw PMA by dipping technique (Fig. 8.7). It was found that the treatment can not only improve the appearance quality of grapes but also have a good preservation effect and extend the shelf life of the grapes. It is also indicated that after the high-Mw PMA film in the in vitro simulated stomach is hydrolyzed for 3.0 h, a considerable amount of L-malic acid is released from the high-Mw PMA film. Therefore, this study demonstrates that coating grapes with the high-Mw PMA film is an effective treatment to maintain the high quality and good texture of the grapes during the storage life and extend the shelf time and marketing period. Furthermore, the released L-malic acid can be absorbed to produce ATP by humans. Thus, the high-Mw PMA film as the editable material on the grapes and any other fruits is considered to be a safe and effective method as well as an environmentally friendly technology (Qi et al. 2023).

In addition, malic acid released from PMA is a four-carbon dicarboxylic acid and an intermediate in the tricarboxylic acid (TCA) cycle. It has many applications in the beverage and food industry as an acidulant and flavor enhancer, and in metal cleaning, textile finishing, pharmaceuticals, and agriculture (Chi et al. 2016). Because L-malic acid is an important intermediate in the TCA cycle, L-malic acid metabolism in the TCA cycle can provide humans with a large amount of ATP, which is very helpful for health.

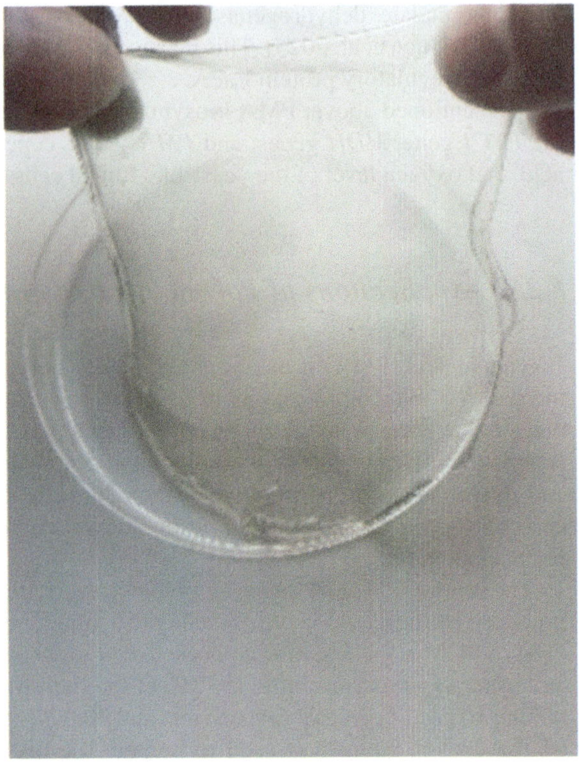

Fig. 8.6 The prepared high-Mw PMA film

Fig. 8.7 The grapes coated with the high-Mw PMA film

Fig. 8.8 Molecular structure of fumaric acid

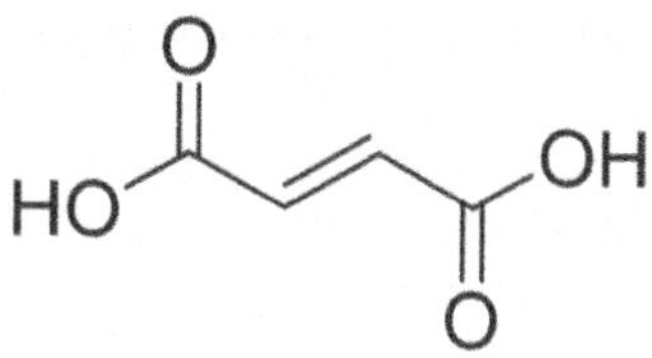

8.3 Production, Biosynthesis, Regulation, and Applications of Fumaric Acid

Fumaric acid (FA), the simplest unsaturated dicarboxylic acid, has been identified as an important platform chemical (Fig. 8.8).

Now, FA's demand is mainly met by traditional chemical reactions with petroleum consumption, leading to strong catalytic toxicity, CO_2 emissions, and dependence on unsustainable petroleum industries. In contrast, FA biosynthesis has many advantages over chemical ones, including mild conditions, multiple substrates, and environmentally friendly reactions. Therefore, it is of great importance to explore efficient and sustainable approaches for FA biosynthesis.

Rhizopus spp., especially *Rhizopus oryzae* and *Rhizopus arrhizus*, have been identified as excellent fumarate producers. These filamentous fungi usually grow as suspended mycelia and clumps, which is detrimental to FA biosynthesis due to blocked stirring and dissolved oxygen and nutrient transfer inside their cells. Additionally, ethanol and lactic acid are the main by-products produced by *R. oryzae* and *R. arrhizus* during FA biosynthesis, leading to reduced conversion of glucose to FA. Furthermore, genetic modification and metabolic engineering of the genomes in *Rhizopus* spp. are difficult to achieve due to unclear genetic information, filamentous morphology, and the lack of effective genetic editing tools. It was reported that *A. aubasidani* DH177 strain could accumulate calcium fumarate in the presence of calcium carbonate and glucose (Wang et al. 2018). *A. aubasidani* DH177 strain usually grows as a single cell during FA biosynthesis, which is considered to have many significant advantages over *Rhizopus* spp. Moreover, direct genetic modification and genome editing in *A. aubasidani* can be achieved easily and successfully with the available excellent genetic editing tools as stated in Chap. 5 (Wei et al. 2023).

8.3.1 Overproduction of Fumarate by Metabolic Engineering

In 2018, Wang et al. (2018) reported that the yeast-like fungus *A. aubasidani* DH177 isolated from the leaves of *Weigela florida* in China can accumulate 64.7% (w/w) oil in its cells, produce 22.4 g/L cell biomass, and secrete 32.3 g/L fumarate during a 5-Liter batch fermentation. However, it is shown that calcium gluconate is the main by-product produced by the DH177 strain during FA biosynthesis, which can accumulate 37.7 g/L gluconate and 21.8 g/L fumarate in the supernatant of the culture (Wei et al. 2023). It has been well known that the biosynthesis of gluconic acid, for which glucose oxidase (Gox) encoded by the *GOX* gene is the key enzyme, must be blocked in order to obtain pure fumarate. After deletion of the *GOX* gene, there is no accumulation of gluconate from the disruptant *Δgox* strain, and the fumarate titer produced by the disruptant *Δgox1* strain increases to 60.5 g/L. In addition, pure fumarate crystals and dimethyl fumarate can be obtained easily from the supernatant of the culture of the *Δgox1* strain, indicating that the exocellular product produced by the *Δgox1* strain is indeed fumaric acid, and the *Δgox1* strain has great potential for fumarate production.

In addition, it was found that the genome (accession number JADDKP000000000) of *A. aubasidani* DH177 is 28.62 Mpb in size, contains 10,215 genes, and has a GC content of 50.32% (Wei et al. 2023). Therefore, it is a haploid yeast-like fungal strain.

The *Δgox1* strain obtained can accumulate 1.2 g/L liamocin; therefore, the key *PKS1* gene encoding polyketone synthase involved in liamocin biosynthesis was removed from the genomic DNA of the *Δgox1* strain to block biosynthesis and secretion of liamocin in the *ΔgoxΔpks11* strain. Then, the *PYC* gene encoding pyruvate carboxylase was overexpressed in the *ΔgoxΔpks11* strain, resulting in the e-PYC1 strain. Finally, fumarate produced by the e-PYC1 strain can reach 82.5 g/L at flask level. After that, the culture conditions for the e-PYC1 strain are optimized and the fumarate titer reaches 88.1 g/L in the shake flasks. Under the optimal conditions, the e-PYC1 strain can produce 93.9 g/L of fumarate from glucose during a 10-Liter fed-batch fermentation (Wei et al. 2023). Therefore, the fumarate titer produced by *A. aubasidani* e-PYC1 strain has reached the highest level compared with those produced by any other yeast and fungal strains obtained so far (Table 8.2).

In addition to glucose from starch and cellulose in lignocellulosic materials, xylose from lignocellulosic materials is also an inexpensive, natural, and the most

Table 8.2 Production of fumaric acid from different strains of *Rhizopus* spp. and *A. aubasidani*

Strain	Substrate	Fermenter	Titer (g/L)	Yield (g/g)	References
R. oryzae	Corn straw	Shake flask	27.8	0.35	Xu et al. (2010a, b)
R. oryzae	Glucose	Stirred tank	56.2	0.54	Fu et al. (2010a, b)
R. oryzae	Brewery wastewater	Shake flask	31.3	/	Das and Brar (2014)
R. oryzae	Apple pomace	Shake flask	25.2	/	Das et al. (2015)
R. oryzae	Glycerol	Stirred tank	25.5	/	Huang et al. (2015)
R. oryzae	Glucose	Bubble column	37.2	0.53	Zhou et al. (2002)
R. oryzae	Glucose	Shake flask	57.4	/	Huang et al. (2010)
R. oryzae	Glucose	Stirred tank	41.1	/	Huang et al. (2010)
R. oryzae	Glucose	Stirred tank	52.7	/	Fu et al. (2010)
R. oryzae	Glucose	Stirred tank	32.1	0.45	Kang et al. (2010)
R. oryzae	Cornstarch	Shake flask	44.1	0.44	Deng et al. (2012)
R. oryzae	Raw corn powder	Shake flask	32.2	0.39	Deng et al. (2012)

(continued)

Table 8.2 (continued)

Strain	Substrate	Fermenter	Titer (g/L)	Yield (g/g)	References
R. oryzae	Glucose	Stirred tank	49.4	0.56	Yu et al. (2012)
R. arrhizus	Xylose	Shake flask	45.3	/	Liu et al. (2015)
R. arrhizus	Glucose and xylose	Shake flask	46.8	/	Liu et al. (2017)
R. arrhizus	Crude glycerol and glucose	Shake flask	22.8	/	Zhou et al. (2014)
R. arrhizus	Food waste	Shake flask	32.7	/	Liu et al. (2016)
A. aubasidani DH177	Glucose	Shak flask	21.8 g/L		Wang et al. (2018)
A. aubasidani GC1	Glucose	Shake flask	60.5 g/L		Wei et al. (2023)
A. aubasidani	Glucose	Shake flask	88.1 g/L		Wei et al. (2023)
e-PYC	Glucose	Fed-batch	93.9 g/L	0.63	Wei et al. (2023)
A. aubasidani TKL-4	Xylose	Shake flask	82.1 g/L		Wang et al. (2024)
A. aubasidani TKL-4	Corncob-derived xylose	Shake flask	77.3 g/L		Wang et al. (2024)
A. aubasidani TKL-4	Xylose	Fermenter	88.5 g/L	0.81	Wang et al. (2024)
A. aubasidani TKL-4	Corncob-derived xylose	Fermenter	84.7 g/l	0.79	Wang et al. (2024)

abundant renewable resource in nature, as stated above. So, it is crucial to know how to efficiently and rapidly transform xylose into intermediates such as pyruvate in the glycolysis pathway and in turn convert pyruvate into fermentation products. Based on the xylose metabolism in eucaryotic cells (Fig. 8.9), in the isomerase pathway, xylose is isomerized into xylulose without needing any cofactor under catalysis of isomerase (Xi). In contrast, in the oxidoreductase pathway, first, xylose is reduced to xylitol using NADPH as a coenzyme under catalysis of xylose reductase (Xr), and then xylitol is converted into xylulose using NAD^+ under catalysis of xylitol dehydrogenase (Xdh). However, this pathway has many drawbacks, such as the cofactor imbalance between Xr (NADPH-dependent) and Xdh (NAD^+-dependent), accumulation of the intermediate byproduct xylitol and reduced supply of NADPH for any other biosynthesis. Xylulose formed from the two pathways is phosphorylated into xylulose-5-phosphate under the catalysis of xylulose kinase (Xk). Finally, the formed xylulose-5-phosphate will be converted into fructose-6-phosphate and glyceraldehyde-3-phospahte under the catalysis of transketolase (Tkl). The obtained fructose-6-phosphate and glyceraldehyde-3-phosphate will enter the central glycolysis pathway via the pentose phosphate pathway (PPP) to form pyruvate for fumarate biosynthesis (Fig. 8.9). It was found that the mutant Δ*gox1* of *A. aubasidani* DH177 has a high ability to utilize xylose and corncob-derived xylose to produce fumarate. In order to further efficiently and rapidly transform xylose into fumarate, abolishment of the *XR* gene encoding Xr and overexpression of the *XI* gene encoding Xi, the

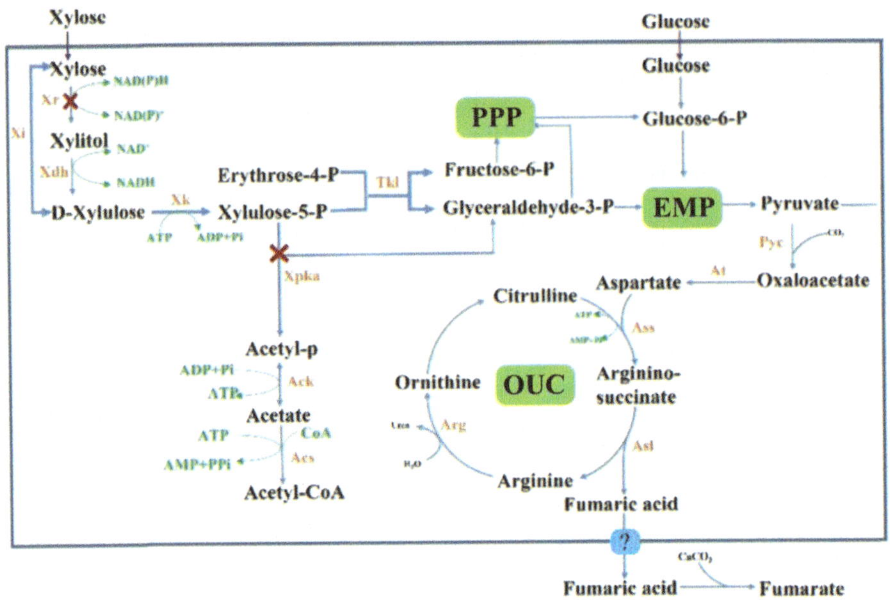

Fig. 8.9 The pathway for how xylose is efficiently converted into fumarate. X: deletion

XK gene encoding Xk, and the *TKL* gene coding for Tkl makes the strain TKL-44 produce 73.1 g/L of fumarate from xylose. At the same time, the transcriptional levels of the key *ASS* gene for argininosuccinate synthase and the *ASL* gene responsible for argininosuccinate lyase in the ornithine-urea cycle (OUC) in the strain TKL-44 (Fig. 8.9) are also obviously enhanced. After optimization of the fumarate production medium, the strain TKL-44 can yield 82.1 g/L of fumarate from xylose while it can also accumulate 77.3 g/L of fumarate from corncob-derived xylose in the culture. It is also strange to note that the TKL-44 strain can produce more fumarate from xylose and corncob-derived xylose than from glucose under the same conditions. During a 10-Liter fermentation, the TKL-44 strain can produce 88.5 g/L of fumarate from xylose, the yield was 0.81 g/g of xylose, and the productivity was 0.52 g/h/L. Meanwhile, it can yield 85.6 g/L of fumarate from corncob-derived xylose, and the yield is 0.8 g/L of xylose. However, the TKL-44 strain only produces 67.1 g/L of fumarate from glucose, and the yield is 0.6 g/g of glucose under the same conditions. These demonstrate that the TKL-44 strain constructed indeed can actively transform xylose and corncob-derived xylose into fumarate (Wang et al. 2024) and the lignocellulosic hydrolysate from corncob is also suitable for green fumarate production. At the same time, it must be stressed that a large amount of corncob from agriculture industry is available in northern China during the harvest period each year.

The fumarate production from different raw materials by different strains of *Rhizopus* spp. and *A. aubasidani* is summarized in Table 8.2. From Table 8.2, it can be seen that the yields produced by the engineered TKL-44 strain of *A. aubasidani* grown in xylose and corncob-derived xylose are the highest. This means that synthetic biology can play a significant role in enhanced fumarate production by the engineered strain.

8.3.2 Elucidation of Biosynthesis Pathway of Fumaric Acid by Genome Editing and Synthetic Biology

It is generally accepted that the oxidative TCA cycle in mitochondria, the reductive TCA pathway in the cytoplasm, amino acid metabolism, purine metabolism, and the ornithine urea cycle (OUC) are involved in FA biosynthesis in different species of fungi and yeasts. The various key genes related to these pathways were removed from the genomic DNA of the *Δgox1* strain obtained above to confirm the main biosynthesis pathway of FA in *A. aubasidani*. It was found that fumarate titers decrease to 9.5 g/L, 14.8 g/L, 16.9 g/L, and 33.5 g/L after the *ASL* gene encoding argininosuccinate lyase catalyzing lysis of argininosuccinate to produce L-arginine and fumaric acid, the *CPS1* gene encoding carbamoyl synthase I catalyzing formation of carbamoyl phosphate from CO_2, *CPS2L* gene encoding carbamoyl phosphate synthase II large subunit, and *CPS2S* gene governing carbamoyl phosphate synthase II small unit related to OUC pathway (Fig. 8.10) were deleted, respectively. FA biosynthesis is then restored after complementation of the corresponding genes in the disruptants. However, there is no significant decrease in fumarate titer from the disruptant mutants of other genes related to the TCA cycle (the key *FUM* gene encoding fumarase, Fum in mitochondria, and the key *SFC* gene encoding succinate-fumarate carrier from mitochondria to cytoplasm), the glyoxylate cycle (the key *ICL1* gene encoding iso-citrate lyase 1 and *ICL2* gene encoding isocitrate lyase 2), the reductive pathway (the key *MDH1* gene encoding malate dehydrogenase, Mdh, and the key *PYC1* gene encoding pyruvate carboxylase, Pyc), the TTP metabolism (the key *FAA* gene encoding fumarylacetoacetase), and the purine metabolism (the key *ADSL* gene for adenylosuccinate lyase) (Guo et al. 2020; Wei et al. 2023). In addition, the genome of *A. aubasidani* does not carry the *FUM* gene encoding cytoplasmic fumarase. This means that only the OUC, rather than the reductive TCA pathway in the cytoplasm and any other pathways, is involved in FA biosynthesis in *A. aubasidani* (Fig. 8.10), which is completely different from that of *R. oryzae* in which fumarate mainly comes from the reductive pathway in the cytoplasm and the oxidative pathway in the TCA cycle.

It can be seen from Fig. 8.10 that the Asl activity, which catalyzes the release of fumaric acid from argininosuccinate in the OUC pathway, is responsible for only fumarate production by *A. aubasidani*, and Asl in the OUC can serve as a bridge between carbon metabolism and nitrogen metabolism in the fungal cells. In the presence of $CaCO_3$, the synthesized fumaric acid can be efficiently transformed into Ca^{2+}-fumarate so that the bridge between carbon metabolism and nitrogen metabolism in the fungal cells can be disconnected. That is why accumulation of a high level of fumarate can occur by *A. aubasidani* in the presence of $CaCO_3$. Meanwhile, HCO_3^{2-} released from $CaCO_3$ may also be involved in the first reaction by the Cps,

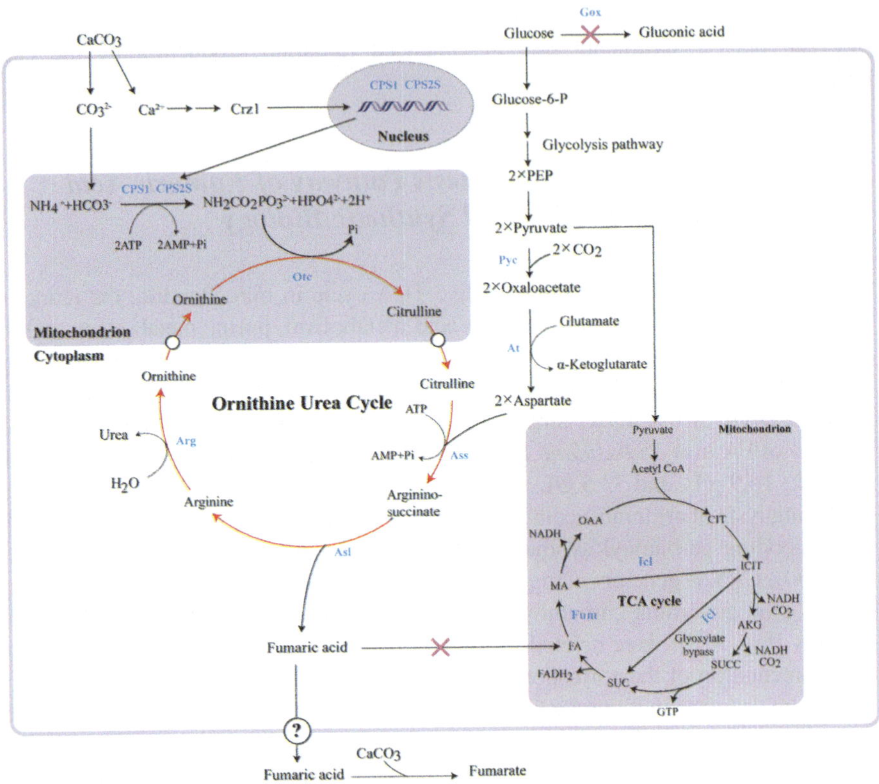

Fig. 8.10 Fumaric acid biosynthesis pathways in the parent *Δgox* strain. Pyc pyruvate carboxylase, At aspartate aminotransferase, Cps carbamoyl phosphate synthase, Otc ornithine transcarboxylase, Ass argininosuccinate synthase, Asl argininosuccinate lyase, Arg arginase, Sfc succinate-fumarate carrier, Icl isocitrate lyase, Fum fumarase. L-Cit L-citrulline, L-Orn L-ornithine, L-Arg L-arginine, AspC aspartate aminotransferase, Arg arginase, Otc ornithine transcarboxylase, Ort1 ornithine-citrulline transporter, ArgE acetylornithine deacetylase, ArgA N-acetylglutamate synthase, ArgB N-acetyl-glutamate kinase, ArgC N-acetyl-glutamyl-phosphate reductase, ArgD N-acetyl-l-ornithine aminotransferase, ArgJ L-glutamate N-acetyltransferase, L-Arg L-arginine, CaM calmodulin, CN calcineurin, Crz1 the key transcriptional factor in the calcium signaling pathway, Gox glucose oxidase

which is a rate-limiting reaction in the OUC pathway (Fig. 8.10). Therefore, according to Fig. 8.10, one mol of glucose theoretically can produce 2 mols of oxaloacetate by fixation of 2 mols of CO_2 through the glycolysis pathway under catalysis of Pyc. Although OUC fixes 1 mol of CO_2 from CO_3^{2-} released from the added $CaCO_3$, 1 mol of urea containing one carbon is lost in OUC, leading to one net carbon fixation in the cycle. Therefore, the OUC-mediated fumaric acid biosynthesis pathway in *A. aubasidani* is a green and eco-friendly process for the global sustainable development of fumaric acid production and carbon neutrality.

It has been reported that L-ornithine, one of the most important intermediates of the OUC, can come from the acetyl-glutamate cycle (AGC) (Wei et al. 2024; Myo

Myo Tint et al. 2025) (Fig. 11.1). In order to confirm this, the key *ARGBC* gene encoding both glutamate kinase and N-acetyl-gamma-glutamyl-phosphate reductase involved in the AGC was disrupted from the genomic DNA of the *Δgox1* strain to investigate whether L-ornithine concentration can affect FA biosynthesis in OUC. It was found that the intracellular L-ornithine concentration and fumarate titer produced by the obtained *ΔgoxΔargbc1* mutant decreased by 59.34% and 43.21%, respectively, indicating that deletion of the key *ARGBC* gene indeed negatively affects L-ornithine biosynthesis and FA production. In contrast, the intracellular L-ornithine concentration increases significantly, and FA biosynthesis is then restored after complementation of the native *ARGB* gene in the *ΔgoxΔargbc1* mutant. In addition, informatic analysis shows that there is a lack of L-arginine binding site in the amino acid sequence of the ArgB from the *Δgox1* strain compared with that of the corresponding N-acetyl-glutamate kinase encoded by the key *ARGBC* gene in any other yeast strains. For example, there are L-threonine (T340) and L-arginine (R336) residues in the N-acetyl-glutamate kinase of the *Δgox1* strain, while there are L-threonine (T340) and L-lysine (K336) residues in the N-acetyl-glutamate kinase from any other fungal strains, especially *S. cerevisiae*, sensitive to the L-arginine feedback inhibition, meaning that the ArgB from the *Δgox1* strain will not be inhibited by feedback inhibition of L-arginine. Indeed, our results show that the specific activity of ArgB and biosynthesis of L-ornithine and FA are not inhibited by different concentrations of added L-arginine, indicating that there is indeed no feedback inhibition of L-arginine from OUC on the ArgB in the *Δgox1* strain, and AGC can provide OUC with a large amount of L-ornithine for FA biosynthesis. These results indicate that the L-ornithine synthesized from the AGC and AGC itself can play a key role in FA biosynthesis by providing sufficient intermediates for the OUC (Fig. 8.10), which may be one of the reasons why FA biosynthesis only comes from the OUC pathway rather than from any other pathways in *A. aubasidani Δgox1* strain (Wei et al. 2023, 2024). Therefore, FA biosynthesis in *A. aubasidani* is completely different from that in *Rhizopus* spp.

8.3.3 Clarification of Regulation Mechanisms of Fumaric Acid Biosynthesis by Genome Editing and Synthetic Biology

The results in Table 8.2 show that the fumarate titer can reach 63.6 g/L when the *Δgox1* strain is grown in the FA production medium with $CaCO_3$, while there is no fumarate accumulation in the absence of $CaCO_3$ in the FA production medium. In addition, the transcriptional levels of the genes related to FA biosynthesis are also significantly increased in the presence of $CaCO_3$. It is therefore speculated that FA biosynthesis in the *Δgox1* strain is activated and greatly enhanced by the calcium signaling pathway as mentioned in the last subsection. Deletion of the key *CRZ1* gene encoding the key transcriptional factor in the calcium signaling pathway results in a significant decrease in fumarate titer to 32.4 g/L by the mutant and makes the transcriptional levels of the genes related to FA biosynthesis decrease too, while fumarate titer and the transcriptional levels of these genes are restored after the native *CRZ1* gene complementation in the mutant. When the transformants carrying the fused *CRZ1-GFP* gene grow in the presence of $CaCO_3$, the expressed Crz1-Gfps are localized in the nucleus, while in the absence of $CaCO_3$, the expressed Crz1-Gfps are distributed throughout the cells (Wei et al. 2023). Thus, like PMA biosynthesis as stated in the last subsection, it is confirmed that FA biosynthesis is activated and greatly enhanced by the calcium signaling pathway, too, and that is why a huge amount of PMA and FA biosynthesis can occur only in the presence of $CaCO_3$ (Fig. 8.10).

The genomic DNA of the fumarate-producing GC1 strain also carries the *NSDD* gene encoding a GATA-type zinc finger transcription factor, the global regulator NsdD, as mentioned above. Deletion of the *NSDD* gene makes the disruptants produce less fumarate and causes decreased expression of the genes for fumarate biosynthesis. Therefore, the NsdD also can be a positive regulator for fumarate biosynthesis and regulation (Hansali et al. 2026).

8.3.4 Applications of Fumaric Acid

Because FA is an (E)-2-butenedioic acid with a carbon-carbon double bond and two carboxylic acid groups and has excellent characteristics such as acidity, non-toxic, and active chemical properties, it is regarded as one of the most important platform chemicals in chemical industries and has been widely used in food, chemical, and pharmaceutical industries. For example, fumaric acid can act as an initial material for polymerization and esterification reactions. Fumaric acid can be used as a sour agent in food additives and in animal feed, as it can decrease methane emissions from rumen animals and increase feed efficiency for pigs and poultry. In addition, fumaric acid can be reacted with long-chain alcohols to produce various industrial products such as unsaturated polyester resins and plasticizers. Most importantly, dimethyl fumarate as a drug can act on dermatosis and cancer (Myo Myo Tint et al. 2025). The fumarate salts also can be transformed into L-aspartate by the action of aspartase that catalyzes reductive amination of fumaric acid in the presence of ammonia or into L-malate by fumarase (Kawabata et al. 1995), respectively.

8.4 Production, Biosynthesis, Regulation, and Applications of Gluconic Acid

Gluconic acid (GA) (Pentahydroxycaproic acid) (Fig. 8.11) is one of the important organic acids produced by using the specific oxidation of the aldehyde group at C_1 in β-D-glucose to a carboxyl group and is a common additive used in pharmaceutical, textile, building, and food industries. Gluconic acid can easily bind di- and trivalent metals such as Ca^{2+} to form highly stable chemical compounds. The industrial producers of gluconate are now the strains of *Penicillium* spp., *Aspergillus niger*, or *Gluconobacter suboxidans*. However, *Penicillium* spp., and *A. niger* are difficult to

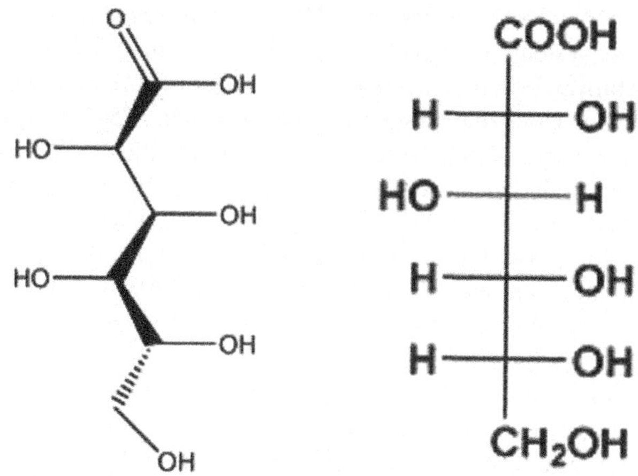

Fig. 8.11 Chemical structure of gluconic acid (GA)

use in the continuous gluconic acid fermentation, while *G. suboxidans* produces a relatively large amount of by-products such as keto-acids during gluconic acid fermentation and is inhibited by high glucose and product concentrations. Furthermore, *Penicillium* spp. and *A. niger* easily form mycelia during agitation in the fermenter, which usually increases the viscosity of the culture and hinders the aeration, oxygen, and nutrient transfer and homogenization of the culture as a result. Therefore, it is important to seek for the better strain to replace *Penicillium* spp., *A. niger*, and *G. suboxidans* to produce gluconic acid. Finally, it was found that some strains of *A. pullulans* are much better than *Penicillium* spp., *A. niger*, and *G. suboxidans* for gluconic acid production. For example, they can be easily used in the continuous cultivation from high concentration of glucose for high gluconate production because of their high osmotolerant properties. Furthermore, its genomes can be more easily edited for enhanced gluconic acid production than those of *Penicillium* spp. *A. niger*, and *G. suboxidans.* It has a broad substrate spectrum (glucose, glycerol, fructose, sucrose, inulin, cellulose, xylan, and xylose) because it can produce various extracellular enzymes as stated in Chap. 6. In addition, it can be easily grown in bioreactors as most of cells are yeast-like fungal cells.

8.4.1 Overproduction of Gluconate by Metabolic Engineering

It has been reported that *A. pullulans* isolate 70 isolated from wildflowers in forests can produce gluconic acid up to 260 g/L in the continuous stirred tank reactors in defined media with a residence time of about 26 h (Anastassiadis et al. 2003; Anastassiadis and Rehm 2006a). The same strain can produce 272–308 g/L of gluconic acid under the continuous cultivation of its free-growing cells in the first bioreactor at the residence time between 19.5 and 24 h and yield 350–370 g/L of gluconic acid in the second bioreactor at a total residence time of 30.8–37 h (Anastassiadis and Rehm 2006b).

A total of 186 strains of *Aureobasidium* spp. were isolated from the mangrove ecosystems in Wenchang and Dongzhaigang of Hainan province and Taoyuan of Taiwan province, China. Three strains of *Aureobasidium* spp. (P25 and P28 strains from Dongzhaigang of Hainan province, T5–2-2 strain from Taoyuan of Taiwan province) with high-level production of calcium gluconate were obtained. The calcium gluconate concentrations produced by the yeast-like fungal strains P25, P28, and T5–2-2 are 99.0 g/L, 98.0 g/L, and 97.0 g/L, respectively. At the same time, exopolysaccharide concentrations of 1.2 g/L, 3.3 g/L, and 3.5 g/L are also reached in the cultures, respectively (Zhao et al. 2019).

After optimization of the medium for calcium gluconate production and cell growth by *A. pullulans* P25, it was found that 140.0 g/L glucose, 7.5 g/L corn steep liquor, and 65.0 g/L $CaCO_3$ are the most suitable medium for calcium gluconate production. After cultivation at the flask level for 5 days, *A. pullulans* P25 can produce 121.3 g/L of calcium gluconate at a flask level. During a 10-Liter batch fermentation, 152.5 g/L of calcium gluconate in the culture and 12.59 g/L of cell dry weight are obtained within 96 h, leaving 4.5 g/L of reducing sugar in the fermented medium (Zhao et al. 2019).

Genome size, GC content, and protein-coding genes of *A. pullulans* P25 are 25.4 Mpb, 49.97%, and 6673, respectively. After the key *GOX1* gene encoding glucose oxidase in the genome of *A. pullulans* P25 was cloned and characterized, it was found that the size of the *GOX1* gene (the accession number: KY673530.1) is 1818 bp without any intron, encoding a protein with 605 amino acids with 65. 0 kDa and a 16 amino acid signal peptide. 55.6% of Gox1 is estimated to be secreted outside the cell wall, and 44.4% of the Gox1 is predicted to be localized in the cells. Knock-out of the gene causes no calcium gluconate production by the disruptant *Δgox*11 and no *GOX1* gene expression. Compared to the specific Gox1 activity in the wild type of *A. pullulans* P25, 96.3% of the specific activity in the *Δgox*11 is lost. In contrast, when the native *GOX1* gene is overexpressed in the P25 strain, the obtained GA-66 strain can produce 200.2 g/L of calcium gluconate from 180.0 g/L of glucose and 65.0 g/L of $CaCO_3$ during a 10-Liter batch fermentation. The productivity of calcium gluconate is 3.10 g/L/h, while that of calcium gluconate by the P25 strain is only 1.59 g/L/h. Compared to that produced by the P25 strain, 34.2% of the specific glucose oxidase activity produced by the GA-66 strain is increased. By measuring the intracellular, extracellular, and supernatant Gox activity of the culture, it was found that most of the Gox1 activity is localized on the cell surface, and a small amount of the Gox1 activity is detected in periplasm of the P25 strains (Zhao et al. 2019). This means that glucose in the medium can be easily and directly transformed into gluconic acid by the P25 strain.

The high levels of dissolved oxygen concentration in *A. pullulans* cultures by increased agitation rate or total air pressure up to 4 bar can greatly improve gluconic acid productivity in the pure glucose medium (Fernandes et al. 2021), showing that the dissolved oxygen in the culture can play an important role in gluconic acid production.

Currently, pure glucose is gradually being replaced by agro-industrial by-products for microbial gluconic acid production to achieve high titers with a more economical and environmentally friendly bioprocess. For example, banana must, corn starch, concentrated cellulosic hydrolysate, sugarcane and beet molasses, sugarcane juice, corn stover hydrolysate, breadfruit hydrolysate, and sugarcane molasses which contain 43% sugars (mainly sucrose, followed by glucose and fructose) also have been used for gluconic acid production. For example, the highest titer of gluconic acid of 75.0 g/L from sugarcane molasses is obtained when *A. pullulans* NCYC 4012 is cultivated in the airlift bioreactor (Fernandes et al. 2021).

A large-scale bioproduction of gluconic acid can be achieved by solid-state fermentation, immobilization of the producer, aerobic submerged fermentation, and continuous fermentation (Pal et al. 2016).

The genome size of *A. hainanensis* P6 strain is 25.4 Mpb, the G + C content of its genome is 50.7%, and the genome carries 6098 protein-coding genes (Jia et al. 2019). A protein deduced from the *GOX1* gene of *A. hainanensis* P6 strain has 1824 bp encoding a protein with 606 amino acids. Deleting the *GOX1* gene allows the disruptant GOK11 to completely lose the ability to produce both gluconic acid and Gox1 activity, whereas overexpression of the native *GOX1* gene renders the overexpressing strain GOEX88 to produce 160.5 g/L of Ca^{2+}-gluconate from 160.0 g/L of glucose and 65.0 g/L of $CaCO_3$ and Gox1 activity of 1438.6 U/mg of protein, much higher than those (118.7 g/L of Ca^{2+}-gluconate and Gox1 activity of 1100.0 U/mg of protein) produced by its parent P6 strain. During a 10-Liter batch fermentation, the overexpressing strain GOEX88 grown in the medium containing 160.0 g/L of glucose and 65.0 g/L of $CaCO_3$ can produce 186.8 g/L of Ca^{2+}-gluconate, and most of the produced Gox1 activity is located in the yeast cell wall (Ma et al. 2018).

The production of gluconic acid by different microbial strains is shown in Table 8.3.

Table 8.3 The production of gluconic acid by different microbial strains

Microorganism	Ca^{2+}-gluconate (g/L)	References
A. niger	96.16	Purane et al. (2011)
A. niger ORS-4 410	110.94	Singh and Singh (2006)
A. pullulans isolate 70	260.0	Anastassiadis et al. (2003)
A. terreus strain A 156	92.1	Dowdells et al. (2010)
Z. mobilis	112.7	
G. oxydans	125.0	Jiang et al. (2017)
A. pullulans P25	99.0	Zhao et al. (2019)
A. pullulans P25	152.5	Zhao et al. (2019)
A. pullulans GA-6	200.2	Zhao et al. (2019)
A. pullulans NCYC 4012	75.0	Fernandes et al. (2021)
A. hainanensis P6	160.5	Ma et al. (2018)
A. hainanensis GOEX8	186.8	Ma et al. (2018)

8.4.2 *Elucidation of Biosynthesis Pathway of Gluconic Acid by Genome Editing and Synthetic Biology*

As gluconic acid (GA) is pentahydroxycaproic acid, production of gluconic acid from glucose by *A. pullulans* proceeds via a dehydrogenation reaction catalyzed by glucose oxidase (Gox1) by utilizing molecular oxygen as an electron acceptor. For example, deletion of the *GOX1* gene encoding Gox1 can totally stop gluconic acid production from glucose in *A. aubasidani* as mentioned above (Wei et al. 2023). Removal of the *GOX1* genes from the genomic DNAs in *A. pullulans* P25 and *A. hainanensis* P6 strain causes no calcium gluconate production from glucose, no Gox1 activity, and no *GOX1* gene expression by the disruptant *Δgox1*. In contrast, overexpression of the native *GOX1* gene in the obtained GA-66 strain of *A. pullulans* P25 and the GOEX88 strain of *A. hainanensis* P6 can produce 200.2 g/L and 160.5 g/L of Ca^{2+}-gluconate, respectively (Table 8.3) (Zhao et al. 2019; Ma et al. 2018). This enzyme Gox1 is primarily located in the cell wall (Ma et al. 2018; Fig. 8.11). The direct bioconversion of one mole of glucose to one mole of gluconic acid in the medium consumes half a mole of oxygen, establishing oxygen as a principal substrate. Consequently, this Gox1-catalyzed oxidation of glucose is a highly oxygen-intensive process, strongly dependent on the dissolved oxygen concentration (DOC) in the culture. Under the optimal conditions, nearly 100% of the supplied glucose can be converted to gluconic acid.

The genomic sequence indicates that the gluconic acid biosynthesis pathway in *A. pullulans* P25 isolated from the mangrove ecosystem aligns with that of any other fungi (Canete-Rodríguez et al. 2016). The pathway initiates with conversion of D-glucose to D-glucono-δ-lactone by Gox1. It has been confirmed that Gox1 is a flavoprotein containing FAD cofactor, which catalyzes this oxidation while transferring hydrogen to FAD. This reaction is favored under the conditions of high glucose and oxygen concentrations and at a pH of 5.5, which can be easily met during cultivation of *A. pullulans P25, A. hainanensis* P6, and *A. aubasidani*. Subsequently, D-glucono-δ-lactone is converted to D-gluconic acid, either enzymatically by D-glucono-δ-lactonase (Gnl) or via spontaneous hydrolysis under natural conditions. Usually, the spontaneous hydrolysis occurs rapidly at neutral pH, a condition established in the culture by the addition of $CaCO_3$. Consequently, gluconic acid production necessitates high concentrations of both $CaCO_3$ and dissolved oxygen in the culture.

H_2O_2 is also generated during the reactions as mentioned above and in Fig. 8.12. Subsequently, this peroxide is decomposed into oxygen (O_2) and water (H_2O) by the action of catalase (Cat). Indeed, the genomic analysis of the yeast-like fungus P25 strain confirms the presence of a *CAT* gene encoding Cat. Alternatively, D-glucose can be converted to D-gluconic acid through activities of glucokinase (encoded by a *GK* gene), NADPH-dependent glucose-6-phosphate dehydrogenase (encoded by a *GPDH* gene), 6-phosphogluconolactonase (encoded by a *PGNL* gene), and gluconokinase (encoded by a *GNK* gene), constituting part of the pentose phosphate pathway in *A. pullulans* (Fig. 8.12). The Gox1 enzyme deduced from the *GOX1* gene in the genome of strain P25 contains two FAD-binding sites, each binding one FAD

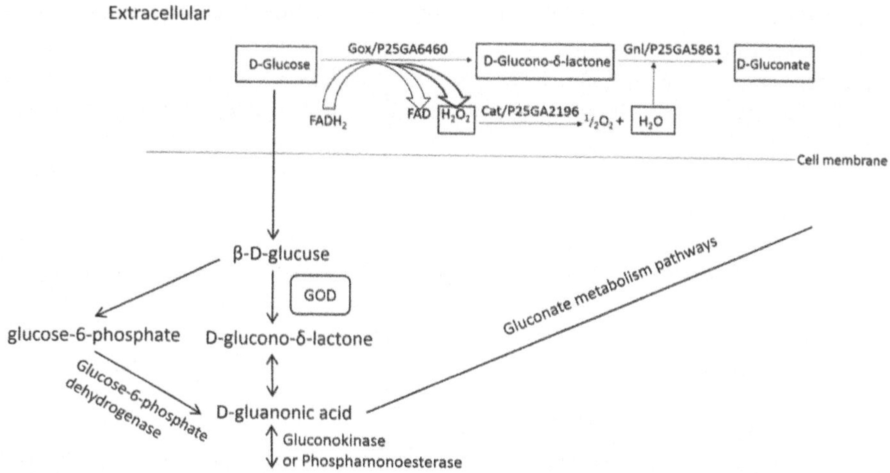

Fig. 8.12 Biosynthesis pathway of gluconic acid in *A. pullulans* P25 strain. *Gox* FAD-dependent glucose oxidase, *Gnl* gluconolactonase, *Cat* catalase

molecule. This flavoprotein harbors a tightly, non-covalently bound FAD cofactor per monomer and catalyzes the dehydration of glucose to D-glucono-δ-lactone while transferring hydrogen to FAD (Fig. 8.12) (Canete-Rodríguez et al. 2016). The resulting $FADH_2$ is then regenerated to FAD by transferring hydrogen to oxygen, yielding H_2O_2 (Fig. 8.12) (Canete-Rodríguez et al. 2016).

8.4.3 Regulation of GA Biosynthesis

It was found that the promoter of the *GOX1* gene contains 5'-HGATAR-3′ motif, indicating that expression of the *GOX1* gene can be repressed by nitrogen sources. So, all the gluconic acid production media must contain a low concentration of nitrogen source. In addition, all the gluconic acid production media must contain a suitable amount of $CaCO_3$ for high Ca^{2+}-gluconate production because Gox1 has the optimal activity at pH 7.0. This suggests that the Ca^{2+} signaling pathway as stated above may also be involved in the regulation of gluconic acid biosynthesis. However, so far, there has not been such evidence to confirm the hypothesis.

8.4.4 Application of Gluconate

Gluconic acid is a multifunctional, non-corrosive, odorless, non-volatile, non-toxic, natural, and readily biodegradable pentahydroxycaproic acid with Generally Recognized as Safe (GRAS) status. It can be a bio-based additive for food, beverages, agricultural products, chemicals, construction materials, pharmaceuticals, and textiles. Gluconic acid and its derivatives, including gluconates, gluconolactones, and Ca^{2+}-gluconate, have also been widely applied in food, beverage, dairy, pharmaceutical, chemical, water treatment, and building construction industries. For instance, Ca^{2+}-gluconate serves as a mineral supplement to prevent human calcium deficiency and can function as a buffer salt in the food industry. In the pharmaceutical industry, Ca^{2+}-gluconate is applied as a calcium source to treat human hypocalcemia via oral or intravenous administration. Consequently, Ca^{2+}-gluconate is a key derivative of gluconic acid, commercially available as tablets, powder, and liquid for dietary supplements to prevent human calcium deficiency (Canete-Rodríguez et al. 2016). At present, the production cost for Ca^{2+}-gluconate is approximately US$8.50/kg. Additionally, Ca^{2+}-gluconate is used as a concrete additive to control setting time and enhance strength and water resistance for any building construction.

As gluconic acid has antimicrobial activity against different bacterial pathogens, it is also widely applied to tanning, textile industry, and fermented tea as an important beverage.

8.5 Advantages and Disadvantages of Added $CaCO_3$ During Organic Acid Production

It can be clearly observed from biosynthesis of PMA, FA, and GA that $CaCO_3$, the neutralizing agent, must be added to the cultures in order to enhance biosynthesis of PMA, FA, and GA. The pH, CO_2, and Ca^{2+} from the added $CaCO_3$ have many functions for the fungal producers and the enhanced biosynthesis of PMA, FA and GA. (1) It serves not only as a neutralizing agent to keep the pH of the culture at approximately 6.5, but also as a source of CO_2 and Ca^{2+}, which have many functions in fungal cells. In the absence of $CaCO_3$, the pH of the culture will decrease to approximately 2–3 as the concentration of the organic acid increases. This will halt fungal cell growth and cellular metabolism and damage fungal cell integrity due to the production of reactive oxygen species (ROS). The use of any other neutralizing agents, such as Na_2CO_3 and NaOH, will release sodium ions into the culture, which can adversely affect cellular metabolism and inhibit organic acids formation. (2) During the organic acid biosynthesis, $CaCO_3$ gradually dissolves, releasing CO_2 for pyruvate carboxylation to produce oxaloacetate under the catalysis of pyruvate carboxylase, as well as liberating HCO_3 for CO_2 fixation under the catalysis of carbamoyl phosphate synthetase in the ornithine-urea cycle (OUC) pathway (Fig. 8.10). (3) The released Ca^{2+} can stimulate the activity of pyruvate carboxylase and other enzymes. Consequently, an increased amount of pyruvate from glycolysis is converted into oxaloacetate, thereby enhancing the oxidative TCA cycle in the mitochondria. This process promotes the production of malic acid, fumaric acid, and ATP. (4) In addition, the addition of $CaCO_3$ can inhibit the production of by-product exopolysaccharides while enhancing the desired production of PMA by

Aureobasidium spp. (5) Synthesized organic acids in the fungal culture broth, in the form of calcium salts, can reduce the likelihood of feedback inhibition exerted by free fumaric acid and malic acid on the reductive tricarboxylic acid (TCA) cycle in the cytoplasm and the oxidative TCA cycle in the mitochondria. Furthermore, the formed calcium salts facilitate the in situ crystallization of malate with $CaCO_3$ in the cultures and the purification of the formed calcium salts in the downstream process. (6) The addition of $CaCO_3$ can mitigate the acidic stress caused by synthesized organic acids and reactive oxygen species (ROS), thereby alleviating stress on fungal cell growth and cellular metabolism and the damage to cell integrity. (7) Ca^{2+} released from $CaCO_3$ can stimulate the activity of the Ca^{2+} signaling pathway by activating a Ca^{2+} sensory protein, calmodulin, and a Ca^{2+}-calmodulin-dependent phosphatase and calcineurin for enhanced biosynthesis of the organic acids. In addition, calcium plays a crucial role in the development of fungal cells, glucose metabolism, maintaining cell membrane integrity, stress responses, and the growth and remodeling of the cell wall. (8) In the presence of Ca^{2+}, the produced fumaric acid and malic acid can be converted into calcium fumarate and calcium malate, disrupting the link between the oxidative urea cycle (OUC) or reductive TCA cycle and the mitochondrial TCA cycle, thereby enhancing the accumulation of fumarate and malate in the culture. (9) In the presence of $CaCO_3$, glucose consumption in many fungal cells is stimulated, resulting in enhanced organic acid production. Therefore, the addition of $CaCO_3$ can play an important role in the biosynthesis of organic acids in the fungi, as evidenced above.

However, the addition of $CaCO_3$ can also display numerous challenges in the fermentation industry and downstream processes. For example, the addition of $CaCO_3$ is not an effective method for controlling pH, as dissolution of $CaCO_3$ is a kinetically controlled equilibrium reaction (Fig. 8.10), resulting in pH fluctuations throughout the fermentation process (Wei et al. 2023). This limitation may restrict the commercial applications of $CaCO_3$. In addition, due to the low aqueous solubility of products such as calcium malate and calcium fumarate, along with their aggregation with filamentous fungal cells, the addition of $CaCO_3$ can lead to significant foaming, increased broth viscosity, limited oxygen and nutrient mass transfer, increased the power input required to keep the culture well mixed, complicated product recovery, and a broth heating step required to solubilize the calcium fumarate and separate it from the biomass and other solids in the culture. These factors are detrimental to equipment, increase mechanical maintenance burdens, and contribute to the formation of calcium organic acids. Furthermore, to obtain free organic acids, calcium must be removed from PMA, calcium malate, calcium fumarate, calcium lactate, and similar compounds using H_2SO_4. This process results in the production of a significant amount of waste $CaSO_4$, leading to environmental pollution and increased downstream processing costs. When a large amount of $CaCO_3$ is added to the culture, the residual $CaCO_3$ will impede the cells' utilization, resulting in the accumulation of a significant amount of commercially useless $CaSO_4$ during the culture and separation processes. In addition, an excess of the water-insoluble $CaCO_3$ buffer may reduce oxygen availability by impeding mass transfer. Thus, effectively removing the residual $CaCO_3$ from the fermenter poses a significant challenge. Therefore, it is still a big challenge how to avoid addion of $CaCO_3$ during free organic acids production by the yeast-like fungal strains. One of the methods is to isolate high organic acid producing yeasts which can actively resist low pH and ROS. Another method is to avoid influence of the Ca^{2+} signaling pathway by metabolic engineering and synthetic biology. The third method is to keep the constant pH around 6.5 of medium by using sterile NaOH solution during fermentation.

8.6 Conclusions and Perspectives

Although the *PMAs* gene encoding PMA synthetase has been demonstrated to be a key gene for PMA biosynthesis, the functions of each domain have not been confirmed at the molecular level so far. In addition, *A. melanogenum* OUC strain can produce high-molecular-weight PMA, which has many applications in biomaterial, agricultural, and medical industries. It is still completely unclear why and how the PMA synthetase in *A. melanogenum* OUC can synthesize such high-molecular-weight PMA. It has been known that filamentous fungi synthesize fumarate via the reductive and oxidative TCA cycle, why does *A. aubasidani* synthesize fumarate through OUC?

It is crucial to avoid the addition of $CaCO_3$ during high-level organic acid production in industry. One way may be to screen for more native acid-tolerant yeasts from low pH environments and to develop genetically modified yeasts with low pH tolerance to achieve large-scale green production of free organic acids without the use of a neutralizer, $CaCO_3$. Another way may be to escape the influence of the Ca^{2+} signaling pathway and further deeply understand the physiological functions of Ca^{2+} and CO_2 released from $CaCO_3$ at the molecular level.

References

Anastassiadis S, Rehm HJ (2006a) Continuous gluconic acid production by Aureobasidium pullulans with and without biomass retention. Electron J Biotechnol 9:494–504

Anastassiadis S, Rehm HJ (2006b) Continuous gluconic acid production by the yeast-like *Aureobasidium pullulans* in a cascading operation of two bioreactors. Appl Microbiol Biotechnol 73:541–548

Anastassiadis S, Aivasidis A, Wandrey C (2003) Continuous gluconic acid production by isolated yeast-like mould strains of *Aureobasidium pullulans*. Appl Microbiol Biotechnol 61:110–117

Canete-Rodríguez AM, Santos-Duenas IM, Jiménez-Hornero JE, Ehrenreich A, Liebl W, García-García I (2016) Gluconic acid: properties, production methods and applications—an excellent opportunity for agro-industrial by-products and waste bio-valorization. Process Biochem 51:1891–1903

Cao W, Chen X, Luo J, Yin J, Qiao C, Wan Y (2016) High molecular weight β-poly (L-malic acid) produced by a. pullulans with Ca^{2+} added repeated batch culture. Intern J Biol Macromole 85:192–199

Cao W, Wang Y, Shen F, Luo J, Yin J, Qiao C, Wan Y (2019) Efficient β-poly (L-malic acid) production from Jerusalem artichoke by *Aureobasidium pullulans* ipe-1 immobilized in luffa sponge matrices. Bioresour Technol 288:1–7

Cao W, Cao W, Shen F, Luo J, Yin J, Qiao C, Wan Y (2020) Membrane-assisted β-poly (L-malic acid) production from bagasse hydrolysates by *Aureobasidium pullulans* ipe-1. Bioresour Technol 295:1–9

Cheng C, Zhou Y, Lin M, Wei P, Yang ST (2017) Polymalic acid fermentation by *Aureobasidium pullulans* for malic acid production from soybean hull and soy molasses: fermentation kinetics and economic analysis. Bioresour Technol 223:166–174

Chi Z, Wang ZP, Wang GY, Khan I, Chi ZM (2016) Microbial biosynthesis and secretion of L-malic acid and its applications. Crit Rev Biotechnol 36(1):99–107

Chi Z, Wei X, Ge N, Jiang H, Liu GL, Chi ZM (2024) NsdD, a GATA-type transcription factor is involved in regulation and biosynthesis of macromolecules melanin, pullulan, and polymalate in *Aureobasidium melanogenum*. Intern J Biol Macromole 268:131820

Das RK, Brar SK (2014) Enhanced fumaric acid production from brewery wastewater and insight into the morphology of *Rhizopus oryzae* 1526. Appl Biochem Biotechnol 172(6):2974–2988

Das RK, Brar SK, Verma M (2015) A fermentative approach towards optimizing directed biosynthesis of fumaric acid by *Rhizopus oryzae* 1526 utilizing apple industry waste biomass. Fung Biol 119(12):1279–1290

Deng YF, Li S, Xu Q, Gao M, Huang H (2012) Production of fumaric acid by simultaneous saccharification and fermentation of starchy materials with 2-deoxyglucose-resistant mutant strains of *Rhizopus oryzae*. Bioresour Technol 107:363–367

Dowdells C, Jones RL, Mattey M, Bencina M, Legisa M, Mousdale DM (2010) Gluconic acid production by *Aspergillus terreus*. Lett Appl Microbiol 51(3):252

Fernandes S, Belo I, Lopes M (2021) Highly aerated cultures boost gluconic acid production by the yeast-like fungus *Aureobasidium pullulans*. Biochem Eng J 175:108133

Feng J, Li TF, Zhang Z, Chen J, Zhao T, Zou X (2019) Efficient production of polymalic acid from xylose mother liquor, an environmental waste from the xylitol industry by a T-DNA-based mutant of *Aureobasidium pullulans*. Appl Microbiol Biotechnol 103:6519–6527

Fu YQ, Li S, Chen Y, Xu Q, Huang H, Sheng XY (2010a) Enhancement of fumaric acid production by *Rhizopus oryzae* using a two-stage dissolved oxygen control strategy. Appl Biochem Biotechnol 162(4):1031–1038

Fu YQ, Xu Q, Li S (2010b) Strain improvement of Rhizopus oryzae for over-production of fumaric acid by reducing ethanol synthesis pathway. Korean J Chem Eng 27(1):183–186

Fu YQ, Xu Q, Shuang L, Yao C, He H (2010) Strain improvement of Rhizopus oryzae for over-production of fumaric acid by reducing ethanol synthesis pathway. Kor J Chem Eng 27:183–6

Geng W, Yang C, Gu Y, Liu R, Guo W, Wang X, Song C, Wang S (2014) Cloning of ε-polyL-lysine (ε-PL) synthetase gene from a newly isolated ε-PL-producing *Streptomyces albulus* NK660 and its heterologous expression in *Streptomyces lividans*. Microb Biotechnol 7:155–164

Guan J, Wang W, Zhang K, Shi X, Yang Q, Song J (2025) Role of AplaeA in the regulation of spore production and polymalic acid synthesis in *Aureobasidium pullulans*. Intern J Biol Macromole 304:140781

Guo F, Wu M, Dai Z, Zhang S, Xin F (2020) Current advances on biological production of fumaric acid. Biochem Eng J 153:107397

Hansali K, Wang P, Wei X, Liu GL, Chi Z, Chi ZM (2026) Dual role of the tra scriptional factor NsdD in reguration of fumarate and melanin biosnthesis in *Aureobasidium abasidani* GC1 strain. Fung Biol 130: 101973

Huang L, Wei PL, Zang R (2010) High-throughput screening of high-yield colonies of *Rhizopus oryzae* for enhanced production of fumaric acid. Ann Microbiol 60(2):287–292

Huang D, Wang R, Du WJ, Wang G, Xia M (2015) Activation of glycerol metabolic pathway by evolutionary engineering of *Rhizopus oryzae* to strengthen the fumaric acid biosynthesis from crude glycerol. Bioresour Technol 196:263–272

Jia SL, Ma Y, Chi Z, Liu GL, Hu Z, Chi ZM (2019) Genome sequencing of a yeast-like fungal strain P6, a novel species of *Aureobasidium* spp.: insights into its taxonomy, evolution, and biotechnological potentials. Ann Microbiol 69:1475–1488

Jiang Y, Liu K, Zhang H, Wang Y, Yuan Q, Su N, Bao J, Fang X (2017) Gluconic acid production from potato waste by *Gluconobacter oxydans* using sequential hydrolysis and fermentation. Acs Sust Chem Eng 5(7):6116

Kang SW, Lee H, Kim D (2010) Strain development and medium optimization for fumaric acid production. Biotechnol Bioprocess Eng 15(5):761–769

Kang XX, Jia SL, Wei X, Zhang M, Liu GL, Hu Z, Chi Z, Chi ZM (2022) Liamocins biosynthesis, its regulation in *Aureobasidium* spp., and their bioactivities. Crit Rev Biotechnol 42(1):93–105

Kawabata N, Hatanaka H, Odaka H (1995) Continuous production of L-aspartic acid from ammonium fumarate using immobilized cells by capture on the surface of nonwo ven cloth coated with a pyridinium-type polymer. J Ferment Bioeng 79:317–322

Leathers TD, Manitchotpisit P (2013) Production of poly(b-L-malic acid) (PMA) from agricultural biomass substrates by *Aureobasidium pullulans*. Biotechnol Lett 35:83–89

Li B, Li B, Wang P, Feng Y, Xu X, Zhang Y, Zou X (2023) Bio-refinery of xylose processing wastes for green polymalic acid production and l-malic acid recovery by engineered Aureobasidium pullulans in a non-waste-disposal system. Chem Eng J 454:140533

Li B, He J, Zuo K, Xu X, Zou X (2024) Engineering the by-products pathway in *Aureobasidium pullulans* for highly purified polymalic acid fermentation with concurrent recovery of L-malic acid. Bioresour Technol 414:131578

Liu H, Wang WN, Deng L, Wang F, Tan T (2015) High production of fumaric acid from xylose by newly selected strain *Rhizopus arrhizus* RH 7-13-9#. Bioresour Technol 186:348–350

Liu H, Ma J, Wang M, Wang W, Deng L, Nie K, Yue X, Wang F, Tan T (2016) Food waste fermentation to fumaric acid by *Rhizopus arrhizus* RH7-13. Appl Biochem Biotechnol 180(8):1524–1533

Liu H, Hu HR, Jin YH, Yue X, Deng L, Wang F, Tan T (2017) Co-fermentation of a mixture of glucose and xylose to fumaric acid by *Rhizopus arrhizus* RH 7-13-9. Bioresour Technol 233:30–33

Ma Y, Wang GY, Liu GL, Wang ZP, Chi ZM (2013) Overproduction of poly (β-malic acid) (PMA) from glucose by a novel *Aureobasidium* sp. P6 strain isolated from mangrove system. Appl Microbiol Biotechnol 97:8931–8939

Ma Y, Chi Z, Li YF, Jiang H, Liu GL, Hu Z, Chi ZM (2018) Cloning, deletion, and overexpression of a glucose oxidase gene in *Aureobasidium* sp. P6 for Ca^{2+}-gluconic acid overproduction. Ann Microbiol 68:871–879

Manitchotpisit P, Skory CD, Peterson SW (2012) Poly (β-L-malic acid) production by diverse phylogenetic clades of *Aureobasidium pullulans*. J Indu Microbiol Biotechnol 39:125–132

Myo Myo Tint K, Wei X, Wang P, Liu GL, Zhang M, Chi ZM, Chi Z (2025) Biotechnological application of *Aureobasidium* spp. as a promising chassis for biosynthesis of ornithine-urea cycle-derived bioproducts. Crit Rev Biotechnol 45(3):591–605

Nakajima-Kambe T, Hirotani N, Nakahara T (1996) Poly (β-malic acid) production by the non-growing cells of *Aureobasidium* sp. strain A-91. J Ferment Bioeng 82:411–413

Pal P, Kumar R, Banerjee S (2016) Manufacture of gluconic acid: a review towards process intensification for green production. Chem Eng Proc: Proc Intensificat 104:160–171

Purane N, Shinde M, Sharma S, Mahajani S (2011) Effect of nutrient additives on gluconic acid production by *A. niger* during submerged fermentation. J Pure Appl Microbiol 5(1):235–239

Qi CY, Jia SL, Liu GL, Chen L, Wei X, Hu Z, Chi ZM, Chi Z (2021) Polymalate (PMA) biosynthesis and its molecular regulation in *Aureobasidium* spp. Int J Biol Macromol 174:512–518

Qi CY, Chi Z, Liu GL, Hu Z, Chi ZM (2022) A novel very high molecular weight polymalate is synthesized by the whole genome duplicated strain *Aureobasidium melanogenum* OUC. Int J Biol Macromol 202:608–619

Qi CY, Chi Z, Liu GL, Wang P, Chi Z-M (2023) A new high molecular weight polymalate coating film on grape. Indu Crops Prod 202:116994

Qin Z, Feng J, Li Y, Zheng Y, Moore C, Yang ST (2024) Engineering the reductive tricarboxylic acid pathway in *Aureobasidium pullulans* for enhanced biosynthesis of poly-L-malic acid. Bioresour Technol 393:130122

Singh OV, Singh RP (2006) Bioconversion of grape must into modulated gluconic acid production by *aspergillus Niger* ORS-4.410. J Appl Microbiol 100(5):1114–1122

Song X, Wang Y, Wang P, Pu G, Zou X (2020) GATA-type transcriptional factor Gat1 regulates nitrogen uptake and polymalic acid biosynthesis in polyextremotolerant fungus *Aureobasidium pullulans*. Environ Microbiol 22(1):229–242

Wang YK, Chi Z, Zhou HX, Liu GL, Chi ZM (2015) Enhanced production of Ca^{2+} -polymalate (PMA) with high molecular mass by *Aureobasidium pullulans* var. pullulans MCW. Microb Cell Factories 14:115

Wang K, Chi Z, Liu GL, Qi CY, Jiang H, Hu Z, Chi ZM (2020) A novel PMA synthetase is the key enzyme for polymalate biosynthesis and its gene is regulated by a calcium signaling pathway in *Aureobasidium melanogenum* ATCC62921. Intern J Biol Macromole 156:1053–1063

Wang P, Chen H, Wei X, Liu GL, Chi Z, Jiang B, Chi ZM (2024) Efficient calcium fumarate overproduction from xylose and corncob-derived xylose by engineered strains of *Aureobasidium pullulans* var. *aubasidani* DH177. Microb Cell Factories 23:327

Wang GY, Bai T, Miao Z, Ning W, Liang W (2018) Simultaneous production of single cell oil and fumaric acid by a newly isolated yeast *Aureobasidium pullulans* var. *aubasidani* DH177. Bioproc Biosyst Eng 41:1707–1716

Wei PL, Cheng C, Lin M, Zhou YP, Yang ST (2017) Production of poly(malic acid) from sugarcane juice in fermentation by *Aureobasidium pullulans*: kinetics and process economics. Bioresour Technol 224:581–589

Wei X, Zhang M, Wang GY, Liu GL, Chi ZM, Chi Z (2023) The ornithine-urea cycle involves fumaric acid biosynthesis in *Aureobasidium pullulans var. aubasidani*, a green and eco-friendly process for fumaric acid production. Synth Syst Biotechnol 8:33–45

Wei X, Zhao SF, Liu GL, Chi Z, Chi ZM (2024) The role of the acetyl-glutamate cycle in fumarate biosynthesis through Lornithine supply via the ornithine-urea cycle in *Aureobasidium pullulans var. aubasidani*. Food Biosci 61:104853

Xu Q, Huang L, Wei PL, Zang R (2010a) High-throughput screening of high-yield colonies of *Rhizopus oryzae* for enhanced production of fumaric acid. Ann Microbiol 60(2):287–292

Xu Q, Li S, Fu Y, Tai C, Huang H (2010b) Two-stage utilization of corn straw by *Rhizopus oryzae* for fumaric acid production. Bioresour Technol 101(15):62624

Yang Y, Dai L, Li F, Xu X, Zou X (2026) Identification of a thermotolerant strain of *Aureobasidium melanogenum* DA22 and functional analysis of Spt23 as a key transcriptional regulator in hightemperature polymalic acid fermentation. Biores Technol 457:134956

Yegina S, Saha BC, Kennedy GJ, Kennedy GJ, Leathers TD (2019) Valorization of egg shell as a detoxifying and buffering agent for efficient polymalic acid production by *Aureobasidium pullulans* NRRL Y2311-1 from barley straw hydrolysate. Bioresour Technol 278:130–137

Yu SZ, Huang D, Wen JP, Li S, Chen Y, Jia X (2012) Metabolic profiling of a Rhizopus oryzae fumaric acid production mutant generated by femtosecond laser irradiation. Bioresour Technol 114:610–615

Zan ZQ, Zou X (2013) Efficient production of polymalic acid from raw sweet potato hydrolysate with immobilized cells of *Aureobasidium pullulans* CCTCC M2012223 in aerobic fibrous bed bioreactor. J Chem Technol Biotechnol 88:1822–1827

Zeng W, Zhang B, Li M, Ding S, Chen G, Liang Z (2019) Development and benefit evaluation of fermentation strategies for poly(malic acid) production from malt syrup by *Aureobasidium melanogenum* GXZ-6. Bioresour Technol 274:479–487

Zeng W, Zhang B, Jiang L, Liu Y, Ding S, Chen G, Liang Z (2020) Poly (malic acid) production from liquefied corn starch by simultaneous saccharification and fermentation with a novel isolated *Aureobasidium pullulans* GXL-1 strain and its techno-economic analysis. Bioresour Technol 304:1–9

Zeng D, Zhang Y, Ma X, Li J, Yin F, Li D, Bie W (2024) Biosynthesis of poly(β-L-malic acid) from rubberwood enzymatic hydrolysates in co-fermentation by *Aureobasidium pullulans*. Intern J Biol Macromole 257:128605

Zhang H, Cai J, Dong J, Zhang D, Huang L, Xu Z, Cen P (2011) High level production of poly (β-L-malic acid) with a new isolated *Aureobasidium pullulans* strain. Appl Microbiol Biotechnol 92:295–303

Zhao SF, Jiang H, Chi Z, Liu GL, Chi ZM, Chen TJ, Yang G, Hu Z (2019) Genome sequencing of *Aureobasidium pullulans* P25 and overexpression of a glucose oxidase gene for hyperproduction of Ca^{2+}-gluconic acid. Antonie Leeuwenhoek 112:669–678

Zhou Y, Du J, Tsao GT (2002) Comparison of fumaric acid production by *Rhizopus oryzae* using different neutralizing agents. Bioprocess Biosyst Eng 25(3):179–181

Zhou YQ, Nie K, Zhang X, Liu S, Wang M, Deng L, Wang F (2014) Production of fumaric acid from biodiesel-derived crude glycerol by *Rhizopus arrhizus*. Bioresour Technol 163:48–53

Zou X, Yang J, Tian X, Guo M, Li Z, Li Y (2016a) Production of polymalic acid and malic acid from xylose and corncob hydrolysate by a novel *Aureobasidium pullulans* YJ 6–11 strain. Process Biochem 51(1):16–23

Zou X, Yang J, Tian J (2016b) Production of polymalic acid and malic acid from xylose and corncob hydrolysate by a novel *Aureobasidium pullulans* YJ 6-11 strain. Process Biochem 51:16–23

Zou X, Cheng C, Feng J, Song X, Lin M, Yang ST (2019) Biosynthesis of polymalic acid in fermentation: advances and prospects for industrial application. Crit Rev Biotechnol 39:408–421

Chapter 9
Production, Biosynthesis, Regulation, and Applications of Oils by *Aureobasidium* spp.

Abstract Oils produced by *Aureobasidium* spp. include extracellular liamocin (heavy oils) and intracellular lipids. The tail group 3,5-dihydroxydecanoic acid of liamocin is synthesized from acetyl-CoA and malonyl-CoA by the highly reducing Pks encoded by the *HR-PKS* gene, and head groups mannitol and arabitol are synthesized by MPDH and MtDH, and ArDH, respectively. 3,5-Dihydroxydecanoic acid is reacted with mannitol or arabitol to form liamocin under the catalysis of Est1. The PPTase activates the ACP domain in the HR-Pks by phosphopantetheinylation, and the specific transcriptional activator Ga11 stimulates expression of the *HR-PKS* gene. The Gltp, the glycolipid transporter, and the Mdr, the ABC transporter, take part in the transport of liamocins from Golgi to plasma membrane and secretion of the produced liamocins into medium, respectively. Liamocin biosynthesis is regulated by many signaling pathway. The biosynthesis pathway of intracellular lipids in eucaryotes has been clearly elucidated. Liamocin (heavy oils) and intracellular lipids have been overproduced by metabolic engineering and have many applications in various sectors of biotechnology.

Keywords Liamocin · Intracellular lipids · Biodiesel · Biosynthesis · Regulation

9.1 Introduction

Oils produced by different strains of *Aureobasidium* spp. include extracellular liamocins and intracellular lipids. These oils have been produced on a large scale and applied in various fields. Although the chemical structure of liamocins was identified in 2014, it is only recently that the liamocin synthetic pathway, secretion, regulation, and corresponding genes have been elucidated through genome editing and synthetic biology, as discussed in Chap. 5. Furthermore, liamocins exhibit anticancer and antibacterial activities, and massoia lactone, released from liamocins, shows multiple biological activities, particularly antifungal activity. In contrast, the synthesis and regulation of intracellular lipids in fungi and yeasts have been well understood for a long time. Lipids extracted from the oleaginous yeast *Aureobasidium*

Z. Chi et al., *Biology and Synthetic Biotechnology of Aureobasidium spp.*,
https://doi.org/10.1007/978-981-92-2607-8_9

spp. can be converted into biodiesel, alkane, and glycerol and have multiple applications in various sectors of biotechnology. Therefore, it is of significant how to greatly improve liamocins and intracellular lipids biosynthesis using metabolic engineering and synthetic biology.

9.2 Production, Biosynthesis, Regulation, and Applications of Liamocins

Liamocins (also known as heavy oils or polyol lipids) are novel biosurfactants synthesized and secreted by *Aureobasidium* spp. Their chemical structure consists of a single, partially acetylated polyol head group and polyester tails. These tails contain one to six 3,5-dihydroxydecanoic ester groups, which may or may not feature 3-O-acetylation in the first acyl group (Fig. 9.1; Tiso et al. 2024), and the headgroup of liamocins consists of different polyols, such as mannitol, arabitol, glycerol, and others.

Liamocins possess several commercially valuable properties. First, their surface tension values fall within the range of commercial surfactants. Additionally, liamocins exhibit antiproliferative effects against certain human cancer cell lines and antimicrobial activity against numerous bacterial pathogens, especially *Streptococcus* spp. Furthermore, 5-hydroxy-2-decenoic acid lactone (massoia lactone), released

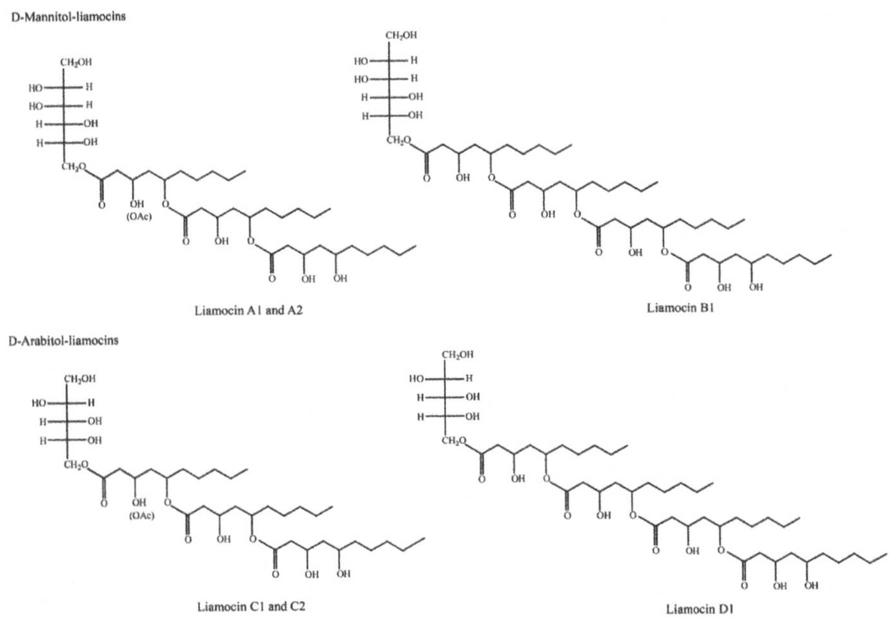

Fig. 9.1 The chemical structure of liamocins produced by *A. melanogenum*

upon chemical and enzymatic hydrolysis of liamocin, possesses a unique aroma and potent antifungal activity against various fungal pathogens affecting plants, vegetables, fruits, and animal and human skin. This combination of properties endows massoia lactone with significant application potential in the perfume, cosmetics, food, agricultural, and pharmaceutical industries. Currently, most research on liamocins focuses on the isolation of native producing strains from different environments, their chemical characterization, and direct applications. Only the authors of this book (Xue et al. 2020; Kang et al. 2022; Wang et al. 2024, 2025) are conducting comprehensive studies on liamocin overproduction, its synthetic pathway, transport, secretion, regulation, and applications by using metabolic engineering, genomic editing, and synthetic biology (Xue et al. 2020; Zhang et al. 2023).

9.2.1 Overproduction of Liamocins *by Metabolic Engineering*

The extracellular liamocins concentration and intracellular lipids content produced by *A. melanogenum* 6-1-2 isolated from the Hainan mangrove ecosystem in China are 25.8 g/L and 28.4% (w/w), respectively, when it is grown in a medium containing 12.0% (w/v) glucose. GC-MS analysis found that the intracellular lipids contain mainly $C_{16:0}$, $C_{16:1}$, $C_{18:0}$, $C_{18:1}$, and $C_{18:2}$ fatty acids, which are similar to the compositions of fatty acids from the intracellular lipid particles of all the oleaginous yeasts, fungi, and plants. While the liamocins produced by *A. melanogenum* 6-1-2 are composed of a single mannitol or arabitol headgroup partially *O*-acylated with three or four 3,5-dihydroxydecanoic ester groups (tails) (Fig. 9.1). In the produced mixtures of liamocins, the arabitol-liamocin accounts for 40.7% and the mannitol-liamocin accounts for 59.3% (Xue et al. 2020).

The yeast-like fungal strain HN6-1, isolated from the Yellow Sea in Qingdao, China, can produce 25.3 g/L of liamocin from glucose. The optimization of nitrogen concentration in the liamocin production medium found that corn steep liquor, which is rich in vitamins and other nutrients, is important for the high production of liamocin by the yeast-like fungal strain. Under the optimal conditions, the secreted liamocin titer by the yeast-like fungal strain HN6-1 grown in the medium containing 0.045% (w/v) corn steep liquor reaches the highest amount (29.5 g/L). Then, the *PYC1* gene encoding pyruvate carboxylase in the glycolytic pathway is actively overexpressed in the yeast-like fungal strain HN6-1 and the transformant strain Y766 with high yield of liamocins is gained and can produce 39.5 g/L liamocins (3.1 g/g of cell dry weight) and its cell dry weight is 12.7 g/L. Overexpression of the *ACL1* gene encoding ATP-citric acid lyase in the transformant strain Y766 makes the new transformant strain A155 reduce the amount of citric acid in the cells and further produce 42.7 g/L of liamocin (3.3 g/g of cell dry weight) (Wang et al. 2025).

Beyond conventional carbon sources like glucose, xylose, starch, cellulose, and xylan, inulin—a polymer of β-(2,1)-linked D-fructosyl residues capped by a glucose unit—has garnered significant attention as a renewable, non-food substrate for microbial conversion into valuable products (Chi et al. 2011) as mentioned above.

Research indicates that inulin is well-suited for single-cell oil production by oleaginous yeasts, as outlined in the following. Inulin is hydrolyzed primarily into fructose, with a minor amount of glucose. Fructose directly contributes to mannitol biosynthesis, a key head group component of liamocins (Fig. 9.1). Thus, enhanced mannitol accumulation is expected to promote liamocin production. To exploit this pathway, the *INU1* gene encoding exo-inulinase from *K. marxianus* was overexpressed in the host *A. melanogenum* strain A155, generating a newly engineered transformant U33. The inulinase activity of the newly engineered transformant U33 reaches 80.7 U/mL, 4.7-fold higher than that of the A155 strain (15.2 U/mL). Consequently, the flask-level liamocin titer increases from 42.7 g/L to 50.7 g/L within 168 h of cultivation. Furthermore, compared to the parent A155 strain, the newly engineered transformant U33 shows higher intracellular mannitol levels and significantly elevated transcriptional levels of the *INU1*, *ACL1*, and the key liamocin/mannitol biosynthesis genes (including *MPDH* and *MtDH* genes; Figs. 9.5 and 9.7). In a 10-L batch fermentation, the newly engineered transformant U33 efficiently converted inulin, yielding 55.2 g/L liamocins, 84.1 U/mL inulinase activity, and 15.2 g/L cell mass. This corresponds to a product yield of 0.34 g/g inulin and a productivity of 0.46 g/L/h, with 99.7% of the supplied inulin being utilized for cell growth, enzyme production, and liamocin synthesis. Massoia lactone released from the produced liamocins exhibits strong inhibitory effects against pathogenic bacteria (*Enterobacter cloacae*) and pathogenic yeasts (*Candida albicans*) (Wang et al. 2025). These results demonstrate that enhancing inulinase activity in the newly engineered transformant U33 strain improves fructose utilization, thereby boosting mannitol and liamocin biosynthesis. This strategy establishes a novel and efficient method for high-level liamocin production from inulin, a non-grain and plant-derived raw material, and highlights its potential applications.

Pyruvate carboxylase (Pyc) is a tetrameric enzyme that catalyzes the carboxylation of pyruvate to oxaloacetate. This enzyme is a key component of the reductive TCA pathway in the fungal cytoplasm. The *PYC1* gene from the marine-derived, oleaginous yeast-like fungus *A. melanogenum* 9-1—a high producer of liamocins and intracellular lipids—was overexpressed. Comparative analysis of their parent strain 9-1 and the transformants confirms that Pyc1 enhances biosynthesis of citric acid, a key precursor for both intracellular lipid and extracellular liamocin biosynthesis. Among the transformants, strain M399 exhibited high liamocin yield, producing 35.3 g/L of extracellular liamocins (equivalent to 2.1 g/g cell dry weight) and intracellular lipids at 26.6% (w/w) of its cell mass, which is 17.0 g/L. During a 10-Liter batch fermentation in the optimized medium, the transformant M399 produces 43.04 g/L of extracellular liamocins, 17.7 g/L of cell dry weight, and 23.0 g/L of intracellular lipids within 156 h of the cultivation (Tang et al. 2018).

ATP-citrate lyase (Acl) catalyzes the formation of oxaloacetic acid and acetyl CoA (precursors of fatty acids biosynthesis) from citric acid in the fungal cytoplasm. The gene *ACL1* encoding ATP-citrate lyase of the 9-1 strain mentioned above was cloned and characterized. Then, the *ACL1* gene was overexpressed successfully in the transformant M399 strain as mentioned above. The newly engineered transformant strain PA11 produces 43.1 g/L of extracellular liamocins

(3.79 g/g of cell dry weight) and 31.7% (w/w) of the intracellular lipid content. The citric acid concentration produced by the newly engineered transformant strain PA11 is reduced. This suggests that Acl also plays an important role in both liamocins biosynthesis and intracellular lipids. It was found that the Pyc also can reduce the amount of citric acid left in the fermented medium and promote conversion of citric acid into liamocins and intracellular lipids, leading to enhanced accumulation of intracellular lipids and secretion of extracellular liamocin.

The analysis showed that the main component of liamocins after alkaline hydrolysis is 5-hydroxy-2-decenoic acid lactone (massoia lactone, 92.1% of the produced liamocin). In contrast, the intracellular lipids primarily consist of $C_{16:0}$ (palmitic), $C_{16:1}$ (palmitoleic), $C_{18:0}$ (stearic), $C_{18:1}$ (oleic), and $C_{18:2}$ (linoleic) fatty acids. This lipid profile is similar to those found in common animal lipids and vegetable oils from plants, suggesting that the intracellular lipids from the newly engineered transformant strain PA11 are a good feedstock for biodiesel synthesis. Indeed, fatty acid methyl esters (FAME) synthesized from these lipids exhibit excellent combustion performance, as shown at the end of this chapter (Tang et al. 2018).

Nitrogen source optimization for higher liamocin production showed that the liamocin titer produced by *A. melanogenum* 9-1 grown in the medium with 0.6 g/L $NaNO_3$ reaches 30.9 g/L, compared to only 26.8 g/L with an equivalent concentration of NH_4NO_3. Thus, using $NaNO_3$ as the nitrogen source can enhance liamocin accumulation in the culture by 15.4% relative to NH_4NO_3 (Zhang et al. 2021a, b, c). This may be due to the fact that $NaNO_3$ as nitrogen source is better than NH_4NO_3, which has nitrogen catabolite repression on liamocin biosynthesis. However, the detailed reason for this is still awaited to be clarified.

When grown in a glucose medium with a high C/N ratio, *A. melanogenum* 9-1 produces 26.6 g/L of liamocin. However, liamocin biosynthesis is repressed by high concentrations of glucose available in the production medium. To enhance liamocin titer, it is therefore crucial to alleviate this glucose repression. It is well documented that in fungi, CreA (Mig1 in *S. cerevisiae*), encoded by the *CREA* gene, is a key global transcriptional repressor in the glucose repression system as mentioned above. CreA is a C_2H_2-type zinc-finger protein that binds to the conserved 5′-SYGGRG-3′ motifs in the promoters of many glucose-repressed genes. Removal of the *CREA* gene from the genomic DNA of *A. melanogenum* 9-1 results in increased liamocin production across all the obtained mutant strains compared to their parent strain 9-1. Notably, the glucose-derepressed mutant *Δcrea55* among them produces 36.5 g/L of liamocin, whereas its parent strain 9-1 produces only 26.6 g/L under the identical conditions. Furthermore, transcriptional analysis confirms that removal of the *CREA* gene also derepresses expression of some liamocin biosynthesis-related genes and all tested sugar transporter genes in the glucose-derepressed mutant *Δcrea55* so that liamocin biosynthesis is obviously enhanced in the glucose-derepressed disruptant *Δcrea55* (Wang et al. 2022a, b).

It has been established that Msn2, the global transcriptional activator encoded by the *MSN2* gene as mentioned above, regulated by the HOG1 and cAMP-PKA signaling pathways, is also involved in modulating liamocin biosynthesis (Kang et al. 2022). Accordingly, the *MSN2* gene is expressed in the glucose-derepressed mutant

Δcrea55 obtained above to further promote liamocin production. Results show that the resulting transformant M600 produces 41.4 g/L of liamocins, whereas the glucose-derepressed mutant *Δcrea55* produces only 36.5 g/L under the identical conditions.

Xue et al. (2020) demonstrated that the specific transcriptional activator Gal1 activates expression of the target genes, such as *HR-PKS1* (involved in 3,5-dihydroxydecanoic acid biosynthesis) and *EST1* (responsible for forming ester bonds between mannitol/arabitol and 3,5-dihydroxydecanoic acid in liamocin molecules) (Fig. 9.6). To further enhance liamocin biosynthesis, the *GAL1* gene encoding Gal1 is overexpressed in the transformant M600, yielding the engineered G400 strain, which secrets 46.7 g/L of liamocins. When cultured in a 10-liter fermenter, the engineered G400 strain achieves a liamocin titer of 49.7 g/L within 156 h of the fermentation. This confirms that the engineered G400 strain constructed by metabolic engineering and synthetic biology is capable of very high-level liamocin production (Wang et al. 2022a, b).

Analysis of the relative transcriptional levels of liamocin biosynthesis and sugar transport-related genes in the transformant M600 and the engineered strain G400 reveals a significant upregulation of them. Specifically, expression of all the liamocin biosynthesis-related genes was greatly enhanced. Transcriptional levels of sugar transport genes and the glucose kinase gene are also increased in both strains. Notably, expression of *HR-PKS1* gene and *EST1* gene was significantly higher in G400 than in M600, confirming that these two key genes are strongly regulated by the specific transcriptional activator Gal1 (Wang et al. 2022a, b). Therefore, this represents a new strategy for efficient liamocin biosynthesis and production and provides a theoretical basis for constructing high-yield liamocin fungal strains through metabolic engineering and synthetic biology in the future.

Biosynthesis of the 3,5-dihydroxydecanoic ester tail group of liamocins, along with intracellular fatty acids and hydroxy fatty acids, begins with the simple building blocks acetyl-CoA and malonyl-CoA, catalyzed by highly reduced polyketide synthases (HR-Pkss) (Xue et al. 2018, 2020). Therefore, to achieve high-level liamocin synthesis, it is essential to increase the availability of these precursor pools and to eliminate the competing pathways that consume acetyl-CoA for the biosynthesis of other intermediates. In the PDH bypass pathway (Fig. 9.2), pyruvate is first decarboxylated. The resulting acetaldehyde is then oxidized by acetaldehyde dehydrogenase (Ald) to acetate, which is subsequently converted to acetyl-CoA by acetyl-CoA synthetase (Acs). This reaction consumes one ATP and releases one CO_2 molecule (Fig. 9.2). This pathway serves as the primary route for acetyl-CoA production in the cytoplasm of fungal cells. Furthermore, Fig. 9.2 indicates that the phosphoketolase pathway, involving phosphoketolase (Pk) together with phosphotransacetylase (Pta) or acetate kinase, provides an attractive alternative for cytosolic acetyl-CoA production in the fungal and yeast cells (Van Rossum et al. 2016). Unlike some routes, it can be clearly seen from Fig. 9.2 that this pathway converts sugar phosphates to acetyl-CoA without carbon loss (Bogorad et al. 2013). Therefore, to support high liamocin biosynthesis, both the PDH bypass and

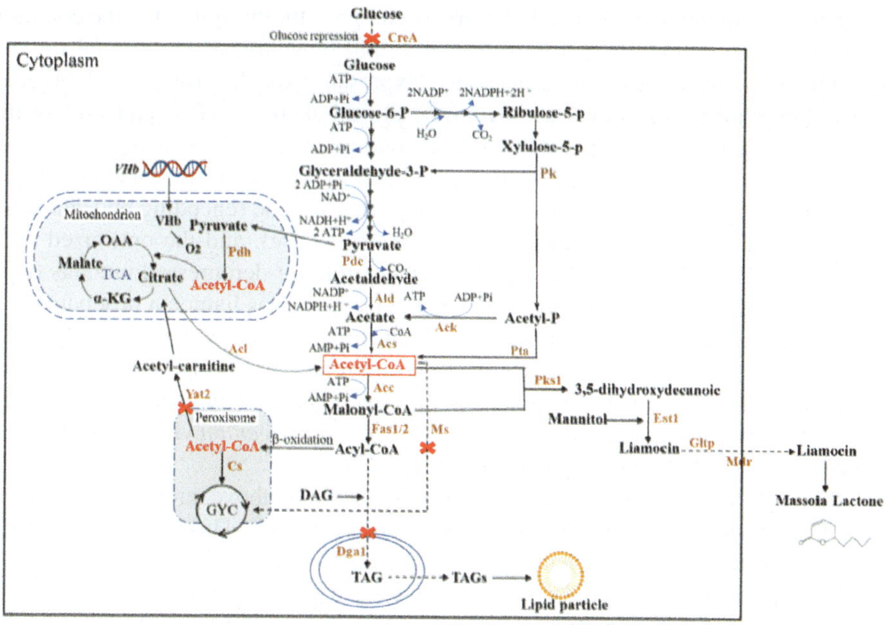

Fig. 9.2 Genetic modification strategy pathway for high liamocin synthesis in *A. melanogenum* 9-1. *VHb* Vitreoscilla hemoglobin, *CreA* glucose repressor, *Pk* phosphoketolase, *Pta* phosphotransacetylase, *Ald* acetaldehyde dehydrogenase, *Ack* acetate kinase, *Acs* acetyl-CoA synthetase, *Dga1* diacylglycerol transferase 1, *Acc* acetyl-CoA carboxylase, Ms malate synthase; Pks1 polyketide synthase, Cs citrate synthase, Est1 esterase 1, Yat2 carnitine acyl-transferase 2, Gltp glycolipid transporter, Fas1/2 fatty acid synthase 1/2, Mdr ABC transporter, Acl ATP-citrate lyase, Pdh pyruvate dehydrogenase, Pdc pyruvate decarboxylase, DAG diacyl glycerol, TAG triacyl glycerol, GYC glyoxylate cycle

phosphoketolase pathways should be enhanced. Additionally, it is found that in the engineered strains of the high liamocin-producing yeast-like fungus *A. melanogenum*, both endogenous and heterologous acetyl-CoA supply pathways can be directly leveraged to increase acetyl-CoA levels and boost liamocin production.

To reduce acetyl-CoA consumption, the competing pathways for this precursor should be eliminated in the engineered strain of *A. melanogenum*. As high liamocin production requires cultivation in a high-glucose medium, its biosynthesis is subject to glucose repression, as previously indicated. Therefore, glucose repression must also be alleviated in the production host of *A. melanogenum*. Furthermore, it can be clearly seen from Fig. 9.2 that the conversion of acetyl-CoA to malonyl-CoA consumes ATP (Xue et al. 2018). Therefore, to support efficient liamocin biosynthesis, it is necessary to enhance ATP supply. This can be achieved by overexpressing the codon-optimized gene for Vitreoscilla hemoglobin (VHb), an oxygen-binding protein that promotes ATP generation in the TCA cycle. We found that reducing acetyl-CoA consumption in mitochondria, peroxisomes, and lipid biosynthesis (Fig. 9.2) did not significantly enhance liamocin production in the engineered strain of

A. melanogenum 9-1 and instead impairs its cell growth, implying that the common function of mitochondria, peroxisomes, and lipid biosynthesis is essential for cell growth of *A. melanogenum.* Conversely, expressing the heterologous *PTA* gene (encoding phosphotransacetylase) from the phosphoketolase (PK) pathway, or the native *ALD* gene (encoding acetaldehyde dehydrogenase) and the *ACS* gene (encoding acetyl-CoA synthetase) genes from the PDH bypass pathway, leads to reduced liamocin biosynthesis. In contrast, expressing the *PK* gene (encoding phosphoketolase), the *PDC* gene (encoding pyruvate decarboxylase), and the optimized *VHb* gene (encoding Vitreoscilla hemoglobin) in the glucose-derepressed mutants (e.g., the disruptant Δ*crea55* mentioned above) greatly increases liamocin production.

Ultimately, the metabolic engineering and synthetic biology-derived strain V333 produces 55.4 g/L of liamocin in flask cultivation and yields 60.2 g/L in a 10-liter batch fermentation at pH 3.0 within 156 h. This represents another novel and efficient strategy for liamocin overproduction by *A. melanogenum* via metabolic engineering and synthetic biology (Zhang et al. 2022).

Meanwhile, the engineered strain V333 obtained above grows optimally at pH 7.0, but produces the highest liamocin titer at pH 3.0. Therefore, a two-phase cultivation strategy was implemented: the engineered strain V333 is first grown at a constant pH of 7.0 for 48 h for active and high cell growth, followed by a shift to a constant pH of 3.0 for continued cultivation and limocin production. Under these conditions, the engineered V333 achieves a liamocin titer of 70.9 g/L. Its cell mass reaches 31.6 g/L, and only 0.16 g/L of glucose remained in the fermented culture at the end of the novel two-phase cultivation process (Wang et al. 2024). This represents the highest liamocin titer reported for *A. melanogenum* and *A. pullulans* to date. This finding is consistent with the established principle that an optimal liamocin synthesis strategy involves a cell growth phase at neutral pH, followed by a liamocin production phase induced by a manual pH shift to an acidic condition (Maria Saur et al. 2019).

As described above, liamocins typically exist as a mixture of mannitol-liamocin and arabitol-liamocin (Fig. 9.1). Studies indicate that these differing head groups confer their distinct functions and bioactivities (Xue et al. 2020). To produce a homogeneous variant, the mannitol phosphate dehydrogenase (*MPDH*) and mannitol dehydrogenase (*MtDH*) genes were sequentially knocked out in the high-yielding and engineered yeast-like fungus V333 mentioned above. The resulting engineered strain, Δ*pt-422*, produces liamocin exclusively with an arabinitol head group (A-liamocin), achieving a flask-level extracellular concentration of 11.7 g/L of A-liamocin. In contrast, the parental V333 strain naturally produces a mixture of mannitol- and arabinitol-headed liamocins (Fig. 9.1) (Xue et al. 2020), with a total liamocin titer of 47.4 g/L. Alternatively, the liamocin head group formation can be controlled through media composition. When fructose is supplied as the sole carbon source, strain V333 produces liamocin predominantly with a mannitol head group (M-liamocin). This approach yields 39.3 g/L M-liamocin at a flask level and is successfully scaled to 46.6 g/L in a 10-Liter fermenter (Song et al. 2026). Thus, the head group of liamocins can be effectively edited through both synthetic biology strategies or optimization of the culture medium.

Table 9.1 Liamocin production by different strains of *Aureobasidium* spp.

Fungi	Strain	Substrates	Titer (g/L)	References
	NRRL 58544	Glucose	5.4	Leathers et al. (2010)
	NRRL 50380	5% sucrose	0.5–6.0	Price et al. (2017)
	NRRL 62042	Glucose	8.6	Manitchotpisit et al. (2014)
	RSU29	Glucose	22	Maria Saur et al. (2019)
A. melanogenum	AS37	Sucrose	5.6	Wang et al. (2022a, b)
	SK25	Xylose	7.0–9.9	Saika et al. (2020)
	6-1-2	Glucose	25.75	Xue et al. (2020)
	M399	Glucose	43.04	Tang et al. (2018)
	G400	Glucose	46.7	Wang et al. (2022a, b)
	9–1	Glucose	26.8	Wang et al. (2022a, b)
	V333	Glucose	60.2	Zhang et al. (2023)
	V333	Glucose	70.9	Wang et al. (2024)
	U33	Inulin	55.2	Wang et al. (2025)
	HN6-1	Glucose	25.3	Tang et al. (2018)
	V333	Glucose	60.2 g/L	Zhang et al. (2022)
	V333	Glucose	70.9	Wang et al. (2024)
	V333	Fructose	46.6	Song et al. (2026)
A. thailandense	NRRL58543	Glucose	>1.0	Rattanahadsadin et al. (2024)

All native strains of *A. pullulans*, *A. melanogenum*, and *A. thailandense* reported to date secrete less than 26.8 g/L of liamocin across various carbon sources (Table 9.1). For instance, a screening of 30 *Aureobasidium* spp. strains (including *A. pullulans*, *A. melanogenum*, and *A. thailandense*) isolated from tropical environments in Thailand reveals that liamocin titers ranging from 0.5 to 10.6 g/L are produced by them (Rattanahadsadin et al. 2024). Consequently, the engineered strain V333 again is demonstrated to produce the highest liamocin titer among all the reported native and genetically modified strains of *Aureobasidium* spp.

9.2.2 Elucidation of Biosynthesis Pathway of Liamocins by Genome Editing and Synthetic Biology

As outlined above, many native and engineered strains of *Aureobasidium* spp. can synthesize and secrete substantial amounts of liamocin into the culture. The elucidation of its biosynthesis, transport, secretion, and regulation is therefore critical (Xue et al. 2020; Kang et al. 2022). Defining this core pathway in *Aureobasidium* spp. through molecular biology will facilitate the metabolic engineering of these novel polyol lipids and may provide insights into the evolutionary origin of liamocin biosynthesis in eukaryotes. To this end, the core biosynthetic pathway was investigated in the strain *A. melanogenum* 6-1-2 as mentioned in Table 9.1 in this laboratory.

Fig. 9.3 Conserved domains of amino acids deduced from the *HR-PKS* genes

The study has strongly demonstrated that a highly reducing polyketide synthase (HR-Pks1 encoded by a *HR-PKS1* gene), mannitol 1-phosphate dehydrogenase (Mpdh1 encoded by a *MPDH1* gene), mannitol dehydrogenase (Mtdh1 encoded by a *MtDH1* gene), arabitol dehydrogenase (Ardh1 encoded by a *ArDH1* gene), and an esterase (Est1 encoded by a *EST1* gene) are intimately involved in extracellular liamocin biosynthesis in this strain. The HR-Pks catalyzes the formation of the 3,5-dihydroxydecanoic acid moiety of liamocin. This enzyme comprises acyl carrier protein (ACP), acyl transferase (AT), dehydratase (DH), enoyl reductase (ER), ketoreductase (KR), and ketosynthase (KS) domains (Fig. 9.3). Sequence analysis identified two phosphopantetheine attachment motifs (LGVDSL and LGMDSL) within the ACP domain, where the serine residues are predicted to bind the thiol-containing cofactor. This indicates that the HR-Pks1 requires activation via phosphopantetheine addition by a PPTase (Fig. 9.3).

Next, it is essential to understand the biosynthesis of the 3,5-dihydroxydecanoic acid tail group in extracellular liamocins. Based on the domain organization and structure of the HR-Pks1 (accession number: XM_013570770.1, Fig. 9.3), a biosynthetic pathway is proposed. It is hypothesized that, in the presence of acetyl-CoA and malonyl-CoA, all catalytic domains (ACP, AT, DH, ER, KR, KS) of the HR-Pks1 act twice to form the intermediate **ACP–S–CO–$(CH_2)_4$–CH_3**. Subsequently, a subset of domains (ACP, KS, AT, and KR) acts twice more to synthesize the final C_{10}-product, **ACP–S–CO–$(CH_2)_2$–CHOH–CH_2–CHOH–$(CH_2)_3$–CH_3** (Fig. 9.4).

Furthermore, the total amount of intracellular lipids in the *HR-PKS1* gene deletion mutant (Δ*pks120*) does not differ significantly from that in the parent strain 6-1-2. Lipid particles were also isolated and purified from the Δ*pks120* mutant. GC-MS analysis revealed that 98.4% of the fatty acids in these purified lipid particles consists of $C_{16:0}$, $C_{18:0}$, $C_{18:1}$, and $C_{18:2}$ fatty acids—the compositions nearly identical to that of the parent strain 6-1-2. These results suggest that the biosynthesis of the extracellular liamocin tail group (3,5-dihydroxydecanoic acid) is independent of the general intracellular lipid synthesis mechanism and that the intracellular lipid biosynthesis mechanism is not governed by the HR-Pks1 for 3,5-dihydroxydecanoic acid biosynthesis.

Our findings indicate that mannitol phosphate dehydrogenase (Mpdh1), encoded by the *MPDH1* gene, and mannitol dehydrogenase (Mtdh1), encoded by the *MTDH1* gene, are involved in mannitol biosynthesis. Similarly, arabitol dehydrogenase (Ardh1), encoded by the *ARDH1* gene, participates in arabitol biosynthesis in strain 6-1-2. Consequently, the liamocin produced by the Δ*pt88* mutant (lacking both the *MPDH1* gene and the *MTDH1* gene) contains only arabitol as the head group of the produced liamocin, while the liamocin yielded by the Δ*A221* mutant (with a removal

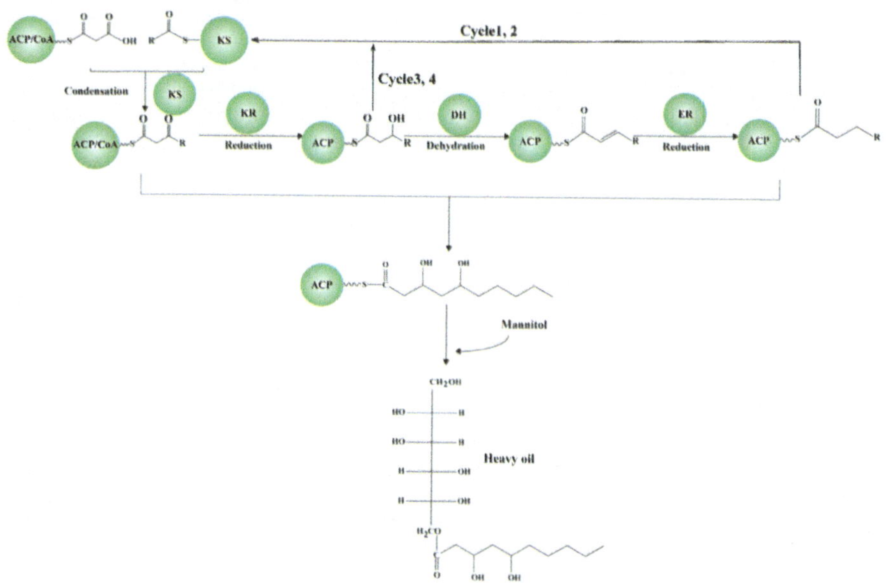

Fig. 9.4 The relationship between the HR-Pks1 and 3,5-dihydroxydecanoic acid biosynthesis

of the *ARDH1* gene) contains only mannitol as the head group of the produced liamocin. Importantly, liamocin production is completely abolished only when all three genes (*MPDH1*, *MTDH1*, and *ARDH1*) are abolished, as seen in the Δ*pta133* mutant (Xue et al. 2020).

Furthermore, the esterase Est1, encoded by the *EST1* gene, is responsible for forming the ester bond between **$ACP–S–CO–(CH_2)_2–CHOH–CH_2–CHOH–(CH_2)_3–CH_3$** (Fig. 9.4) and a single mannitol or arabitol head group of liamocins. Given that the polyketide synthase HR-Pks1 lacks a thioesterase (TE) domain, we propose that Est1 directly catalyzes this esterification step, as shown in Fig. 9.5.

However, the mechanism by which one 3,5-dihydroxydecanoic acid molecule forms an ester linkage with another one still remains completely unknown. So, which esterase catalyzes the mechanism and which gene in the genomes of *Aureobasidium* spp. encodes the esterase have not been mined so far. Furthermore, it is also completely unclear which environmental, physiological, and genetic factors determine the number of 3,5-dihydroxydecanoic acid molecules linked to liamocin.

Furthermore, the *EST1*, *HR-PKS1*, and *GAL1* genes are co-localized within a single gene cluster (Fig. 9.6). Gal1, encoded by the *GAL1* gene, functions as a specific transcriptional activator that regulates this cluster, including the *HR-PKS1* and *EST1* genes. Therefore, removal of the *GAL1* gene also causes the cessation of liamocin biosynthesis in the corresponding mutant.

Research on liamocin transport and secretion indicates that the *GLTP1* gene (encoding a glycolipid transfer protein) and the *MDR11* gene (encoding an ABC transporter) are involved in exporting these compounds into the extracellular

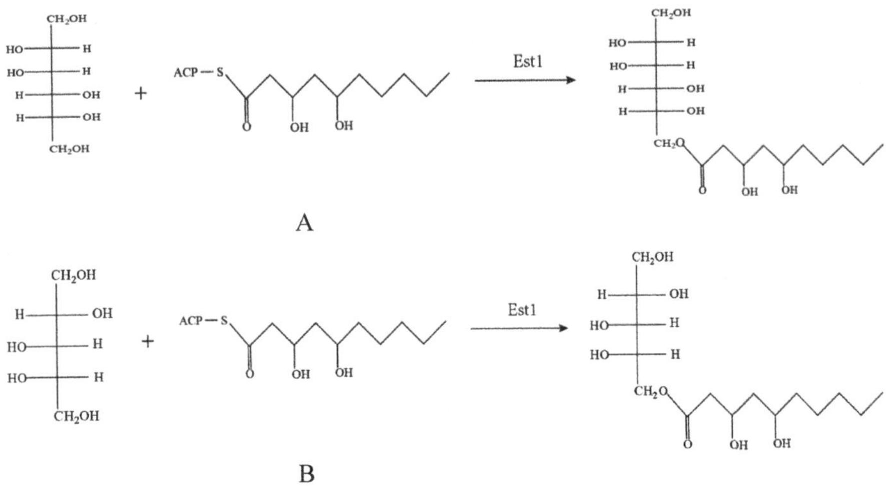

Fig. 9.5 The relationship between Est1 and 3,5-dihydroxydecanoic acid biosynthesis. Biosynthesis of mannitol-liamocin (**A**) and arabitol-liamocin (**B**)

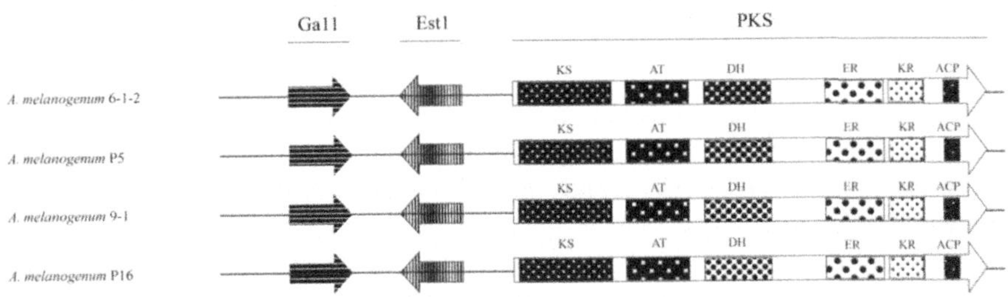

Fig. 9.6 The gene clusters for liamocin biosynthesis in *Aureobasidium* spp.

medium. The glycolipid transfer protein, Gltp1, is a cytosolic protein belonging to the intermembrane transporter family. In vitro studies have established that such proteins can transfer β-linked glycolipids between bilayer membranes. The Gltp1 likely functions as a cellular modulator or sensor of glycolipid levels. Abolishment of the *GLTP1* gene produces the mutant Δ*gltp22*. This mutant exhibits a reduced capacity for liamocin secretion, resulting in heavy intracellular accumulation of the product in the yeast-like fungal cells. This accumulation correlates with observable cellular swelling (Fig. 9.7) and a significantly higher intracellular lipid content (37.0%) compared to that in the cells of the parent strain 6-1-2, which displays normal morphology and a lower lipid content (28.4%). Genetic complementation of the *GLTP1* gene in Δ*gltp22* mutant restores the normal phenotype, yielding the transformant BG177. The BG177 transformant produces increased extracellular liamocin and an intracellular lipid level similar to that of the parent strain.

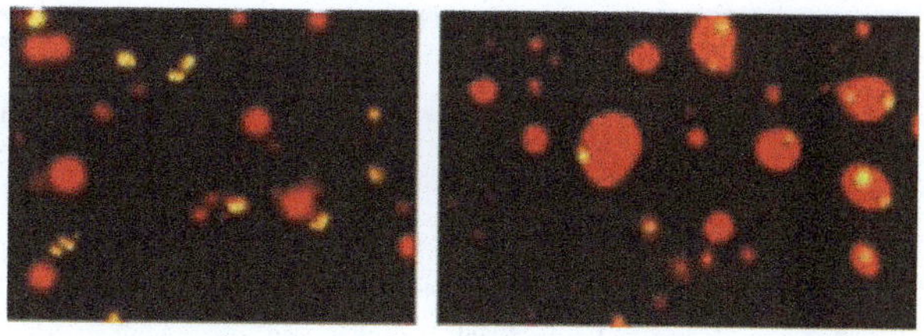

Fig. 9.7 The lipid particles (yellow ones) and liamocin particles (red ones) in the cells of the wild type strain 6-1-2 (**A**) and the disruptant *Δgltp2* in which the *GLTP* gene is totally removed (**B**)

Notably, the structural similarity of liamocins to glycolipids and polyol lipids (Tiso et al. 2024) supports the hypothesis that the Gltp1 facilitates liamocin transport within the cells of strain 6-1-2 used in this study.

Removal of the *MDR1* gene mentioned above reveals that liamocin secretion is impaired in a *Δmdr16*6 mutant which genomic DNA lacked the *MDR11* gene. A large quantity of liamocins accumulates intracellularly in the *Δmdr16*6 mutant cells, although some secretion into the culture still occurs. This suggests that other transmembrane transporters may also contribute to liamocin export in strain 6-1-2.

In summary, combined with the findings above, the complete process of core liamocin biosynthesis, regulation, transport, and secretion in the yeast-like fungal cells is elucidated (Fig. 9.8). Specifically, the highly reducing polyketide synthase (HR-Pks1), which carries ACP, AT, DH, ER, KR, and KS domains, synthesizes the 3,5-dihydroxydecanoic acid moiety of extracellular liamocins from acetyl-CoA and malonyl-CoA. Mannitol phosphate dehydrogenase (Mpdh1) and mannitol dehydrogenase (Mtdh1) are involved in mannitol biosynthesis, while arabinitol dehydrogenase (Ardh1) mediates arabitol biosynthesis. The esterase Est1 then catalyzes ester bond formation between 3,5-dihydroxydecanoic acid and mannitol or arabitol. Furthermore, a phosphopantetheinyl transferase (PPTase) activates the acyl carrier protein (ACP) domain of the HR-Pks1 by the post-translational modification, and the specific transcriptional activator Gal1 upregulates expression of the *HR-PKS1* gene and *EST1* gene in the gene cluster. Finally, the glycolipid transfer protein Gltp1 and the ABC transporter Mdr1 are proposed to facilitate liamocin transport from the Golgi to the plasma membrane and subsequent secretion of the synthesized liamocin into the extracellular medium, respectively (Fig. 9.8) (Xue et al. 2020). However, as stated above, many questions about the detailed, complete process of liamocin biosynthesis still remain to be answered.

9.2.3 Clarification of Regulation Mechanisms of Liamocin Biosynthesis by Genome Editing and Synthetic Biology

The regulatory genes governing liamocin biosynthesis were also investigated in the wild-type strain 6-1-2 (Xue et al. 2020). As described previously, both a phosphopantetheinyl transferase (PPTase) and the specific transcriptional activator Gal11

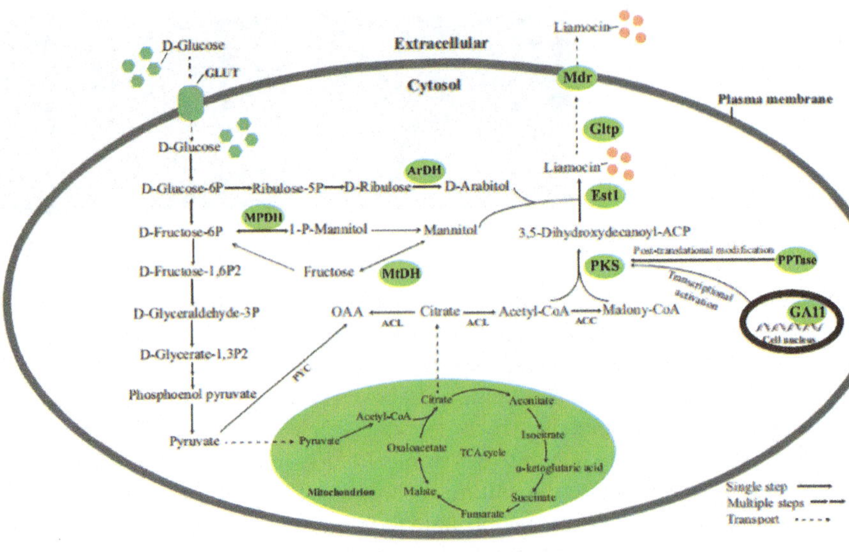

Fig. 9.8 The core pathway for liamocins biosynthesis, regulation, transport, and secretion in *A. melanogenum* L1-1. *OAA* oxaloacetate, *AKG* α-ketoglutaric acid, *PYC* pyruvate carboxylase

positively influence this biosynthetic pathway. The PPTase from strain 6-1-2 contains the two conserved domains, VGTDIV and FCYKEAYIKL, identifying it as an Sfp-type enzyme essential for phosphopantetheinylation. This reaction also activates the acyl carrier protein (ACP) domains of any other polyketide synthases (Pkss), non-ribosomal peptide synthases (NRPSs), and fatty acid synthases (Fass), as shown in the following. Consistent with this essential function, disruption of the *PPTase* gene (the mutant Δ*P122*) abolishes all liamocin production. Conversely, genetic complementation of the native gene in strain BPPT6 restores liamocin biosynthesis to its parent strain levels. These results indicate that the PPTase performs an indispensable post-translational modification on the ACP domain of the core HR-Pks1 in strain 6-1-2 (Fig. 9.6).

Sequence analysis indicates that the Gal1 protein encoded by the *GAL1* gene harbors a $Cys(X)_2Cys(X)_4His(X)_4Cys$ motif at its C-terminus, a characteristic feature of a novel DNA/RNA-binding zinc finger protein (Fig. 9.6). To investigate Gal1's role in regulating liamocin biosynthesis, we constructed a *GAL1*gene deletion mutant (Δ*gal1–12*). This mutant failed to produce any liamocins. However, reintroduction of the native *GAL1* gene into the disruptant Δ*gal1-1* restores liamocin biosynthesis and secretion in the complemented strain BGA33. Quantitative real-time PCR (qPCR) analysis reveals that removal of the *GA11* gene drastically reduces transcription of the *HR-PKS1* gene for 3,5-dihydroxydecanoic acid biosynthesis, with its expression level remaining at only 0.47% of the parent strain level. This result suggests that Gal1 indeed acts as a strong transcriptional activator of the HR-PKS1 gene, thereby directly regulating liamocin biosynthesis (Fig. 9.6).

Many studies have shown that the global transcriptional activator Msn2/4 can recognize the promoter sequences containing the motifs AGGGG or CCCCT, thereby regulating the expression of the target genes. Analysis of the genomic DNA of *A. melanogenum* strain 6-1-2 reveals that the promoters of its liamocin biosynthetic-related genes indeed contain such stress-responsive elements with the core sequences AGGGG or CCCCT. To investigate its role, the single-copy *MSN2* gene in the genomic DNA of *A. melanogenum* 6-1-2 was knocked out and subsequently complemented. The extracellular liamocin yield in the knockout strain decreases by 92.3%, while this phenotype is restored in the complementary strain. Furthermore, the results from real-time quantitative PCR (RT-qPCR) analysis and subcellular localization of a Msn2-GFP fusion protein demonstrate that when *A. melanogenum* 6-1-2 was cultured in the liamocin production medium with high-glucose and low-nitrogen, Msn2 was localized to the nucleus. This nuclear localization correlates with its positive regulatory role in the expression of the liamocin biosynthesis-related genes and high liamocin titers in the culture.

Studies have shown that the cAMP-PKA signaling pathway modulates protein kinase A (Pka) activity, which in turn phosphorylates or dephosphorylates the transcription factor Msn2 to regulate the downstream target gene expression. To investigate this pathway in *A. melanogenum* 6-1-2, we knocked out and subsequently complemented its *BCY1* gene, which encodes the regulatory subunit of Pka in the cAMP-PKA signaling pathway. The results demonstrate that the *BCY1* gene obviously affects extracellular liamocin production. Specifically, liamocin yield in the *BCY1* knockout mutant decreases by 96.7%. Further analysis of transcription levels for the *TPK1/2* genes (encoding catalytic PKA subunits) and key liamocin biosynthesis-related genes, combined with the Msn2 subcellular localization studies, reveals a regulatory mechanism by Msn2. When the PKA activity in the cAMP-PKA signaling pathway is low, Msn2 is dephosphorylated, accumulates in the nucleus, and positively regulates the liamocin biosynthesis-related genes and promotes liamocin biosynthesis. Conversely, the high PKA activity in the cAMP-PKA signaling pathway leads to Msn2 phosphorylation, its sequestration in the cytoplasm, and subsequent negative regulation of these liamocin biosynthesis-related genes and reduces liamocin biosynthesis.

In summary, extracellular liamocin biosynthesis in strain 6-1-2 is regulated by the cAMP-PKA signaling pathway through the global transcription factor Msn2 (Zhang et al. 2021a).

Within the high-osmolarity glycerol (HOG) signaling pathway as mentioned above, Msn2 also functions as a key transcriptional factor downstream of the Hog1 cascade. Analysis of liamocin biosynthesis-related gene transcription in the *HOG1* knockout mutant (Δ*hog266*) and the complementary strain (BHOG22) reveals that expression of the *HOG1* gene synergistically regulates expression of both the *MSN2* gene and the liamocin biosynthesis-related genes. Specifically, removal of the *HOG1* gene decreases liamocin biosynthesis in the Δ*hog266* mutant by 98.1%. This liamocin phenotype is restored upon genetic complementation of the native *HOG1* gene, as observed in the BHOG22 strain.

Combined with the Msn2 protein subcellular localization data, finally, it is concluded from these results that Hog1 likely forms a part of a transcriptional complex regulating the *MSN2* gene transcription. This regulation modulates the cellular level of the Msn2 protein in the yeast-like fungal cells of *A. melanogenum*, thereby affecting expression of the liamocin biosynthesis-related genes and liamocin production.

Msn2 activity is regulated not only by the cAMP-PKA and HOG pathways, but also may be regulated by the TOR1 and SNF1 pathways. To investigate their roles in this context, the key *TOR1* and *SNF1* genes in *A. melanogenum* strain 6-1-2 were abolished. However, extracellular liamocin production by the corresponding Δ*tor11* or Δ*snf11* mutants does not differ significantly from that of their parent strains. Consistent with this phenotypic observation, the real-time quantitative PCR analysis shows that disrupting the *TOR1* or *SNF1* gene had no significant effect on the transcriptional levels of either the liamocin biosynthesis-related genes or the *MSN2* gene. Thus, the *TOR1* and *SNF1* genes in the TOR1 and SNF1 signaling pathways may not be involved in regulating Msn2 activity.

Therefore, in *A. melanogenum* strain 6-1-2, liamocin biosynthesis is primarily regulated by the transcriptional activator Msn2 via the cAMP-PKA and the HOG signaling pathways. These findings establish a foundation for further research into the regulatory mechanisms of liamocin biosynthesis, as well as for further enhancing liamocin production using synthetic biology and metabolic engineering approaches.

It is well established that fungal autogenous pH activation, which stems from ammonia or organic acid production by themselves, profoundly influences various physiological processes. These include the biosynthesis of secreted enzymes and extracellular secondary metabolites (such as toxins), as well as cellular growth and development. Depending on the ambient pH shift induced by these self-produced compounds, this phenomenon can be classified into two distinct types: autogenous alkaline activation (triggered by ammonia) and autogenous acid activation (triggered by organic acids like citric, oxalic, gluconic, or fumaric acid). Usually, ambient pH signals are sensed, transmitted, and responded to by a dedicated pH signaling pathway. In most fungi, this pathway involves several key components besides the central transcriptional activator PacC1. Typically, three upstream components—PalH, PalI, and PalF—are primarily responsible for sensing and responding to ambient pH signals. Three additional components—PalA, PalB, and PalC—are involved in the proteolytic processing and activation of PacC1. During the fungal autogenous alkaline activation (triggered by ammonium production), the plasma membrane sensor PalH detects environmental alkalinity and initiates the signaling process. This triggers the proteolytic processing of the full-length activator PacC72 into a transient 53-kDa intermediate (PacC53). Subsequently, PacC53 undergoes further, pH-independent processing by the proteasome, yielding the fully functional PacC27 protein. Finally, PacC27 binds to the pH-responsive 5′-GCCARG-3′ consensus sequence in the target gene promoters, inducing the expression of alkaline-expressed genes (such as its own *PACC1* gene and other PacC-regulated genes) while repressing the acid-expressed genes. In contrast, the molecular mechanisms underlying the fungal autogenous acid activation remain largely unclear. It is not yet understood how ambient acidic pH signals are sensed, transduced, and responded to, or how PacC1 is processed and activated under these acid conditions. To date, it is only known that the *PACC1* gene in *A. melanogenum* itself is acid-expressed (Zhang et al. 2022).

Previous studies have shown that the medium pH decreases from 6.5 to 3.0 during liamocin production by *Aureobasidium* spp. (Maria Saur et al. 2019), suggesting that low environmental pH promotes liamocin biosynthesis via the pH signaling pathway. Our results confirm that this low pH is primarily caused by secreted citric acid, which also serves as a precursor of acetyl-CoA for liamocin synthesis (Fig. 9.8). Assays of citric acid in the culture, along with deletion, complementation, and overexpression of the key *CEXA* gene (encoding a citrate exporter), demonstrate that pH reduction is directly attributable to citric acid secretion. Furthermore, deletion, complementation, and overexpression of the *ACL* gene (encoding ATP-citrate lyase), combined with experiments using different initial pH values, indicate that the acidic environment created by citric acid is optimal for its own catabolism, for liamocin biosynthesis, and for expression of the key pH-responsive transcription factor gene *PACC1*.

Accordingly, the *PACC1* gene is an acid-expressed gene. Deletion, complementation, and overexpression of this gene demonstrate that expression of the key liamocin biosynthesis-related gene cluster (*GA11-EST1-HR-PKS1*) as indicated in (Fig. 9.6) is driven by the transcription factor PacC under low-pH conditions. This explains why liamocin biosynthesis is induced at low pH in the presence of citric acid (Maria Saur et al. 2019; Wang et al. 2024). A proposed model for the enhanced liamocin biosynthesis via the autogenous acid activation—involving the pH signaling pathway—is presented in Fig. 9.9. Collectively, our results show that the high-liamocin-producing strain 9-1 possesses all the components of the pH signaling pathway; the genes encoding these components are acid-expressed and exhibit the optimal expression at pH 3.0. Under the low ambient pH (~3.0) induced by secreted citric acid, only the protein **PalA** and **PacC** appear to regulate liamocin biosynthesis. PalA functions as a negative regulator, suppressing *PACC* gene expression and thereby inhibiting liamocin production. This suppression likely occurs through PalA binding to PacC71, which reduces the transcriptional activity of PacC and downregulates *PACC1* gene transcription. Consequently, expression of the key biosynthesis-related and regulatory genes for liamocin is reduced, ultimately lowering liamocin yield. In contrast, the full-length PacC71 form remains unprocessed under acidic conditions and serves as the transcriptional activator that promotes liamocin biosynthesis (Zhang et al. 2023). This brief regulatory process is summarized in Fig. 9.9. However, the detailed mechanisms of the autogenous acid activation at molecular level are still awaited to be elucidated by using synthetic biology.

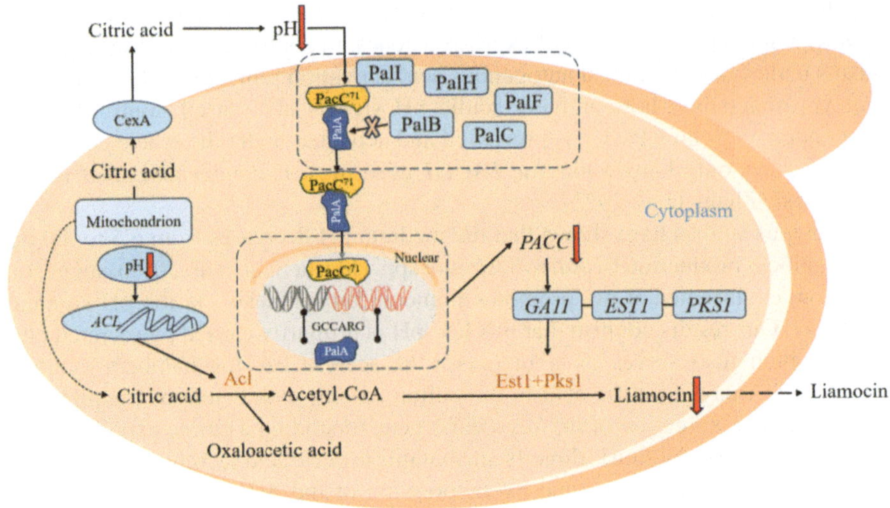

Fig. 9.9 A new regulation mode of liamocin biosynthesis under acid conditions in *A. melanogenum* 9–1. The *GAL1* gene encoding the specific transcriptional activator Gal1. The *EST1* gene encoding the Est1 which catalyzes ester bond formation between 3,5-dihydroxydecanoic acid and mannitol or arabitol (heads) in liamocin molecule. The *PKS1* gene encoding the HR-Pks1 which is responsible for the synthesis of 3,5-dihydroxydecanoic acid tail in liamocin molecule. The *GAL1*, *EST1*, and *PKS*1 genes forming the gene cluster. The Acl is the ATP-citric acid lyase. Any other proteins in this figure have been described in the text

9.2.4 Formation of Massoia Lactone from Liamocins

Liamocins produced by *Aureobasidium* spp. feature polyester tails consisting of one to six 3,5-dihydroxydecanoic acid ester groups. Upon hydrolysis with 2.0 N NaOH, these ester linkages are cleaved, releasing free 3,5-dihydroxydecanoic acids. Subsequent acidification with 5.0 N H_2SO_4 to pH 3.0 converts the released free 3,5-dihydroxydecanoic acids into C-10 massoia lactones (5-hydroxy-2-decenoic acid lactones). The complete reaction scheme is shown in Fig. 9.10. It has been reported that 92.1% of the 43.04 g/L of extracellular liamocins produced by *A. melanogenum* M399 mentioned above can be converted into 5-hydroxy-2-decenoic acid lactone (Tang et al. 2018). This corresponds to a yield of approximately 40.0 g/L of massoia lactone from the produced liamocins. To date, this represents the most efficient method for obtaining massoia lactone from liamocin molecules (Kang et al. 2022). The resulting massoia lactone has numerous industrial applications, as discussed in the following.

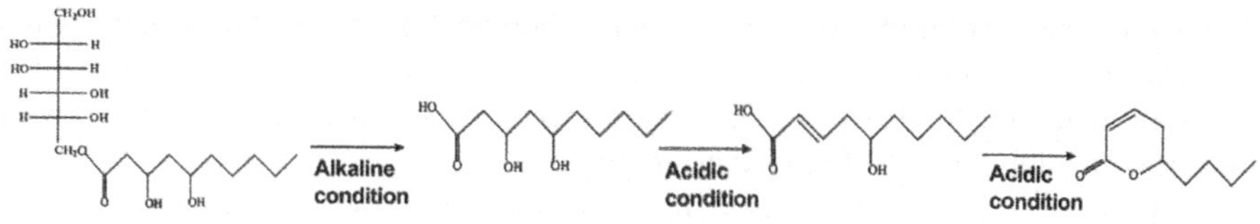

Fig. 9.10 Formation of massoia lactone from the produced liamocins

9.2.5 *Application of Liamocins and Their Massoia Lactone*

Liamocins, produced by *Aureobasidium* spp., exhibit antimicrobial activity. Specifically, they are effective against multidrug-resistant strains of *Streptococcus mutans* (causing dental caries), *Streptococcus sobrinus* (causing dental caries), and *Streptococcus suis* (swine disease). This activity is mediated through the disruption of the bacterial cell membrane, leading to the leakage of small-molecule nucleic acids from the treated bacterial cells. Additionally, liamocins demonstrate anticancer activity against various human cancer cell lines while showing no toxicity to normal cell lines (Manitchotpisit et al. 2011).

Dental caries is a common chronic infectious disease of bacterial origin that significantly threatens the oral and overall systemic health of humans. Its high prevalence and the substantial cost of treatment impose a heavy burden on healthcare systems worldwide. Especially, *S. mutans* plays a crucial role in the development and progression of dental caries. This bacterium contributes to the dental caries through its acidogenic and aciduric properties, as well as its ability to form robust biofilms, particularly in sucrose-rich environments. Current primary methods for controlling *S. mutans* and its biofilm formation involve mechanical removal and applications of broad-spectrum antibiotics or fluorides. However, these approaches are often inadequate, facing challenges such as incomplete biofilm removal and the risk of bacterial antibiotic resistance.

Recent studies have demonstrated that mannitol-liamocin, produced by the engineered strain V333 cultured in the fructose-based media (Wang et al. 2024), can significantly inhibit the growth of *S. mutans* and disrupt its biofilm formation. This inhibition is achieved by increasing the permeability of the bacterial cell membrane and inducing cell necrosis. Furthermore, a mannitol-liamocin nanoemulsion prepared via an ultrasonic-assisted method also effectively combats *S. mutans*. Treatment with this nanoemulsion at a concentration of 2.5 mg/mL of mannitol-liamocin for two hours results in nearly 100% bacterial necrosis (Song et al. 2026). These findings indicate that both mannitol-liamocin and its nanoemulsion formulation containing mannitol-liamocin hold promise as potential therapeutic agents for fighting dental caries in humans.

Liamocins headgroups can be edited to incorporate mannitol, arabitol, xylitol, sorbitol, or any other polyols into liamocins via synthetic biology (Xue et al. 2020). Finally, the bioactivity of liamocins with different polyol headgroups was compared against human 4 T1 breast cancer and B16 melanoma cell lines. The results demonstrate that liamocin bioactivity varies with the polyol headgroup. Arabitol-liamocins show the strongest inhibitory activity against human 4 T1 breast cancer cells, while xylitol-liamocins are the most active against the B16 melanoma cells. Furthermore, the B16 melanoma cells were found to be more sensitive to all tested liamocin variants than the human 4 T1 breast cancer cells.

Skin fungi can cause various diseases, such as dandruff, atopic dermatitis (eczema), ringworm (tinea corporis), athlete's foot (tinea pedis), jock itch (tinea cruris), and nail infections (onychomycosis). Such dermatophyte infections are notably difficult to eradicate completely and often recur at the original site of infection. Therefore, developing novel antifungal agents, particularly those effective against fungal spores, is urgently crucial. As illustrated in Fig. 9.10, massoia lactone can be derived from liamocin produced by the high liamocin-producing *Aureobasidium* spp. The released massoia lactone has proven effective against fungal spores isolated from the diseased human skin. These findings strongly confirm that massoia lactone exhibits potent anti-fungal spore activity by inhibiting spore germination and inducing cellular necrosis in the pathogenic fungal spores (Wang et al. 2022a, b). Importantly, this activity is vital for achieving complete eradication of both skin fungi and their spores, thereby preventing recurrence at the initial infection site due to its potent anti-fungal spore activity by inhibiting spore germination and inducing cellular necrosis in the pathogenic fungal spores.

Trichophyton rubrum is a pathogenic fungus responsible for several superficial infections on human and animal skin, such as tinea corporis (ringworm), tinea pedis (athlete's foot), tinea cruris (jock itch), and onychomycosis (nail infections). The primary antifungal agents currently used against *T. rubrum* are azoles and terbinafine. However, the azoles are typically fungistatic, meaning they only inhibit fungal growth without killing the fungal cells, which can lead to treatment failure or recurrence. The terbinafine, while fungicidal, is associated with potential hepatotoxicity. Consequently, there is an urgent need to develop new antifungal agents against *T. rubrum* infection. A promising candidate for controlling *T. rubrum* infection is massoia lactone, which has demonstrated substantial antifungal activity against this pathogen. Studies indicate that massoia lactone induces the accumulation of reactive oxygen species (ROS) within the *T. rubrum* cells, inhibits ergosterol synthesis in the cell membrane, reduces overall ergosterol content, and disrupts the fungal cell integrity. These combined effects ultimately trigger fungal cell necrosis, effectively inhibiting and eradicating both the hyphal growth and spores of *T. rubrum* on human skin (Wang et al. 2024).

However, the poor water solubility of massoia lactone limits its practical applications. To overcome this limitation, researchers have developed a massoia lactone nanoemulsion. This formulation not only possesses excellent water solubility, but also exhibits favorable stability upon dilution and during storage. Consequently, it has been successfully applied for treating *T. rubrum* infections. The nanoemulsion significantly enhances the aqueous solubility of massoia lactone and demonstrates superior stability and transdermal permeability. Furthermore, the encapsulated massoia lactone exhibits enhanced antifungal efficacy and reduced cytotoxicity compared to the free compound. These advantages indicate that the nanoemulsion formulation holds greater promise than free massoia lactone for practical therapeutic use in the treatment of fungal pathogenic infection (Wang et al. 2024).

In parallel, in vivo efficacy was evaluated by treating *T. rubrum*-infected mice with the massoia lactone nanoemulsion. The infection site on the mice exhibits characteristic symptoms of erythema (redness) and scaling. It was found that the nanoemulsion formulation is effectively localized and accumulated at the infection site. Consequently, a significant therapeutic effect against the *T. rubrum* infection is reached. These results suggest that the massoia lactone nanoemulsion holds promise for the future treatment of dermatophytosis in both animal models and human.

Massoia lactone, derived from liamocins produced by the engineered *A. melanogenum* M399 mentioned above, exhibits high stability and potent antifungal activity against several fungal crop pathogens responsible for significant plant diseases and food safety issues (Tang et al. 2018).

It has well documented that several fungal plant pathogens responsible for widespread crop infections include *Alternaria alternata, Aspergillus* spp., *Fusarium* spp., *Magnaporthe grisea, Penicillium* spp., *Phytophthora nicotianae, Rhizoctonia solani, Sclerotinia sclerotiorum, Ustilaginoidea oryzae,* and *Verticillium dahliae.* These fungal pathogens not only cause various diseases on plants, vegetable and fruits, but also often produce harmful mycotoxins to human. For instance:

(a) *Aspergillus flavus* produces aflatoxin which can cause human lung cancer, causing aflatoxicosis and yellow mold in peanuts.
(b) *Aspergillus niger* produces ochratoxin, which is hazardous and can lead to peanut root rot.
(c) Key *Fusarium* species (*F. graminearum, F. proliferatum, F. oxysporum*) are responsible for crop diseases and produce mycotoxins such as fumonisins, trichothecenes, and zearalenone.
(d) *Penicillium expansum* produces patulin, resulting in blue mold of apples and pears.

The mycotoxins produced by such pathogenic fungi pose a significant threat to animal and human health through the food and feed chain, as many are immunosuppressive, carcinogenic, hepatotoxic, nephrotoxic, or neurotoxic. Consequently, these plant, vegetable, and fruit diseases and associated mycotoxin contamination severely reduce crop quality and cause substantial economic losses in production and storage globally. This issue is further exacerbated by the stability of many mycotoxins during common food processing, meaning contaminated products can enter the food supply and jeopardize public health.

Currently, chemical fungicides are widely used to control fungal pathogens in crops, but their application has led to many challenges, such as significant environmental and ecosystem pollution. Recent studies have shown that treating fungal pathogens with massoia lactone not only effectively inhibits hyphal growth and spore germination but also induces pore formation in the cell membrane, reduces ergosterol content, increases intracellular reactive oxygen species (ROS) levels, and triggers leakage of cellular components. These effects ultimately lead to fungal cellular necrosis and death. Moreover, direct exposure of *Fusarium graminearum* spores to massoia lactone can suppress the development of Fusarium head blight symptoms in the infected wheat plants (Zhang et al. 2021c). Due to its potent antifungal activity against pathogen spores and its multimodal mechanism of action—which may help prevent the emergence of multidrug-resistant strains—massoia lactone represents a promising candidate for development as an effective and environmentally friendly bio-fungicide (Zhang et al. 2021b).

Massoia lactone, when delivered via nanoemulsions, actively inhibits the growth of *A. flavus* cells in liquid culture, as well as spore germination and aflatoxin biosynthesis. The minimum inhibitory concentration (MIC) of massoia lactone is 0.6 mg/mL, while its minimum fungicidal concentration (MFC) is 0.8 mg/mL. The treatment damages the cell walls and membranes of both spores and mycelia, leading to cell necrosis. It also suppresses the expression of the key genes involved in ergosterol and aflatoxin biosynthesis. Furthermore, the nanoemulsion-loaded massoia lactone also effectively inhibits fungal growth, conidial production, and aflatoxin contamination in peanuts. These findings indicate its high potential for controlling *A. flavus* and aflatoxin production in food and feed products (Wang et al. 2024).

Massoia lactone exhibits effective antimicrobial activity, indicating promising potential for applications in the biopharmaceutical industry. Studies on its stability demonstrate that high temperature and UV treatments have minimal impact on its antimicrobial properties. This compound shows activity against nine tested strains: *C. albicans*, *E. cloacae*, *A. baumannii*, *A. hydrophila*, *B. subtilis*, *E. coli*, *S. aureus*, *V. harveyi*, and *S. mammitis*. Among these, *C. albicans* is the most sensitive to massoia lactone. At a concentration of 50.0 mg/mL, massoia lactone produces a clear zone of 28.03 mm, with a minimum inhibitory concentration (MIC) of 2.0 mg/mL. Growth curve experiments confirm that massoia lactone significantly inhibits the yeast proliferation. Furthermore, the treatments damage the cell wall and membrane of *C. albicans*, resulting in leakage of alkaline phosphatase, intracellular proteins, and sugars, as well as enabling propidium iodide (PI) to enter cells and bind to DNA, indicating that its cell integrity is damaged (Wang et al. 2025).

The cultured crabs infected with the pathogenic yeast *Metschnikowia bicuspidata* develop "milky disease," exhibiting weakness, anorexia, opaque and whitish muscle tissue, and milky hemolymph. The widespread occurrence of this disease in aquacultural industries has severely reduced crab yields and economic returns. Currently, there are no safe, stable, or effective control methods available. Studies show that massoia lactone, released from liamocin, strongly inhibits the pathogenic *M. bicuspidate*, with a minimum inhibitory concentration (MIC) of 0.15 mg/mL and a minimum fungicidal concentration (MFC) of 0.34 mg/mL. Even the hydrolysate containing massoia lactone from *A. melanogenum* 9-1 cultures containing liamocins exhibits antifungal activity against this pathogen. Both free massoia lactone and its nanoemulsion formulation can damage the cell membrane, even rupturing entire cells and causing necrosis in the pathogenic yeast *M. bicuspidate*. Therefore, massoia lactone shows promise as a treatment for the milky disease in Chinese mitten crabs by targeting the pathogenic yeast (Zhang et al. 2021c). Recent findings have indicated that the pathogenic *M. bicuspidate* and any other pathogenic fungi also widely infect various farmed aquatic and marine animals (Hansali et al. 2023). Consequently, developing environmentally friendly antifungal agents like massoia lactone is economically important for managing such pathogens in the aquaculture and mariculture ecosystems.

In summary, liamocin and massoia lactone released from liamocin indeed have a wide spectrum of antimicrobial activity; especially, massoia lactone has high antifungal activity against various fungal pathogens of plants, animals, and humans. Therefore, it is of biotechnological significance to scale up the production of liamocin and massoia lactone by large-scale fermentation in industry.

9.3 Production, Biosynthesis, Regulation, and Applications of Intracellular Lipids

The imperative to address the escalating global energy crisis has intensified the development of renewable energy sources. Biodiesel, being renewable, safe, and non-toxic, presents a viable alternative to fossil fuels. Notably, the lipid composition of single-cell oils (SCOs) produced by fungi and yeasts (Fig. 9.11) resembles that of conventional plant oils. This similarity has spurred growing research interest in utilizing SCOs for biodiesel production. By conventional definition, oleaginous microorganisms are those capable of accumulating lipids exceeding 20% of their dry cell weight. Among these, oleaginous yeasts have garnered significant attention due to their superior growth rate and higher lipid accumulation capacity compared to any other oleaginous microbes.

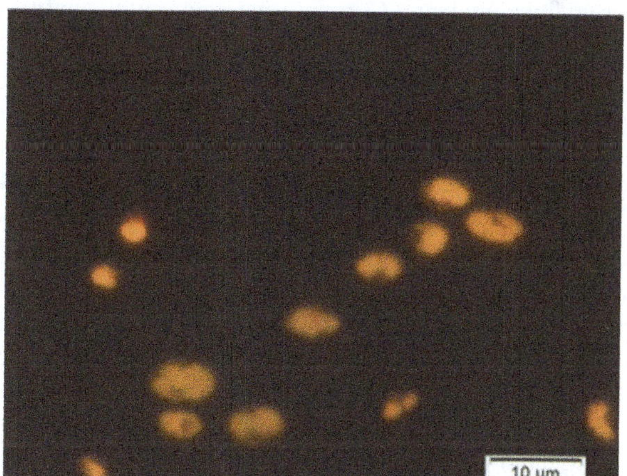

Fig. 9.11 The lipid particles of *A. melanogenum* P10

9.3.1 Overproduction of Intracellular Lipids by Metabolic Engineering

The results in the previous subsection indicate that many *A. melanogenum* strains with high liamocin production also accumulate substantial amounts of intracellular lipids, and such oleaginous yeast-like fungal cells also can be used as the raw biomaterials for the production of biodiesel.

Following isolation and purification from the mangrove ecosystem samples, among a total of 100 yeast strains obtained from the mangrove ecosystem samples, strain P10, identified as *A. melanogenum*, exhibits the highest cellular lipid content. The culture medium for SCO production by strain P10 was optimized using a one-variable-at-a-time approach, and the optimal medium composition, pH, and temperature for its cell growth and intracellular lipid accumulation were fixed (Wang et al. 2014a, b). In a 10-Liter fermentation, the lipid content reaches 66.3% (g oil/100 g dry cell weight) with a cell mass of 13.0 g/L of the culture. The lipid yield was 0.11 g oil/g of sugar consumed, and the lipid productivity was 0.0009 g oil/(g of sugar/h) within 120 h under the optimal conditions (Wang et al. 2014a, b).

The fatty acid composition of the produced lipids was determined to be $C_{16:0}$ (26.7%), $C_{16:1}$ (1.7%), $C_{18:0}$ (6.1%), $C_{18:1}$ (44.5%), and $C_{18:2}$ (21.0%). Combined, $C_{16:0}$, $C_{18:1}$, and $C_{18:2}$ fatty acids account for over 90% of the total fatty acids from the extracted lipids produced by strain P10. This profile suggests that these fatty acids are suitable as feedstock for biodiesel production. Consequently, the lipids extracted from strain P10 cells cultivated in the optimized medium and under the optimal conditions can be successfully converted into biodiesel, which exhibits favorable combustion properties as indicated in the following.

Studies have shown that, like glucose from cellulose and starch, xylose derived from xylan is also a suitable substrate for microbial lipid production. However, the capacity of *A. melanogenum* P10 to accumulate intracellular lipids from xylose remains unclear. Comparative genomic analysis of *A. melanogenum* P10, *Trichoderma reesei*, and *Scheffersomyces stipitis*—all known as the nice xylose-utilizing fungi—reveals that strain P10 harbors more genes encoding xylanases and carbohydrate-active enzymes of the glycoside hydrolase family than the other two fungal species. For instance, the genome of P10 strain contains complete pathways for xylose metabolism via both xylose reductase/xylitol dehydrogenase (Xr encoded by the *XR* gene/Xdh encoded by the *XDH* gene, respectively) and xylose isomerase (Xi encoded by the *XI* gene). It also carries all the genes for downstream metabolic steps of xylose utilization, such as D-xylulose-5-P phosphoketolase (Xpka encoded by the *XPKA* gene) and acetate kinase (Ack encoded by *ACK* gene). After optimizing the medium for lipid production and cell growth of the yeast strain P10 on xylose, the following conditions in a 100 mL medium were identified as optimal: 8.0 g xylose, 0.02 g yeast extract, 0.02 g ammonium sulfate, and a pH of 7.0. The fatty acid profile of the lipids produced by strain P10 cultivated under these conditions comprises $C_{16:0}$ (27.3%), $C_{16:1}$ (2.6%), $C_{18:0}$ (7.7%), $C_{18:1}$ (33.5%), and $C_{18:2}$ (23.1%) fatty acids (Song et al. 2022).

A genetically engineered strain, XI-155, was then constructed from strain P10 by knocking out the xylose reductase/xylitol dehydrogenase (Xr1/Xdh1) pathway and overexpressing the xylose isomerase (Xi1) pathway using synthetic biology tools. In the engineered strain XI-155, the expression levels of the *XR1* and *XDH1* genes are significantly reduced, while the expression of the *XI1* gene and the genes related to lipid biosynthesis is also markedly increased compared to that of the *XI1* gene and the genes related to lipid biosynthesis in its parent strain P10. In a 10-Liter batch fermentation using cheap industrial corncob xylose as the carbon source, strain XI-155 exhibits superior performance. It achieves an intracellular lipid content of 53.7 g per 100 g of cell dry weight, a cell mass of 11.9 g/L, a lipid yield of 0.082 g per g of the consumed corncob xylose, a productivity of 0.059 g/L/h within 108 h of the fermentation, and a corncob xylose utilization rate of 0.70 g/L/h. In contrast, its parent strain P10, under the same fermentation conditions, accumulates only 38.3 g lipid/100 g of cell dry weight, with a cell mass of 11.0 g/L, a yield of 0.059 g/g of corncob xylose, a productivity of 0.036 g/L/h within 120 h, and a corncob xylose utilization rate of 0.62 g/L/h (Song et al. 2022). The lipid synthesis capability of the recombinant strain XI-155 using industrial corncob xylose surpasses that of most of the known oleaginous yeasts.

A variety of low-cost feedstocks, including corncob, corn stover, wheat straw, cassava starch, waste oils, inulin, sugarcane bagasse, and sugarcane molasses, have been utilized for the production of microbial lipids which can be transformed to biodiesel (Table 9.2). Among them, different strains of *Aureobasidium* spp. can utilize corncob xylose, inulin, glucose, soy molasses, and whey powder to accumulate high levels of lipids in their cells.

Table 9.2 Substrates, lipid titer, biomass produced by different yeasts

Yeast strains	Lipid content(w/w%)	Cell mass (g/L)	Substrate	References
Lipomyces starkeyi	47.0	17.2	Corn hydrolysate	Sitepu et al. (2014)
Rhodotorula mucilaginosa TJY15a	52.9	21.8	Hydrolysate of cassava starch	Li et al. (2010)
Rhodosporidium toruloides	70.4	15.6	Inulin	Wang et al. (2014a, b)
A. melanogenum P10	38.3	11.4	Corncob xylose	Song et al. (2022)
A. melanogenum API4	66.2	14.4	Inulin	Wang et al. (2014a, b)
A. melanogenum P10	66.3	13.0	Glucose	Wang et al. (2014a, b)
A. melanogenum XI-155	53.7	11.9	Xylose	Song et al. (2022)
A. namibiae A12	53.3	13.0	Soy molasses	Wang et al. (2021)
A. namibiae A12	51.5	12.5	Whey powder	Wang et al. (2021)

Furthermore, the yeast-like fungal cells of a mutant strain that produces only a small amount of pullulan were found to accumulate more lipid droplets than any other mutants and its parent strain *A. melanogenum* P16, which produces large quantities of pullulan. Lipid extraction and quantitative analysis of the mutant and strain P16 reveal an inverse relationship between pullulan biosynthesis and lipid accumulation: the strain producing less pullulan exhibits higher lipid content and enhances cell growth (Yang et al. 2020). It is well known that citric acid is a key precursor for lipid biosynthesis as shown above. In the oleaginous yeasts, it can be cleaved into acetyl-CoA and oxaloacetate by ATP-citrate lyase (Acl), providing the acetyl-CoA substrate for fatty acid biosynthesis and leading to substantial intracellular lipid accumulation. Notably, the transcriptional levels of the lipid biosynthesis-related genes are also significantly upregulated in the low-pullulan mutants, but downregulated in the high-pullulan mutants (Yang et al. 2020).

The oleaginous yeast *A. namibiae* A12 genome harbors three copies of α-galactosidase gene, five copies of β-galactosidase gene, and one copy of invertase gene. It was found that the enzymes, including β-galactosidase, α-galactosidase, and invertase, produced by *A. namibiae* A12 can efficiently hydrolyze lactose and stachyose in whey powder, as well as galactose-based carbohydrates and raffinose in soy molasses. These obtained hydrolysates are subsequently metabolized by the yeast-like fungus into intracellular lipids, achieving lipid titers of 5.3 g/L from the whey powder and 5.2 g/L from the soy molasses (Table 9.2) (Wang et al. 2021).

9.3.2 *Biosynthesis Pathway of Intracellular Lipids*

It has been demonstrated that the genomes of *A. melanogenum* strains P10 and P16 mentioned above contain all the genes essential for lipid biosynthesis. Like any other oleaginous yeasts, their primary storage lipids are also triacylglycerols (TAGs) and sterol esters (SEs) (Fig. 9.12), which accumulate intracellularly.

Fatty acids for triacylglycerols (TAGs) and sterol esters (SEs) shown in Fig. 9.12 are synthesized by fatty acid synthase (Fas1), a two-subunit enzyme, which commonly exists in plants, animals, humans, fungi, and yeast. A. The β-subunit (Fas1p)

$$\begin{array}{l} CH_2-O-CO-R_1 \\ | \\ CH_2-O-CO-R_2 \\ | \\ CH_2-O-CO-R_3 \end{array}$$

(A)

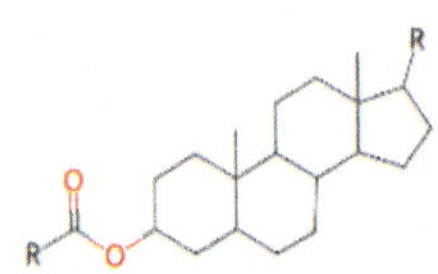

(B)

Fig. 9.12 The chemical structures of triacylglycerol (TAG) (R_1, R_2, and R_3 are individual fatty acid moieties) (**A**) and sterol ester (SE) (R is a fatty acid moiety) (**B**)

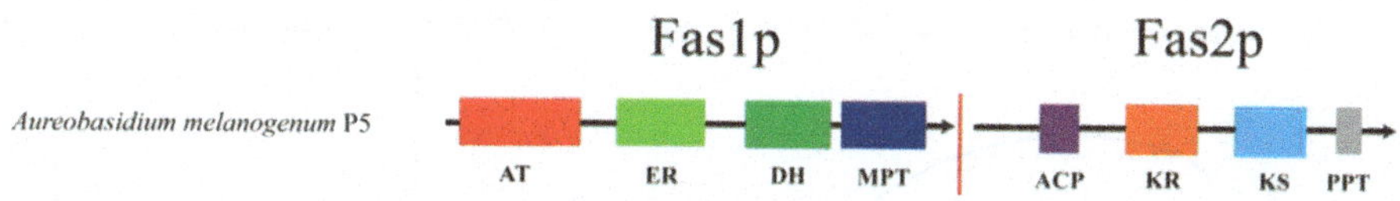

Fig. 9.13 The domains of the Fas1p subunit and the Fas2P subunit for fatty acid biosynthesis in *A. melanogenum*

contains acyltransferase (AT), enoyl-ACP reductase (ER), β-hydroxyacyl-ACP dehydratase (DH), and malonyl/palmitoyl transferase (MPT) domains. B. The α-subunit (Fas2p) contains acyl carrier protein (ACP), β-ketoacyl-ACP reductase (KR), ketosynthase (KS), and phosphopantetheine transferase (PPT) domains (Fig. 9.13). These domains closely resemble those found in the highly reducing polyketide synthases (HR-Pkss) involved in liamocin biosynthesis, as previously described (Fig. 9.6). The fatty acid synthesis process begins with the decarboxylation of malonyl-ACP to form acetyl-ACP. Subsequently, the KS domain catalyzes the condensation of two acetyl-ACP molecules to form β-ketoacyl-ACP. The KR, DH, and ER domains then sequentially reduce the resulting β-ketoacyl-ACP to yield a saturated acyl-ACP. This saturated acyl-ACP is elongated through further additions of acetyl-ACP, a step catalyzed by MPT (Xue et al. 2018). Finally, the acyl group from acyl-ACP is transferred to diacylglycerol (DAG) to form TAG or to a sterol to form SE (Fig. 9.12) via diacylglycerol O-acyltransferase (Fig. 9.14). Figures 9.13 and 9.14 further illustrate that the ACP domain's serine residue must be activated by phosphopantetheine by the post-translational modification like the ACP domain in Pks1 and NRPS, a cofactor attachment catalyzed by PPTase

9.3.3 Regulation of Intracellular Lipid Biosynthesis

Fatty acid biosynthesis is known to be regulated by both glucose and nitrogen repression. This means that the Snf1/Mig1 signaling pathway and the TORC1 signaling pathway, as stated above, must be involved in the regulation of fatty acid biosynthesis. For example, removing the *MIG1* gene from the genomic DNA—which encodes the main glucose repressor protein, Mig1—is key to alleviating glucose repression as stated above. In *Aureobasidium* spp., substantial intracellular lipid accumulation occurs only under the nitrogen-limited medium (Wang et al. 2014a, b; Song et al. 2022), confirming that nitrogen repression also regulates this pathway.

Multiple genes involved in fatty acid biosynthesis are coordinately regulated. This is exemplified by malate esterase (Me1 encoded by the *ME1* gene), ATP-citrate lyase (Acl1 encoded by the *ACL1* gene), acetyl-CoA carboxylase (Acc1 encoded by the *ACC1* gene), and fatty acid synthase (Fass encoded by the *FAS1* gene), which together form a functional enzyme complex. Both pyruvate carboxylase (Pyc1 encoded by the *PYC1* gene) and Acl are key enzymes that drive lipid biosynthesis,

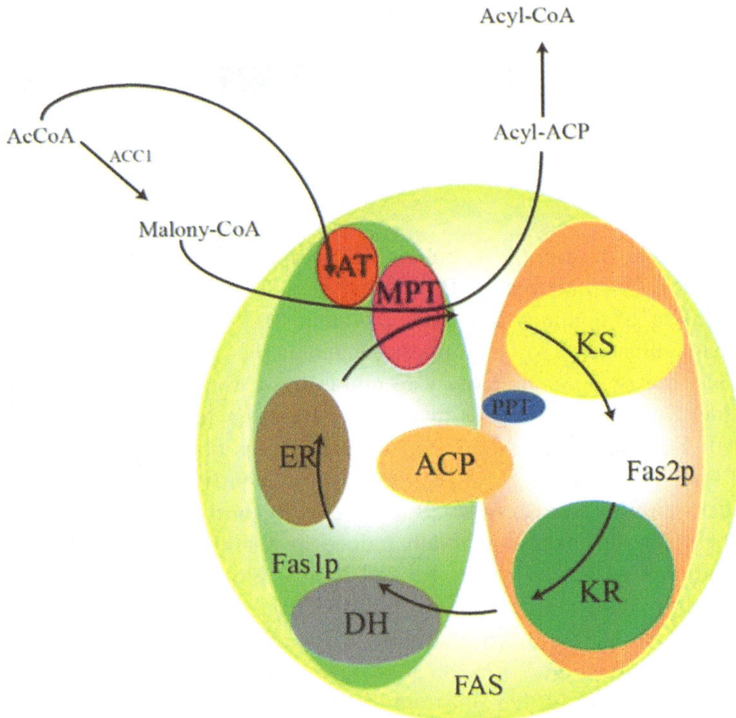

Fig. 9.14 Fatty acid synthase and biosynthesis of fatty acids. *AcCoA*: acetyl-CoA *ACC1*: acetyl-CoA carboxylase, *FAS*: fatty acid synthetase

particularly during citrate-mediated lipid production. Specifically: a. Acl1 cleaves citrate, utilizing ATP and CoA, to continuously supply acetyl-CoA and oxaloacetate—two essential precursors for fatty acid biosynthesis. b. Pyc1 catalyzes the carboxylation of pyruvate to oxaloacetate in the cytoplasm. This oxaloacetate can then condense with acetyl-CoA to form citrate, thereby channeling carbon flux toward lipid overproduction in the yeast-like fungal cells. For instance, heterologous expression of the *PYC1* gene from *Aureobasidium* sp. P6 in *A. melanogenum* P10 significantly enhanced lipid biosynthesis in the yeast-like fungal strain AM533 (Wang et al. 2014a, b). Similarly, expression of the same gene in *A. melanogenum* P5 boosts liamocin production, intracellular lipid accumulation, and cellular biomass in strain AMF156 (Xue et al. 2018).

Furthermore, for fatty acid biosynthesis to proceed, a conserved serine residue within the inactive acyl carrier protein (ACP) domain (Fig. 9.14) must undergo the post-translational modification. This activation is catalyzed by a phosphopantetheinyl transferase (PPTase), which covalently attaches a 4′-phosphopantetheine (P-pant) moiety from coenzyme A to the serine residue, thereby converting the inactive ACP domain into its functional, active form (Xue et al. 2018).

9.3.4 Application of Intracellular Lipids

Oleaginous yeasts and fungi offer distinct advantages for microbial lipid production. Among these, *Aureobasidium* spp. stand out due to a combination of beneficial traits:

(a) Short duplication time, supporting rapid growth.
(b) High cell density (>30.0 g dry cell weight/L) and substantial intracellular lipid accumulation during a large scale of fermentation (>50% of dry cell weight).
(c) Genomic amenability to editing via synthetic biology tools.
(d) Broad substrate utilization, including cellulose, starch, inulin, lactose, maltose, glucose, fructose, glycerol, xylan, xylose, and pectin, as well as the raw materials containing them (Wang et al. 2022a, b).
(e) Scalability for large-scale fermentation (Xue et al. 2018).
(f) Inherent viral resistance (Xue et al. 2018).

The average unsaturation (AU) of the fatty acid methyl esters (FAMEs) in prepared biodiesel can be calculated using the following Eq. 9.1. The values of the fatty acid methyl esters prepared from the lipids produced by the recombinant strain XI-155 obtained above and its parent counterpart P10 are 0.88 and 0.83, respectively.

$$AU = \sum N \times Ci \tag{9.1}$$

Following the chemical conversion of intracellular lipids produced by the recombinant strain XI-155 and its parent counterpart P10 to biodiesel (Fig. 9.15), the

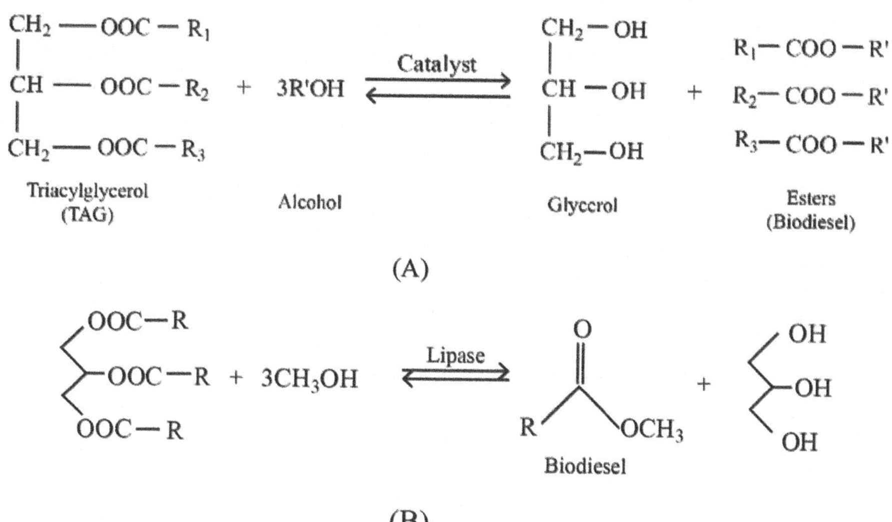

Fig. 9.15 Biodiesel production from intracellular lipids under the catalysis of the chemical catalyst (**A**) and lipases (**B**)

resulting biodiesel properties were calculated using the standardized Eqs. (9.2–9.7) used by the academic cycle as shown below:

$$\text{Viscosity} = -0.6316\text{AU} + 5.2065 \tag{9.2}$$

$$\text{Specific gravity} = 0.0055\text{AU} + 0.8726 \tag{9.3}$$

$$\text{Cloud point} = -13.356\text{AU} + 19.994 \tag{9.4}$$

$$\text{Cetane number} = -6.6684\text{AU} + 62.876 \tag{9.5}$$

$$\text{Iodine number} = 74.373\text{AU} + 12.71 \tag{9.6}$$

$$HHV = 1.7601\text{AU} + 38.534 \tag{9.7}$$

The fuel properties of the biodiesel derived from the intracellular lipids produced by *A. melanogenum* were characterized. For the parent P10 strain: the average unsaturation (AU) = 0.83, viscosity = 4.7 mm^2/s, Gravity = 0.9, cloud point = 8.7 °C, cetane number = 57.3, iodine value = 74.7, and higher heating value (HHV) = 40 MJ/kg. For the engineered XI-155 strain: AU = 0.9, viscosity = 4.6 mm^2/s, Gravity = 0.9, cloud point = 7.9 °C, cetane number = 56.8, iodine value = 56.8, and HHV = 80.2 MJ/kg (Song et al. 2022).

All the prepared biodiesels meet all the property requirements stipulated by both the US standard ASTM D6751 and the EU standard EN 14214 (Viscosity = 3.5–5.0, Gravity = 0.86–0.9; Cenetane = 51 min, Iodine value = 120 max). Notably, the cetane numbers (57.3 and 56.8 min) significantly exceed the standard 51 min, and the iodine values (74.7.3 and 80.2) are markedly lower than the standard 120 max, the thresholds defined by EN12214 (Song et al. 2022). These results indicate that the fuel properties of the prepared biodiesels from the engineered XI-155 strain and its parent strain P10 are superior to the standard specifications, demonstrating their high potential for industrial application of the lipids produced by *A. melanogenum* in the biofuel industry. Consequently, the biodiesel derived from intracellular lipids produced by *A. melanogenum* represents a promising, sustainable biofuel that can contribute to the development of the sustainable biofuel industry and environmental protection, trying to use less unsustainable petroleum oils and coals from fossil resources.

Significantly, more than 73.7% of the lipids extracted from *A. melanogenum* P10 can be converted into biodiesel within just 20 min. Furthermore, the resulting biodiesel demonstrates excellent combustibility (Wang et al. 2014a, b; Fig. 9.16).

In addition to biodiesel production, the medium-chain fatty acids octanoic acid (C_8 fatty acid) and decanoic acid (C_{10} fatty acid) released from intracellular lipids serve as key aromatic components in certain fermented beverages. Furthermore, they can be transformed into value-added products such as fatty alcohols—the key feedstocks for surfactant production—and conjugated linoleic acid. Conjugated linoleic acid exhibits beneficial properties, including anti-carcinogenic,

Fig. 9.16 Burning of the prepared biodiesel from intracellular lipids of the yeast strain P10

anti-atherogenic, anti-diabetic, anti-inflammatory, and anti-obesity effects. Long-chain polyunsaturated fatty acids, another derivative, also possess numerous physiological functions and confer various health benefits (Xue et al. 2018). Recently, many researchers have focused on the directed conversion pathway of biolipids into bioalkanes, similar to gasoline used in the existing fuel industry. It was found that fatty acid photodecarboxylases from *Chlorella variabilis* at blue light, acyl-ACP reductase, and aldehyde-deformylating oxygenase from cyanobacteria can catalyze the transformation of fatty acids from biolipids into alkanes (Sorigué et al. 2017; Hayashi and Arai 2022). If the existing gasoline and diesel from the petroleum industries can be replaced by biodiesel and bioalkanes in the future, this will play a significant role in global economy, security and politics in this world.

9.4 Conclusions and Perspectives

Although the synthetic pathway and regulation of liamocin biosynthesis have been largely elucidated through genome editing and synthetic biology, several key questions remain: (a) How is one 3,5-dihydroxydecanoic acid unit linked to another via an ester bond in the liamocin tail group? (b) What physiological functions does the produced liamocin provide to the fungal producers? (c) What are the specific functions of each domain of the highly reducing polyketide synthase (HR-PKS) in 3,5-dihydroxydecanoic acid biosynthesis at the molecular level?

Furthermore, although massoia lactone exhibits high antifungal activity against various plant fungal pathogens, its ecological impacts require careful evaluation before it can be considered a safe replacement for toxic chemical fungicides. Additionally, a practical and efficient method—whether chemical or biochemical—for converting intracellular free fatty acids into alkane biofuels has yet to be developed to date.

References

Bogorad IW, Lin TS, Liao JC (2013) Synthetic nonoxidative glycolysis enables complete carbon conservation. Nature 502:693–697

Chi ZM, Zhang T, Cao TS, Liu XY, Cui W, Zhao CH (2011) Biotechnological potential of inulin for bioprocesses. Bioresour Technol 102:4295–4303

Hansali K, Zhang ZR, Liu GL, Chi Z, Chi ZM (2023) The pathogenic yeast *Metschnikowia bicuspidata* var. *bicuspidata* in the aquacultured ecosystem and its biocontrol. J Fung 9:1024

Hayashi Y, Arai M (2022) Recent advances in the improvement of cyanobacterial enzymes for bioalkane production. Microb Cell Factories 21:256

Kang XX, Jia SL, Wei X, Zhang M, Liu GL, Hu Z, Chi Z, Chi ZM (2022) Liamocins biosynthesis, its regulation in *Aureobasidium* spp., and their bioactivities. Crit Rev Biotechnol 42(1):93–105

Leathers TD, Price NPJ, Manitchotpisit P (2010) Production of anti-streptococcal liamocins from agricultural biomass by *Aureobasidium pullulans*. World J Microb Biot 32:199–205

Li M, Liu GL, Chi Z, Chi ZM (2010) Single cell oil production from hydrolysate of cassava starch by marine-derived yeast *Rhodotorula mucilaginosa* TJY15a. Biom Bioeng 4:101–107

Manitchotpisit P, Price NP, Leathers TD, Punnapayak H (2011) Heavy oils produced by *Aureobasidium pullulans*. Biotechnol Lett 33:1151–1157

Manitchotpisit P, Watanapoksin R, Price NPJ, Bischoff KM, Tayeh M, Teeraworawit S, Kriwong S, Leathers TD (2014) *Aureobasidium pullulans* as a source of liamocins (heavy oils) with anticancer activity. World J Microbiol Biotechnol 30:2199–2204

Maria Saur K, Brumhard O, Scholz K, Hayen H, Tiso T (2019) A pH shift induces high-titer liamocin production in *Aureobasidium pullulans*. Appl Microbiol Biotechnol 103(12):4741–4752

Price NPJ, Bischoff KM, Leathers TD, Cosse AA, Manitchotpisit P (2017) Polyols, not sugars, determine the structural diversity of anti-streptococcal liamocins produced by *Aureobasidium pullulans* strain NRRL 50380. J Antibiot 70:136–141

Rattanahadsadin N, Lotrakul P, Bankeeree W, Hunsa Punnapayak H, Prasongsuk S (2024) Production and characterization of extracellular liamocins produced from fungal strains of *Aureobasidium* spp. World J Microbiol Biotechnol 40:253

Saika A, Fukuoka T, Mikome S, Kondo Y, Habe H, Morita T (2020) Screening and isolation of the liamocin-producing yeast *Aureobasidium melanogenum* using xylose as the sole carbon source. J Biosci Bioeng 129(4):428–434

Sitepu IR, Garay LA, Sestric R, Levin D, Block DE, Bruce German J, Boundy-Mills KL (2014) Oleaginous yeasts for biodiesel: current and future trends in biology and production. Biotechnol Adv 32:1336–1360

Song SS, Tian BC, Chen H, Chi Z, Liu GL, Chi ZM (2022) Transformation of corncob-derived xylose into intracellular lipid by engineered strain of *Aureobasidium melanogenum* P10 for biodiesel production. Renew Energ 200:1211–1222

Song SS, Wang P, Chi Z, Liu GL, Chi ZM (2026) The production of mannitol-liamocin by the engineered *Aureobasidium melanogenum* V33 and its anti-*Streptococcus mutans* activity. Biochem Eng J 226:110012

Sorigué D, Légeret B, Cuiné S et al (2017) An algal photoenzyme converts fatty acids to hydrocarbons. Science 357:903–907

Tang RR, Chi Z, Jiang H, Liu GL, Xue SJ, Hu Z, Chi ZM (2018) Overexpression of a pyruvate carboxylase gene enhances extracellular liamocin and intracellular lipid biosynthesis by *Aureobasidium melanogenum* M39. Process Biochem 69:64–74

Tiso T, Welsing G, Lipphardt A, Sauer DF, Chi ZM, Blank LM, Hayen H (2024) Proposal for a systematic naming convention for liamocins. J Surfactant Deterg 27(3):1–3

Van Rossum HM, Kozak BU, Pronk JT, Van Maris AJA (2016) Engineering cytosolic acetyl-coenzyme a supply in *Saccharomyces cerevisiae*: pathway stoichiometry, free-energy conservation and redox-cofactor balancing. Metab Eng 36:99–115

Wang ZP, Fu WJ, Xu HM, Chi ZM (2014a) Direct conversion of inulin into cell lipid by an inulinase-producing yeast *Rhodosporidium toruloides* 2F5. Bioresour Technol 161:131–136

Wang CL, Li Y, Xin FH, Liu YY, Chi ZM (2014b) Evaluation of single cell oil from *Aureobasidium pullulans* var. *melanogenum* P10 isolated from mangrove ecosystems for biodiesel production. Process Biochem 49:725–731

Wang ZP, Zhang XY, Ma Y, Ye JR, Jiang J, Wang HY, Chen W (2021) Whole conversion of agro-industrial wastes rich in galactose-based carbohydrates into lipid using oleaginous yeast *Aureobasidium namibiae*. Biotechnol Biofuels 14:181

Wang P, Jia SL, Liu GL, Chi Z, Chi ZM (2022a) *Aureobasidium* spp. and their applications in biotechnology. Process Biochem 116:72–83

Wang Z, Zhang M, Chi Z, Liu GL, Chi ZM (2022b) Liamocin overproduction by the mutants of *Aureobasidium melanogenum* 9-1 for effectively killing spores of the pathogenic fungi from diseased human skin by Massoia lactone. World J Microb Biot 38:107

Wang P, Zhang M, Zhao SF, Zhang ZR, Liu GL, Chi Z, Chi ZM (2024) Liamocins overproduction via the two-pH stage fermentation and anti-*aspergillus flavus* activity of Massoia lactone. Biotechnol J 19:2300675

Wang P, Tiso T, Chi Z, Chi ZM (2025) Efficient transformation of inulin into liamocins by a mutant of *Aureobasidium melanogenum* HN6-1 and antimicrobial activity of the produced Massoia lactone. Int J Biol Macromole 318:145245

Xue SJ, Chi Z, Zhang Y, Li YF, Liu GL, Jiang H, Hu Z, Chi ZM (2018) Fatty acids from oleaginous yeasts and yeast-like fungi and their potential applications. Crit Rev Biotechnol 38(7):1049–1060

Xue SJ, Liu GL, Chi Z, Gao ZC, Hu Z, Chi ZM (2020) Genetic evidences for the core biosynthesis pathway, regulation, transport and secretion of liamocins in yeast-like fungal cells. Biochem J 477:887–903

Yang et al. (2020). Guang Yang G, Liu GL, Wang SJ, Chi ZM, Chi Z (2020) Pullulan biosynthesis in yeast-like fungal cells is regulated by the transcriptional activator Msn2 and cAMP-PKA signaling pathway. Intern J Biol Macromole 157: 591–603

Zhang M, Gao ZC, Chi Z, Wang Z, Liu GL, Li XF,·Hu Z, Chi ZM (2021a) cAMP-PKA and HOG1 signaling pathways regulate liamocin production by different ways via the transcriptional activator Msn2 in *Aureobasidium melanogenum*. Enzy Microb Technol 143:109705

Zhang HQ, Chi Z, Liu GL, Zhang M, Hu Z, Chi ZM (2021b) *Metschnikowia bicuspidate* associated with a milky disease in *Eriocheir sinensis* and its effective treatment by Massoia lactone. Microbiol Res 242:126641

Zhang M, Gao ZC, Chi Z, Wang Z, Liu GL, Li XF, Hu Z, Chi ZM (2021c) Massoia lactone displays strong antifungal property against many crop pathogens and its potential application. Microb Ecol 84:376–390

Zhang M, Wang Z, Chi Z, Liu GL, Chi ZM (2022) Metabolic engineering of *Aureobasidium melanogenum* 9–1 for overproduction of liamocins by enhancing supply of acetyl-CoA and ATP. Microbiol Res 265:127172

Zhang M, Wei X, Wang P, Chi Z, Liu GL, Chi ZM (2023) Liamocin biosynthesis is induced by an autogenous host acid activation in *Aureobasidium melanogenum*. Biotechnol J 23:e2200440

Chapter 10
Production, Biosynthesis, Regulation, and Application of Melanin by *Aureobasidium* spp.

Abstract Melanin synthesized by *A. melanogenum* and accumulated in their cell walls has many physiological functions for the producers and several biological activities. It has been well confirmed that *A. melanogenum* only can produce DHN-melanin polymerized from 1,8-DHN under catalysis of laccase, and 1,8-DHN is synthesized by the non-reducing (NR) Pks1 encoded by the *NR-PKS1* gene and the NR-Pks1 contains one ketosynthase (KS) domain, one acyltransferase (AT) domain, two acyl carrier protein (ACP) domains, one thioesterase (TE) domain, and one cyclase (CYC) domain. It has been demonstrated that the NR-Pks1 is activated via phosphopantetheinylation catalyzed by PPTase. The DHN-melanin biosynthesis is regulated by the melanin-specific transcriptional activator Cmr1, which is in turn regulated by another transcriptional activator Sw4 from cell wall integrity signal pathway, nitrogen catabolite repression, and the dual-function transcriptional factor NsdD. In fact, so far, the detailed chemical structure of DHN-melanin has not been clearly elucidated. Furthermore, DHN-melanin and even some intermediates of 1,8-DHN biosynthesis have many applications.

Keywords DHN-melanin · Biosynthesis · Regulation · Applications

10.1 Introduction

Melanin is a negatively charged, hydrophobic, heterogeneous, and insoluble biopolymer with high molecular weight, typically appearing as brown or black pigment. Its chemical formula and the chemical structure can be represented as $[C_{18}(OR)_3H_7O_4N_2]_n$ (Fig. 10.1). These polymers form graphite-like planar sheets that aggregate in a layered arrangement, resulting in colloidal particles. However, due to melanin's frequent association with proteins, polysaccharides, lipids, and other macromolecules in cell walls, it forms complex compounds that are difficult to isolate, purify, and characterize. As a result, the structural details of melanin molecules and the precise chemical configurations of various melanin types from *Aureobasidium* spp. still remain poorly understood.

Z. Chi et al., *Biology and Synthetic Biotechnology of Aureobasidium spp.*,
https://doi.org/10.1007/978-981-92-2607-8_10

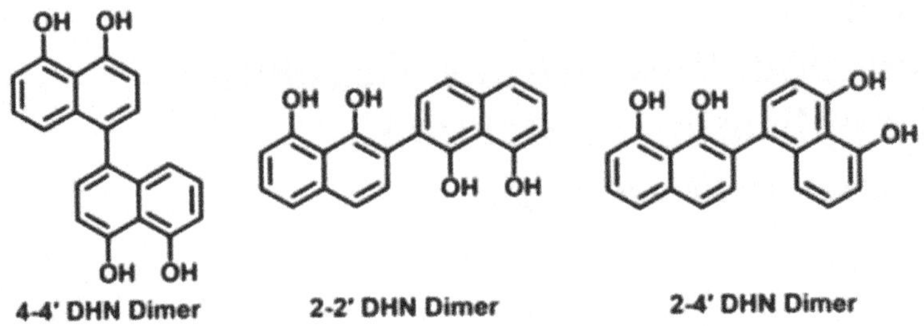

Fig. 10.1 Possible chemical structures of DHN-melanin from 1,8-dihydroxynaphthalene (1,8-DHN)

Melanin also can be chemically synthesized. However, the chemically synthesized melanin is expensive and easily causes environmental pollution. Recently, microbial-based bio-production has gained great attention due to its environmental-friendliness, high manipulability, and widespread availability. Especially, black yeast-based melanin production has many advantages, such as its remarkable metabolic versatility as mentioned above, its easy genomic editing and genetic modification, melanin accumulation in the cell wall, cost-effective culture substrates, well-established fermentation techniques, and easy scale-up production.

Fungal melanin can be classified into five types: (a) eumelanin; (b) 1,8-dihydroxynaphthalene (DHN) melanin; (c) pyomelanin; (d) pheomelanin, and (e) 4-glutaminylhydroxybenzene (GHB) melanin (Langfelder et al. 2003). Studies have confirmed that *Aureobasidium* spp. exclusively produce DHN melanin, which is deposited in the yeast-like fungal cell wall (Fig. 10.2) (Jiang et al. 2017).

It has been known that melanin can be easily dissolved in 0.5 M NaOH solution. Therefore, after lysing fungal cells, all the melanin will be released and dissolved in 0.5 M NaOH. However, melanin is insoluble in 5.0 M HCl. When the melanin solution is mixed with 5.0 M HCl under constant stirring, melanin will precipitate. The precipitate can then be collected via centrifugation.

In addition, Fourier transform infrared (FTIR) spectroscopy can be used to analyze purified melanin. The purified melanin also can be analyzed using a spectrophotometer within the wavelength range of 200–800 nm.

Melanin possesses diverse physicochemical and biological characteristics that enable numerous applications in healthcare, medicine, environmental protection, energy, food industry, and cosmetic industry. Studies have confirmed its antioxidant, anti-radiation, and free radical scavenging properties. Additionally, melanin-producing fungi exhibit enhanced fitness and survival in harsh environments, including extreme temperatures, oxidative stress, drought, toxic heavy metals, and ionizing radiation. Given melanin's critical physiological functions in organisms and its wide industrial utility, it is essential to explore synthetic biotechnological techniques for its overproduction and to elucidate its biosynthesis pathway and regulatory mechanisms.

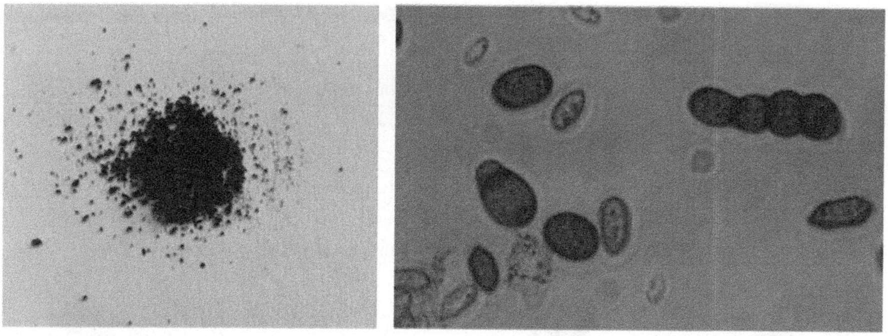

A B

Fig. 10.2 Melanin (**A**) produced by a mutant of *A. aubasidani* (**B**)

10.2 Overproduction of Melanin by Metabolic Engineering

Aureobasidium spp. are commonly referred to as black yeasts due to their ability to synthesize and accumulate abundant melanin within their cell walls (Figs. 10.2 and 10.3). This characteristic indicates that melanin is widely distributed across various *Aureobasidium* strains, regardless of their growth conditions or developmental stages.

The Taklimakan Desert, the largest desert in China and the world's second-largest shifting sand desert, is characterized by extreme aridity, temperature fluctuations, limited organic carbon availability, and intense ultraviolet and infrared radiation. Yeast strains capable of surviving in such harsh conditions likely possess unique stress adaptation mechanisms, including cell wall melanization, carotenoid production, and the synthesis of trehalose and exopolysaccharides under stressed conditions (Jiang et al., 2018, 2019). However, little is known about the yeasts inhabiting the Taklimakan Desert, particularly their stress response mechanisms and metabolites under these extreme conditions.

The yeast strain XJ5-1, isolated from Taklimakan Desert soil, was identified as *A. melanogenum* and exhibits high melanin production in PDA medium. However, its melanin biosynthesis is significantly repressed in the presence of $(NH_4)_2SO_4$. In contrast, *A. melanogenum* P5 strain, isolated from a mangrove ecosystem, does not produce melanin under either $(NH_4)_2SO_4$-supplemented or unsupplemented conditions (Fig. 10.4). Additionally, the cell size of the XJ5-1 strain is substantially larger than that of the P5 strain (Fig. 10.5), potentially reflecting adaptive traits for desert stress tolerance.

Neither the Δ*pks1* mutant (with deleted *NR-PKS1* gene) nor the *Δnpg1* mutant (DP07 strain with deleted *NPG1* gene) can produce melanin. In contrast, their parent strain XJ5-1 and the BP17 strain (Δ*npg1* complemented with the *NPG1* gene) produce significant melanin levels (OD_{400nm} = 10.5 and 14.4, respectively) (Jiang et al. 2017).

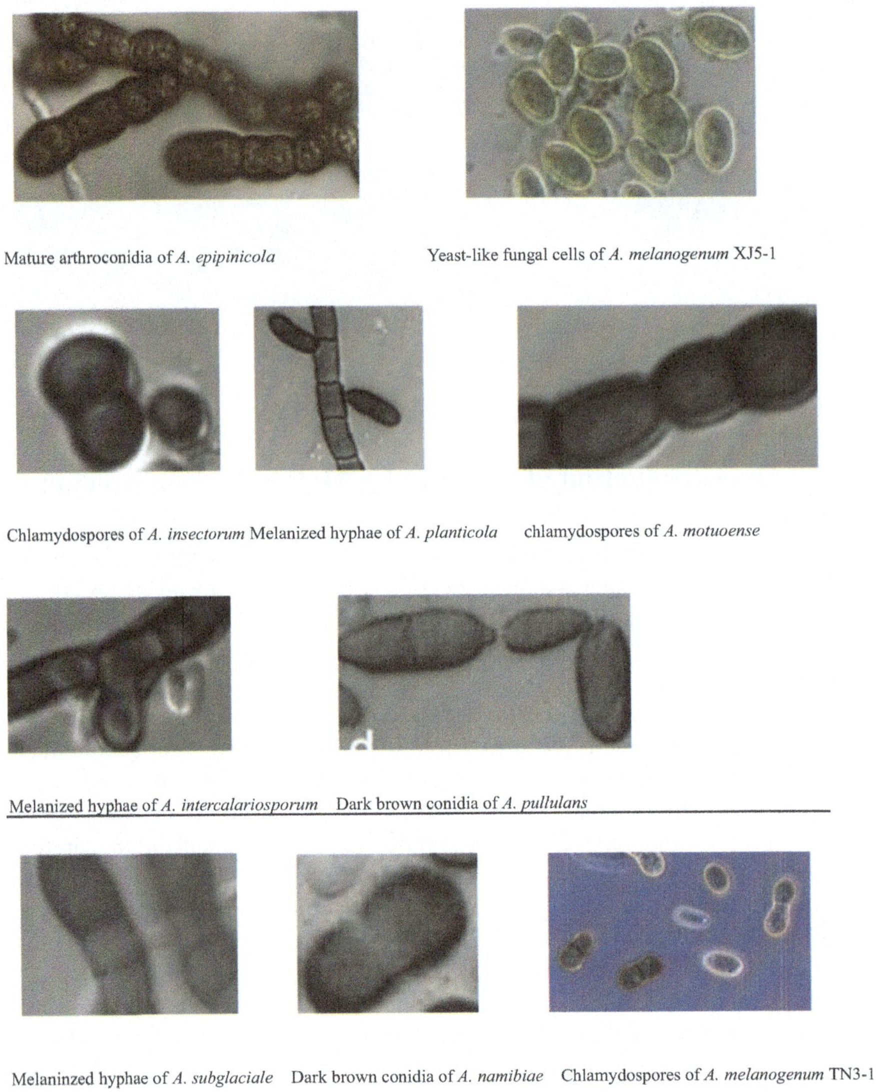

Fig. 10.3 Melanized cells of different strains of *Aureobasidium* spp.

Interestingly, deleting the *NSDD* gene in four *Aureobasidium* strains enhances melanin production in their respective disruptants compared to their parent strains. Specifically, this effect is observed in: (1) PMA-producing *A. melanogenum* ATCC62921, (2) pullulan-producing *A. melanogenum* 13-2, (3) melanin-producing *A. melanogenum* XJ5-1, (4) fumarate-producing *A. aubasidani* GC1 (Chi et al. 2024; Hansali et al. 2026). For example, the Δ*nsdNK1* strain—in which the *NSDD* gene was completely deleted—produces the highest melanin yield ($OD_{400nm} = 10.99$) after 72 h of cultivation, whereas its wild-type strain XJ5–1 and the complemented

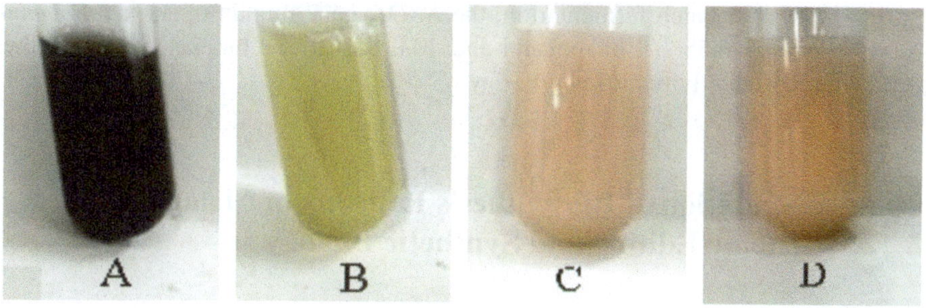

Fig. 10.4 Melanin production (**A**) by *A. melanogenum* XJ5-1 grown in the PDA medium, no melanin production (**B**) by *A. melanogenum* XJ5-1 grown in the PDA medium with 2.0% (w/v) of $(NH_4)_2SO_4$, no melanin production (**C**) by *A. melanogenum* P5 grown in the PDA medium, and no melanin production (**D**) by *A. melanogenum* P5 grown in the PDA medium with 2.0% (w/v) of $(NH_4)_2SO_4$

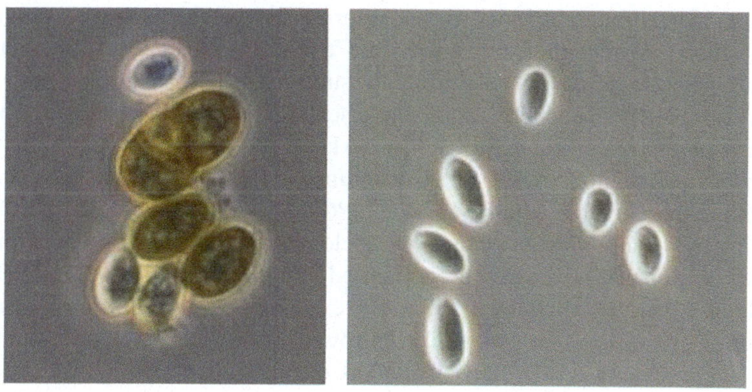

A B

Fig. 10.5 Cell morphologies of *A. melanogenum* XJ5-1 (**A**) and *A. melanogenum* P5 (**B**) grown in the PDA medium

strain CMNSDNE1 (with restored *NSDD* gene) show lower melanin production (OD_{400nm} = 8.26 and 4.78, respectively) under the identical conditions (Chi et al. 2024). Additionally, the transcriptional levels of the melanin biosynthesis-related genes—including the *PKS1* gene (polyketide synthase), the *CMR1* gene (the melanin-specific transcriptional activator), the *T4HNR1* gene (1,3,6,8-tetrahydroxy naphthalene reductase), the *SCD1* gene (scytalone dehydratase), and the *T3HNR1* gene (1,3,8-trihydroxynaphthalene reductase)—are significantly upregulated in the disruptants (Chi et al. 2024; Hansali et al. 2026). These results indicate that the transcription factor NsdD suppresses expression of the melanin-related genes and melanin biosynthesis. In contrast, the transcription factor NsdD promotes expression of the pullulan, polymalate, and fumarate-related genes, and pullulan, polymalate, and fumarate biosynthesis (Chi et al. 2024; Hansali et al. 2026). Furthermore, during a 10-Liter batch fermentation, the *NSDD* gene mutant of *A. melanogenum* XJ-15 produces 6.55 g/L of melanin and 30.0 g/L of cell biomass within 7 days, achieving a productivity of 0.94g/L. day.

10.3 Elucidation of Biosynthesis Pathway of Melanin by Genome Editing and Synthetic Biology

The DHN-melanin (Fig. 10.1) is synthesized through the oxidation of 1,8-dihydroxynaphthalene (1,8-DHN) subunits, catalyzed by laccase. Unlike DOPA-melanin in animals, DHN melanin precursors (acetyl-CoA and malonyl-CoA derived from acetate) are produced intracellularly, leading to their characteristic association with the cell wall. Granular or fibrous DHN-melanin in the cell wall primarily functions as a barrier to modulate the transport of cellular materials.

Fungal DHN-melanin biosynthesis is primarily catalyzed by polyketide synthases (Pks1), while phosphopantetheinyl transferases (PPTases) mediate post-translational modifications of proteins such as PKS, nonribosomal peptide synthetases (NRPS), and fatty acid synthases (FAS) as mentioned in the last two chapters. In *A. melanogenum* XJ5-1, the two key genes: the *PKS1* gene (KT429644), a 6495-bp open reading frame (ORF) encoding a 2165-amino acid Pks1 responsible for melanin precursor synthesis, and the *NPG1* gene (KX094969), a 1076-bp ORF encoding a 340-amino acid PPTase, were cloned and characterized. The promoter analysis reveals that both genes are repressed by nitrogen sources (e.g., $(NH_4)_2SO_4$) and glucose (Fig. 10.4) (Jiang et al. 2017), explaining why the absence of melanin production by *A. melanogenum* XJ5-1 grown in the $(NH_4)_2SO_4$-containing PDA media occurs (Jiang et al., 2016). The Pks1 encoded by the *PKS1* gene comprises one ketosynthase (KS) domain, one acyltransferase (AT) domain, two acyl carrier protein (ACP) domains, one thioesterase (TE) domain, and one cyclase (CYC) domain (Fig. 10.6A). These features confirm its classification as a non-reducing Pks (NR-Pks1) involved in melanin precursor biosynthesis. Notably, this means that no reduction reaction occurs during the initial synthetic step of melanin biosynthesis. Under the catalytic activity of its KS domain and AT domain, each iterative extension cycle incorporates a C_2 unit via decarboxylative thio-Claisen condensation, using five or six malonyl units with the growing chain (initiated by acetyl-CoA). The intermediates remain tethered as thioesters to the 4′-phosphopantetheinyl group on the two ACP domains. The C-terminus of NR-Pks1 contains a TE domain that releases the final product from the ACP domain, along with a Claisen cyclase (CYC domain) that catalyzes C–C bond formation (Vagstad et al. 2012) (Fig. 10.6A and Fig. 10.7).

So, the key intermediate 1,3,6,8-tetrahydroxynaphthalene is synthesized by the NR-Pks1 (Figs 10.6A and 10.7) (Jiang et al. 2017). The PPTase belongs to the Sfp-type family, indicating that the ACP domains of the NR-Pks1 in the yeast strain XJ5-1 require activation via phosphopantetheinylation catalyzed by PPTase (Jiang

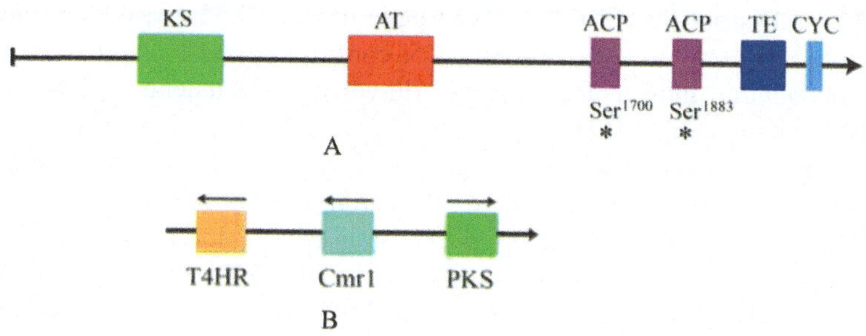

Fig. 10.6 The domain architecture of the NR-Pks1 (**A**), cloned from strain XJ5-1, and its gene cluster (**B**) with *T_4HNR* gene and *CMR1* gene. Ser[1700] and Ser[1883] in the two ACP domains undergo phosphopantetheinylation catalyzed by PPTase

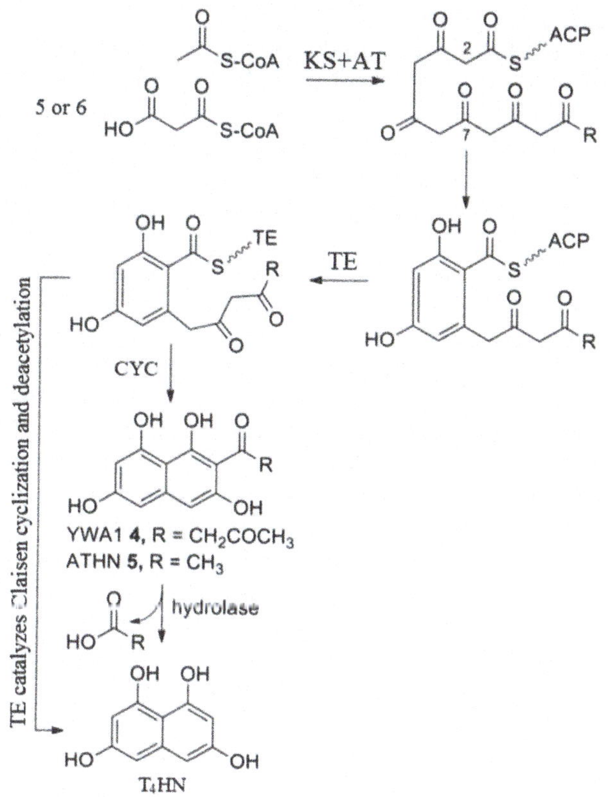

Fig. 10.7 Function of each domain of the NR-Pks1 in biosynthesis of 1,3,6,8-tetrahydroxynaphthalene (T_4HN) (Vagstad et al. 2012)

et al. 2017). Disruption of the *NR-PKS1* gene in strain XJ5-1 results in the loss of melanin biosynthesis in the disruptant K5, confirming that melanin production in *A. melanogenum* indeed follows the dihydroxynaphthalene (DHN) pathway (Fig. 10.7) (Jiang et al. 2017).

The genome of the yeast-like fungal strain XJ5–1 carries several other genes involved in melanin biosynthesis and regulation: (a) The *CMR1* gene (accession: KEQ67497.1), encoding a melanin-specific transcriptional activator Cmr1; (b) The *T4HNR1* gene (accession: KU844847.1), encoding 1,3,6,8-tetrahydroxynaphthalene reductase; (c) The *SCD1* gene (accession: KU844848.1), encoding scytalone dehydratase; (d) The *T3HNR1* gene (accession: KU844849.1), encoding 1,3,8-trihydroxynaphthalene reductase; (e) A vermelone dehydratase gene (ORF size: 513 bp, encoding 170 amino acids), which catalyzes the conversion of vermelone to 1,8-DHN, a key step in melanin biosynthesis and regulation; (f) The *LAC1* gene (ORF size: 1716 bp, encoding 571 amino acids), encoding laccase for 1,8-DHN polymerization (Fig. 10.8).

Furthermore, the *NR-PKS1*, *T4HNR1*, and *CMR1* genes are located within the same gene cluster (Fig. 10.6B; Jiang et al. 2020). Disruption of the *NPG1* gene abolishes melanin biosynthesis in the DP107 disruptant and reduces the *NR-PKS1* gene expression. Upon complementation of the *NPG1* gene in the disruptant DP107, melanin production is restored in the BP17 strain, and the *NR-PKS1* gene and the *NPG1* gene expression significantly increase. These results indicate that the ACP domain of the NR-Pks1 is activated and regulated by PPTase, and that the *NR-PKS1* gene and the *NPG1* gene expression are coordinately regulated (Jiang et al. 2017).

Therefore, the initial biosynthetic step of melanin involves the condensation of acetyl-CoA and malonyl-CoA, derived from acetate and glucose, catalyzed by the non-reducing polyketide synthase (NR-Pks1) to form polyketide (Fig. 10.7). Subsequently, the polyketide is hydrolyzed to yield the intermediate 1,3,6,8-tetrahydroxynaphthalene (T_4HN). T_4HN is then reduced by T4HN reductase to produce scytalone, which undergoes dehydration to form 1,3,8-trihydroxynaphthalene (1,3,8-THN). Through further reduction and dehydration steps, 1,3,8-THN is converted to vermelone and then to 1,8-dihydroxy naphthalene (1,8-DHN). Finally, laccase-mediated polymerization of 1,8-DHN forms DHN-melanin (Figs. 10.1 and 10.8) (Jiang et al. 2020).

Several studies have demonstrated that phenolic compounds (e.g., 1,8-dihydroxynaphthalene, 1,8-DHN) can be oxidized by laccase to form phenoxyl radicals. These radicals undergo either coupling-based polymerization—converting low-molecular-weight phenolics like 1,8-DHN into polymeric pigments such as DHN-melanin—or radical rearrangement (Janusz et al. 2020). The laccase involved in 1,8-DHN polymerization contains four copper atoms bound to conserved histidine residues and catalyzes one-step oxidation of dihydroxy phenols to quinones.

It has been reported that the DHN melanin synthesis pathway (Figs. 10.7 and 10.8) is highly conserved in ascomycetes, including *Aureobasidium* spp. For instance, in *Aspergillus fumigatus* (a phylogenetically close relative of *Aureobasidium* spp.), the gene cluster responsible for DHN melanin synthesis comprises six genes: (a) *PKSP/ALB1* gene (polyketide synthase); (b) *AYG1* (function unknown), (c)

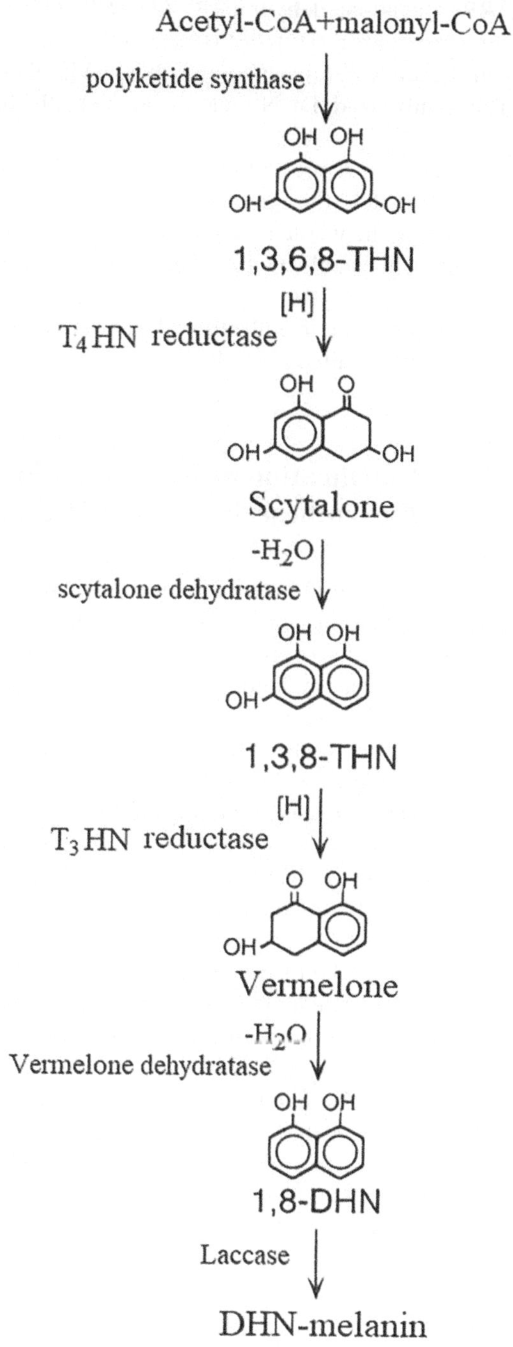

Fig. 10.8 DHN-melanin biosynthesis in *Aureobasidium* spp. The reduction steps are indicated by [H], and the oxidation steps are indicated by [O]. Dehydration steps are labeled -H_2O

ARP1 (scytalone dehydratase); (d) *ARP2* (1,3,8-THN reductase); (e) *ABR1* (multicopper oxidase), (f) *ABR2* (laccase) (Tsai et al. 1999). Among these, the first four genes encode cytoplasmic enzymes, while the last two encode secretory proteins. The synthesized DHN-melanin is typically localized to its cell wall (Qin and Xia 2024).

It has been reported that each intermediate and final product melanin shown in Figs. 10.7 and 10.8 have unique characteristics and special functions. Therefore, elucidating the whole melanin biosynthetic pathway and its regulatory mechanisms in fungi will establish a foundation for metabolic engineering and synthetic biology approaches. This also will enhance production efficiency and enable the creation of customized melanin with tailored material properties. All the work in *Aureobasidium* spp. is being waited to be done.

10.4 Clarification of Regulation Mechanisms of Melanin Biosynthesis by Genome Editing and Synthetic Biology

The Taklimakan Desert, mentioned earlier, is often referred to as the "Dead Sea" due to its extreme environmental conditions. In our previous study (Jiang et al. 2016), it was discovered that *A. melanogenum* XJ-1, isolated from the Taklimakan Desert, produces melanin, which plays a crucial role in its resistance and adaptation to such harsh conditions. Melanin biosynthesis is catalyzed by a polyketide synthase encoded by the *NR-PKS1* gene. However, the molecular regulatory mechanisms of melanin biosynthesis remain poorly understood. Additionally, the relationship between signaling pathways and melanin production in the desert yeasts has not been fully elucidated. Given that melanin is a component of the cell wall, the signaling pathways involved in sensing and responding to environmental stresses—including the filamentation MAPK pathway, the stress-responsive MAPK pathway (HOG1), and the cell wall integrity (CWI) pathway—warrant further investigation. Therefore, to determine whether the main components of the MAPK pathways regulate melanin biosynthesis in *A. melanogenum* XJ5-1, the key genes in the three MAPK signaling pathways were cloned, characterized, and disrupted. These include: (a) The ***HOG1*** gene in the HOG signaling pathway; (b) The ***KSS1*** gene in the filamentation MAPK signaling pathway; (c) The ***SWI4***, ***SLT2***, and ***RLM1*** genes in the cell wall integrity (CWI)-MAPK signaling pathway. Results show that only the colony color of the both disruptants *Δslt2* and *Δswi4* (lacking the *SLT2* gene or *SWI4* gene, respectively) is reduced compared to that of the wild-type strain XJ5-1. No color changes are observed in any other mutants (e.g., Δ*hog1*, Δ*kss1*, Δ*rlm1*). Complementing the *SLT2* and *SWI4* genes in Δ*slt2* and Δ*swi4*, respectively, restore the colony color in the transformants CMΔSLT2 and CMΔSWI4.

Furthermore, the transcript levels of all the genes involved in melanin biosynthesis and the CWI signaling pathway are lower in the disruptants (*Δslt2* and *Δswi4*)

Fig. 10.9 The binding sites of the promoters associated with key melanin biosynthesis and regulatory genes in *A. melanogenum* XJ5-1 were analyzed for interactions with different transcription factors. Only the promoter of the *CMR1* gene (a key gene) can be bound to Swi4, a transcriptional activator from the CWI signaling pathway. Similarly, the promoter of the *NR-PKS1* gene (another key gene) can be specifically bound by the melanin-specific transcriptional activator Cmr1

than those of all the genes involved in melanin biosynthesis and the CWI signaling pathway in the wild-type strain XJ5-1 and the complementation transformants (CMΔ*SLT2* and CMΔ*SWI4*). These results confirm that melanin biosynthesis in the desert yeast is exclusively regulated by the CWI-MAPK signaling pathway (Jiang et al. 2020). This is reasonable because the synthesized melanin is localized in the cell wall.

To investigate how the CWI-MAPK signaling pathway regulates melanin biosynthesis in the desert yeast *A. melanogenum* XJ5-1, the *CMR1* gene was cloned and characterized, which encodes the melanin-specific transcription factor Cmr1. The analysis reveals that only the *CMR1* promoter contains the binding site for Swi4, a transcription factor from the CWI-MAPK signaling pathway (Fig. 10.9), indicating that the *CMR1* gene transcription is regulated by this pathway via Swi4. Deletion of the *CMR1* gene abolishes melanin biosynthesis in the disruptant (*Δcmr1*) and negatively affects expression of the *NR-PKS1* gene. However, when the *CMR1* gene is reintroduced into the *Δcmr1* mutant, the complemented strain (CMΔCMR1) restores melanin production to its wild-type levels and the *NR-PKS1* gene expression is recovered. These results demonstrate that Cmr1 directly activates and regulates expression of the *NR-PKS1* gene, and that their expression is coordinately controlled. Together, the findings confirm that the transcription factor Cmr1 plays a direct role in melanin biosynthesis in *A. melanogenum*.

So, the DNA-binding domain of Cmr1 (Cmr1-DBD) was expressed in *E. coli* and the recombinant protein was purified. Electrophoretic mobility shift assay (EMSA) results demonstrate that the purified Cmr1-DBD specifically binds to the promoter of the *NR-PKS1* gene, with the binding sequence CTCCTTCTCTCC (Fig. 10.9). This finding again strongly supports that the transcription factor Cmr1 regulates the *NR-PKS1* gene expression, consistent with the quantitative real-time RT-PCR data mentioned above. Additionally, it has been well established that Slt2, an MAP kinase in the cell wall integrity (CWI) signaling pathway, regulates the transcriptional activator Swi4, which in turn controls the *CMR1* gene expression. The encoded Cmr1 protein binds specifically to the TTCTCTCCA motif in the *NR-PKS1* promoter, enhancing its transcription along with any other melanin biosynthesis-related genes (Fig. 10.10). Consequently, *A. melanogenum* XJ5-1

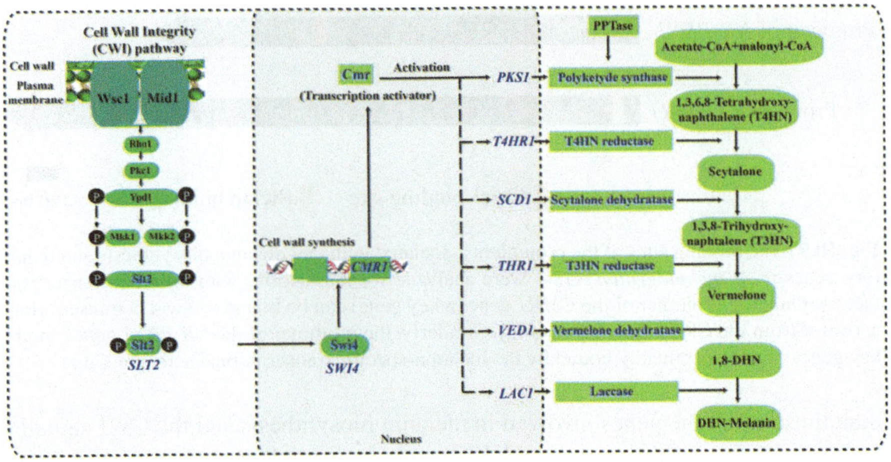

Fig. 10.10 Regulation of melanin biosynthesis by Cmr and Slt2 and Swi4 in the cell wall integrity pathway

produces large amounts of melanin when cultured in the PDA medium lacking $(NH_4)_2SO_4$ (Fig. 10.4). These regulatory mechanisms provide important insights into stress adaptation strategies in the desert yeasts and establish a foundation for further research on their ability to sense, respond to, and thrive in the extreme environments.

As shown above, *A. melanogenum* XJ5-1 grown on the PDA medium synthesizes a large amount of melanin (Fig. 10.4). After the *NSDD* gene, which encodes a GATA-type global transcription factor containing the conserved $C-X_2-C-X_{18}-C--X_2-C$ motif, is completely deleted from the genome of *A. melanogenum* XJ5-1, melanin production by the resulting Δ*nsdD* mutant increases by 133% compared to that by the parental strain XJ5-1. Additionally, the transcriptional levels of certain melanin biosynthesis-related genes are also up-regulated to varying degrees. In contrast, melanin production by the complementation strain NE1 (in which the *NSDD* gene is reintroduced into the Δ*nsdD* mutant) decreases by 57.8% compared to that by the wild-type strain XJ5-1. Moreover, the expression levels of some melanin biosynthesis-related genes in the complementation strain NE1 are significantly down-regulated compared to those in strain XJ5-1 and the Δ*nsdD* mutant. Deletion of the same *NSDD* gene in strain GC1 of *A. aubasidani, A. melanigenum* ATCC62921, and *A. melanogenum* 13-2 also make the corresponding mutants enhance melanin production (Chi et al. 2024; Hansali et al. 2026).

The binding sites of NsdD are located in the promoter sequences (5′-HGATAR-3′) of melanin biosynthesis and regulatory genes (*CMR1*, *NR-PKS1*, *T4HR1*, *SCD1*, and *THR1*) (Figs. 10.8 and 10.10). These results indicate that the GATA-type transcription factor NsdD negatively regulates melanin biosynthesis in *A. melanogenum* XJ5-1 (Chi et al. 2024). In contrast, as demonstrated earlier, the same GATA-type transcription factor NsdD acts as a positive regulator for pullulan, polymalate, and fumarate biosynthesis in other *Aureobasidium* spp. strains. It has been confirmed

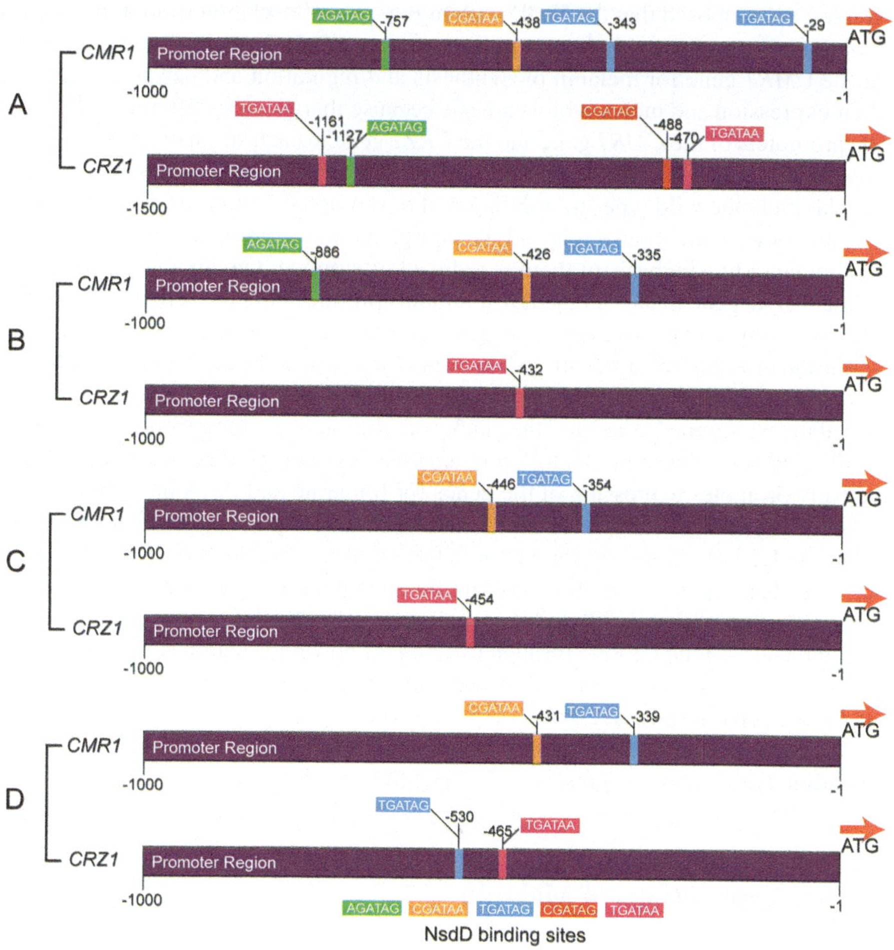

Fig. 10.11 The binding sites (the conserved 5′-HGATAR-3) of the promoters of *CMR1* genes and the *CRZ1* genes in different strains of *Aureobasidium* spp. by NsdD. (**A**) GC1 strain; (**B**) XJ5-1 strain; (**C**) ATCC62921 strain; (**D**) P16 strain mentioned in the text (Hansali et al. 2026)

that expression of the *GFP* gene encoding green fluorescence protein under control of the promoter of the *CMR1* gene in the Δ*nsdD* mutant is greatly enhanced compared to that of the *GFP* gene in its wild type strain GC1 while expression of the *GFP* gene under control of the promoter of the *CRZ1* gene in the Δ*nsdD* mutant is greatly reduced compared to that of the *GFP* gene in its wild type strain GC1. This again strongly demonstrates that the *CMR1* gene specific for melanin biosynthesis is negatively control by the transcriptional activator NsdD while the *CRZ1* gene in the Ca^{2+} signaling pathway is positively controlled by the transcriptional activator NsdD. However, the underlying mechanisms for these dual roles of the transcriptional factor NsdD remain completely unknown (Chi et al. 2024). In our previous studies, we proposed that the NsdD with highly phosphorylation from *A. melanogenum* XJ5-1 may be tightly bound to the conserved 5′-HGATAR-3′ in the promoter of the *CMR1* gene for melanin biosynthesis and regulation and negatively regulate their expression and melanin biosynthesis because there are big differences between the promoters of the *CMR1* gene and the *CRZ1* gene in each strain of *Aureobasidium* spp. (Fig. 10.11). So, the expression of all the genes for melanin biosynthesis and regulation in the wild type strain XJ5-1 and the complementing strain CMNSDNE1 and any other wild type strains and the complementing strains was greatly reduced and melanin biosynthesis in them is reduced or stopped. On contrast, abolishment of the *NSDD* gene makes all the Δ*nsd* mutants produce only the active Cmr1 monomer to stimulate expression of all the genes for melanin biosynthesis and regulation, resulting in enhanced melanin production. Meanwhile, in all the Δ*nsd* mutants, expression of the *CRZ1* gene is reduced and biosynthesis of fumarate, PMA, and pullulan is stopped due to the lack of the active transcriptional activator NsdD. However, in their wild-type strains and the complementing strains, NsdD actively promotes expression of the genes for fumarate and PMA biosynthesis, and regulation in the presence of $CaCO_3$, in turn, enhances production of fumarate and PMA. Indeed, the transcription factor WRKY8 of the fungus *Sclerotinia sclerotiorum* also has a dual role in the transcriptional regulation of the *PRR* genes: repressing expression of the nlp20/nlp24 receptor gene *RLP23* while promoting that of the chitin receptor gene *CERK1* through differential phosphorylation (Ren et al. 2024). In fact, such dual role of transcriptional factor is common in human, plant, fungal, and yeast cells (Ren et al. 2024).

In addition, Fig. 10.4 shows that melanin production is repressed by the nitrogen signaling pathway in the presence of $(NH_4)_2SO_4$ (Rødkær and Færgeman 2014).

10.5 Applications of Melanin

Melanin possesses diverse functions, including anti-radiation, antioxidant, and photoprotective effects. Additionally, it acts as a biosorbent for organic substances, metal ions, and radioactive materials while exhibiting antibacterial and anti-aging properties. Fungal melanin further demonstrates hepatoprotective effects, antitumor activity, and beneficial biological activities regulating blood lipid and blood sugar levels in human. Due to these versatile properties, melanin is a highly valuable pigment with applications in packaging materials, anti-radiation products, food processing, environmental remediation, and biomedical industries (Qin and Xia 2024; Liu et al. 2022).

DHN-melanin is also highly attractive for applications in biomaterials, radiation shielding, and environmental remediation. Because DHN-melanin possesses excellent photoprotection, antiradiation, and antimicrobial activities, it can be applied in the textile industry, spacecraft equipment, and cosmetic industry, thereby protecting human skin from intense UV radiation. Because of efficient electron conduction and the superior capacity for charge storage, it can be used as a bio-semiconductor material and organic electrode, respectively. Because melanin contains carboxyl, phenolic, hydroxyl, and amine functional groups, it can be used to immobilize heavy metals or remove heavy metals in polluted environments. For example, the purified melanin from the *NSDD* gene mutant of *A. melanogenum* XJ5-1 has potent bioactivities, such as strong Fe^{2+} chelation, ABTS· +radical scavenging, DPPH radical scavenging, hydroxyl radical (·OH) scavenging, photoprotective effects and antibacterial activity against *Staphylococcus aureus*, *Salmonella enterica*, *Listeria monocytogenes* and *Bacillus subtilis*.

Melanin in the yeast-like fungal cells has various physiological functions. For instance, the melanized cells of the yeast strain XJ5-1, grown in PDA medium, exhibit greater tolerance to the following stressors compared to its non-melanized cells: (a) UV radiation; (b) Oxidation (200.0 mM H_2O_2); (c) Heat treatment (40 °C); (d) Salt shock (200.0 g/L NaCl); (e) Desiccation; f. Strong acid hydrolysis (6.0 M HCl at 80 °C). The non-melanized controls include: (a) *A. melanogenum* XJ5-1 cells grown in PDA medium with 2.0% (w/v) $(NH_4)_2SO_4$; (b) *A. melanogenum* P5 cells (which cannot produce melanin).

Additionally, the albino mutants derived from the XJ5-1 strain via gene deletion show increased sensitivity to stresses compared to the melanized wild-type XJ5-1 cells grown in the same PDA medium. These findings suggest that melanin enhances the adaptability of the desert-isolated yeast cells to extreme environments, explaining why the wild-type XJ5-1 strain can thrive in the harsh desert conditions (Jiang et al. 2016, 2017).

10.6 Conclusions and Perspectives

Although the key precursor 1,8-DHN of melanin is synthesized by the NR-Pks1 and 1,8-DHN can be linked into high-molecular-weight melanin under catalysis of laccase, it is still completely unknown what the function of each domain of the NR-Pks1 is, how 1,8-DHN is linked into high-molecular-weight melanin, and what the properties and functions of each intermediate produced during 1,8-DHN biosynthesis are. It has been known that the synthesized melanin is accumulated in the cell walls of *Aureobasidium* spp. So, it is crucial to know how to make the accumulated melanin be secreted into the medium to produce a large amount of melanin for biotechnological applications.

References

Chi Z, Wei X, Ge N, Jiang H, Liu GL, Chi ZM (2024) NsdD, a GATA-type transcription factor is involved in regulation and biosynthesis of macromolecules melanin, pullulan, and polymalate in *Aureobasidium melanogenum*. Int J Biol Macromole 268:131820

Hansali K, Wang P, Wei X, Liu GL, Chi Z, Chi ZM (2026) Dual role of the transcriptional factor NsdD in regulation of fumarate and melanin biosynthesis in *Aureobasidium aubasidani* GC1 strain. Fung Biol 130:101973

Janusz G, Pawlik A, Swiderska-Burek U, Polak J, Sulej J, Jarosz-Wilkołazka A, Paszczynski A (2020) Laccase properties, physiological functions, and evolution. Int J Mol Sci 21:966

Jiang H, Liu NN, Liu GL, Chi Z, Wang JM, Zhang LL, Chi ZM (2016) Melanin production by a yeast strain XJ5-1 of *Aureobasidium melanogenum* isolated from the Taklimakan desert and its role in the yeast survival in stress environments. Extremoph 20:567–577

Jiang H, Liu GL, Chi Z, Wang JM, Zhang LL, Chi ZM (2017) Both a PKS and a PPTase are involved in melanin biosynthesis and regulation of *Aureobasidium melanogenum* XJ5-1 isolated from the Taklimakan desert. Gene 602:8–15

Jiang H, Liu GL, Chi Z, Hu Z, Chi ZM (2018) Genetics of trehalose biosynthesis in desert-derived *Aureobasidium melanogenum* and role of trehalose in the adaptation of the yeast to extreme environments. Curr Genet 64:479–491

Jiang H, Chen TJ, Chi Z, Hu Z, Liu GL, Sun Y, Zhang SH, Chi ZM (2019) Macromolecular pullulan produced by *Aureobasidium melanogenum* 13-2 isolated from the Taklimakan desert and its crucial roles in resistance to the stress treatments. Intern J Biol Macromole 135:429–436

Jiang H, Chi Z, Liu GL, Hu Z, Zhao SZ, Chi ZM (2020) Melanin biosynthesis in the desert-derived *Aureobasidium melanogenum* XJ5-1 is controlled mainly by the CWI signal pathway via a transcriptional activator Cmr1. Curr Genet 66:173–185

Langfelder K, Streibel M, Jahn B, Haase G, Brakhage AA (2003) Biosynthesis of fungal melanins and their importance for human pathogenic fungi. Fung Genet Biol 38(2):143–158

Liu R, Meng X, Mo C, Wei X, Ma A (2022) Melanin of fungi: from classification to application. World J Microbiol Biotechnol 38:228

Qin Y, Xia Y (2024) Melanin in fungi: advances in structure, biosynthesis, regulation, and metabolic engineering. Microb Cell Factories 23:334

Ren CX, Chen SY, He YH, Xu YP, Yang J, Cai XZ (2024) Fine-tuning of the dual-role transcription f actor WRKY8 via differential phosphorylation for robust broad-spectrum plant immunity. Plant Commun 5(12):101072

Rødkær SV, Færgeman NJ (2014) Glucose-and nitrogen sensing and regulatory mechanisms in *Saccharomyces cerevisiae*. FEMS Yeast Res 14(5):683–696

Tsai H, Wheeler MH, Chang YC, Kwon-Chung KJ (1999) A developmentally regulated gene cluster involved in conidial pigment biosynthesis in *Aspergillus fumigatus*. J Bacteriol 18(20):6469–6477

Vagstad AL, Hill EA, Labonte JW, Townsend CA (2012) Characterization of a fungal thioesterase having claisen cyclase and deacetylase activities in melanin biosynthesis. Chem Biol 19:1525–1534

Chapter 11
Production, Biosynthesis, Regulation, and Application of L-Ornithine and Its Derivatives by *Aureobasidium* spp.

Abstract L-ornithine is an intermediate in OUC and the end product in AGC. L-ornithine biosynthesis can be enhanced by deletion or low expression of the *OTC* gene in OUC and promotion of AGC via overexpression of the *ArgB/C* gene. L-ornithine can be transformed into N^5-hydroxy-L-ornithine under catalysis of L-ornithine-N5-oxygenase encoded by the gene *SidA*, and N^5-hydroxy-L-ornithine is reacted with anhydromevalonyl-CoA under catalysis of anhydromevalonyl-CoA transferase encoded by the gene *SidF* to form anhydromevalonyl-N^5-hydroxy-L-ornithine, which is subsequently polymerized into siderophores under catalysis of NRPS or cyclized into L-piperazic acid by heterologous piperazate synthase. N^5-hydroxy-L-ornithine also can be decarboxylated to yield putrescine by heterologous L-ornithine decarboxylase. Siderophore biosynthesis is easily repressed by Fe^{2+}, and the derepressed strain can grow in the Fe^{2+} rich medium and synthesize a large amount of siderophores. L-ornithine, siderophore, L-piperazic acid, and putrescine have many applications. Especially, the FsC-powered dual-mode biosensor can be used to achieve ultrasensitive and accurate detection of *H. pylori* in the fecal samples.

Keywords L-ornithine · Siderophores · L-Piperazic acid · Putrescine · Biosynthesis

11.1 Introduction

L-ornithine (L-Orn), a valuable non-protein amino acid, serves as an intermediate in L-arginine (L-Arg) biosynthesis via the ornithine urea cycle (OUC) and as a product of acetyl-glutamate cycle (AGC) (Fig. 8.10 in Chap. 8 and Fig. 11.3). It exhibits diverse physiological and biochemical functions. Under hydrolase catalysis, L-Orn can be converted into N^5-hydroxy-L-ornithine, which is subsequently polymerized into siderophores or cyclized into L-piperazic acid by piperazate synthase. Alternatively, decarboxylation of N^5-hydroxy-L-ornithine by L-ornithine decarboxylase yields putrescine. Thus, L-Orn acts as a key precursor for siderophores,

Z. Chi et al., *Biology and Synthetic Biotechnology of Aureobasidium spp.*,
https://doi.org/10.1007/978-981-92-2607-8_11

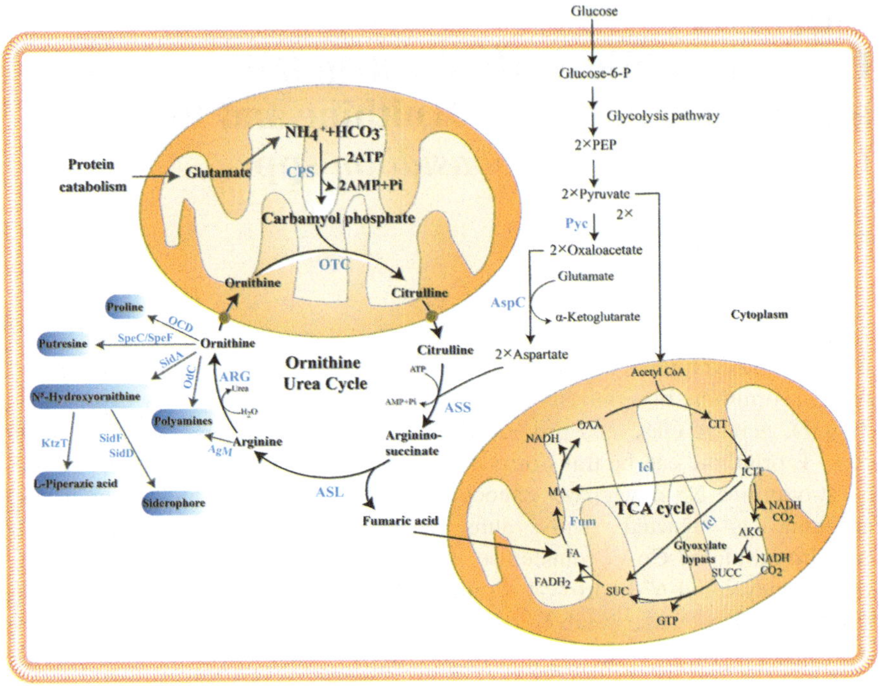

Fig. 11.1 The useful products from the ornithine urea cycle (OUC)

putrescine, and L-piperazic acid biosynthesis. These compounds can be synthesized by both native and engineered strains of *Aureobasidium* spp., with applications spanning healthcare, pharmaceuticals, and chemical industries (Myo Myo Tint et al. 2024).

11.2 Production, Biosynthesis, Regulation, and Application of L-Ornithine by Metabolic Engineering

L-ornithine (L-Orn) is a 2,5-diaminopentanoic acid (Fig. 11.2) with diverse physiological functions in humans. As illustrated in Figs. 8.2 and 8.3 in Chap. 8 and Fig. 11.1, L-Orn is primarily synthesized via two pathways: (a) As an intermediate in the ornithine-urea cycle (OUC); (b) As a product derived from *N*-acetyl-ornithine through the action of *N*-acetyl-ornithine acetyltransferase (ArgE) in the acetyl-glutamate cycle (AGC) of *Aureobasidium* spp. (Fig. 11.3).

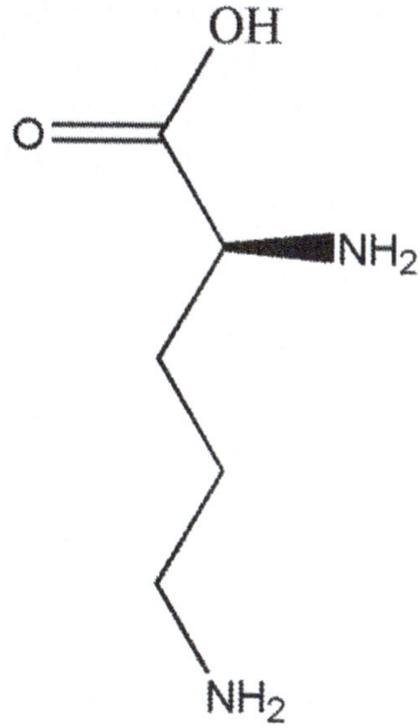

Fig. 11.2 Chemical structure of L-ornithine

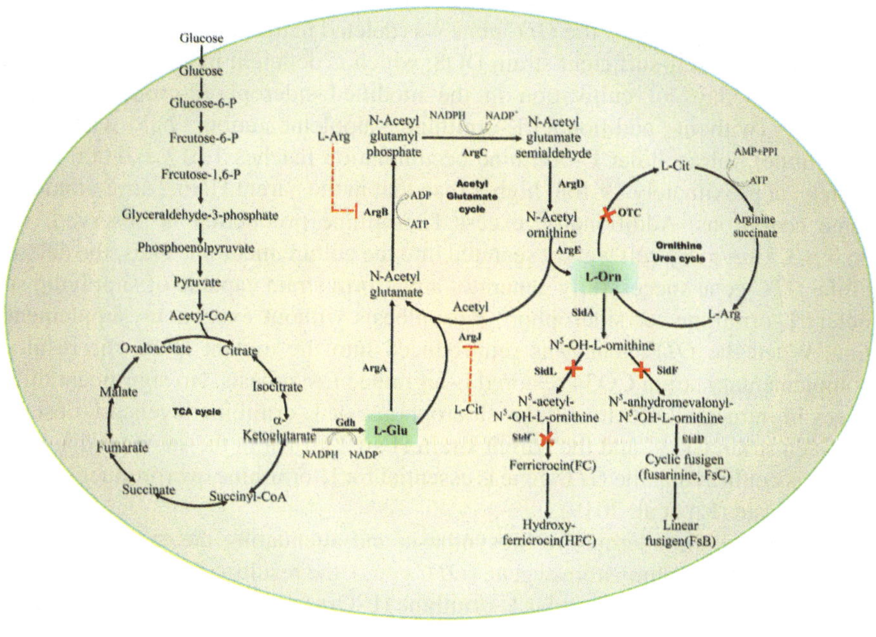

Fig. 11.3 Ornithine metabolic pathways in *A. melanogenum* HN6.2. ×: deletion; T: repression

11.2.1 Overproduction of L-Ornithine by Metabolic Engineering

L-Ornithine (L-Orn) can be biosynthesized by engineered strains of *A. fumigatus*, *A. melanogenum*, *S. cerevisiae*, sake yeast, and *Corynebacterium glutamicum*. Among these, the engineered *C. glutamicum* strain KBJ11 (SNK118ΔargRΔargFΔncgl2228/pXMJ19-CsgapCBsrocG) exhibits the highest L-Orn production (88.3 g/L) (Dong et al. 2020).

Previous studies demonstrated that high-level production of siderophore fusigen (FsC) by *A. melanogenum* HN6.2 isolated from the marine environments requires supplemental L-ornithine (L-Orn) in the medium. However, L-Orn is an uncommon and expensive amino acid, making its supplementation impractical for large-scale fermentation. Thus, enhancing the HN6.2 strain's ability to synthesize sufficient endogenous L-Orn to support FsC production is desirable. Ornithine carbamoyltransferase (Otc), encoded by the *OTC* gene, catalyzes the conversion of L-ornithine to L-citrulline (Fig. 11.3). Consequently, deleting this gene can lead to L-ornithine accumulation in the OUC pathway. Studies have shown that the *OTC* gene deletion or knock downregulation of this gene using specific technique can reduce L-ornithine consumption by blocking its further conversion.

Based on these results, the *OTC* gene was deleted in the HN6.2 strain, generating the L-ornithine self-sufficient strain DO8, which is deficient in arginine biosynthesis. After 5 days of cultivation in the modified siderophore-producing (MSP) medium (without additional L-ornithine supplementation, but with added L-arginine), intracellular L-ornithine accumulation reaches 1.617 ± 0.006 mg/mg protein, approximately 19-fold higher than that in the parent HN6.2 strain under the same conditions. Additionally, excess L-ornithine production is observed, with 36.4 ± 3.5 mg/g dry cell weight secreted into the culture medium. Thus, the deletion of the *OTC* gene successfully generates a platform strain capable of supplying sufficient L-ornithine for siderophore biosynthesis without exogenous supplementation. When the *OTC* gene was reintroduced into the mutant DO8, the resulting complementary strain CO34 restored L-arginine biosynthesis. No significant differences in intracellular/extracellular siderophores or L-ornithine levels are observed between strain CO34 and the parent strain HN6.2 grown in the same media. These findings confirm that the *OTC* gene is essential for L-ornithine overproduction in the HN6.2 strain (Lu et al. 2019).

After blocking siderophore biosynthesis and attenuating the expression of the ornithine carbamoyltransferase gene (*OTC* gene), the resulting D-LCFAO-cre strain produces 2.1 mg of intracellular L-ornithine (L-Orn) per mg of protein (Fig. 11.3). Overexpressing the *SPE1-S* gene (encoding L-ornithine decarboxylase from *S. cerevisiae*) in the D-LCFAO-cre mutant enables the transformant E-SPE1-S to synthesize 3.6 mg of L-Orn per mg of protein. Further overexpression of the *ArgB/C* gene (encoding a bifunctional acetylglutamate kinase/N-acetyl-γ-glutamyl-phosphate reductase from the AGC pathway) in the transformant E-SPE1-S increases L-Orn accumulation to 4.2 mg/mg protein in the new transformant E-SPE1-S-ArgB/C (Kong et al. 2021; Fig. 11.3).

11.2.2 Elucidation of Biosynthesis Pathway and Regulation Mechanisms of L-Ornithine by Genome Editing and Synthetic Biology

From Fig. 11.3, it can be seen that L-Orn is an intermediate in the OUC pathway. Its production can be enhanced by blocking siderophore biosynthesis and attenuating the expression of the ornithine carbamoyltransferase gene (*OTC* gene) as discussed above. L-Orn in the OUC pathway can be supplied from AGC pathway (Fig. 11.3).

11.2.3 Applications of L-Ornithine

As mentioned above, L-ornithine serves as a key precursor in microbial metabolic pathways and an intermediate in the ornithine-urea cycle (OUC). Additionally, it acts as a crucial building block for synthesizing bioactive compounds such as the anesthetic analgesics scopolamine and atropine.

L-ornithine (L-Orn) possesses anti-fatigue properties by enhancing energy metabolism and facilitating the elimination of toxic ammonia in humans and animals. This will reduce plasma ammonia levels after intense exercise, alleviating fatigue-related symptoms, improving sleep quality, and stimulating growth hormone secretion, thereby accelerating recovery from various health conditions. Additionally, L-Orn serves as a dietary supplement in food and beverages, aiding in liver disease treatment and wound healing. Due to these benefits, it has applications in the food, chemical, pharmaceutical, and medical industries. For instance, consuming L-Orn-rich foods and drinks may reduce fatigue. Studies demonstrate its efficacy in preventing and managing liver diseases, promoting wound repair, and supporting cardiovascular health.

11.3 Production, Biosynthesis, Regulation, and Application of Siderophore

Siderophores are low-molecular-weight, natural iron-chelating compounds synthesized by microorganisms such as fungi and yeasts under iron-limited conditions. Notably, fungi and yeasts exclusively produce hydroxamate siderophores (Fig. 11.4). Due to their high affinity for ferric iron and other metal ions, siderophores have been widely applied in diverse fields. Recently, growing interest has focused on

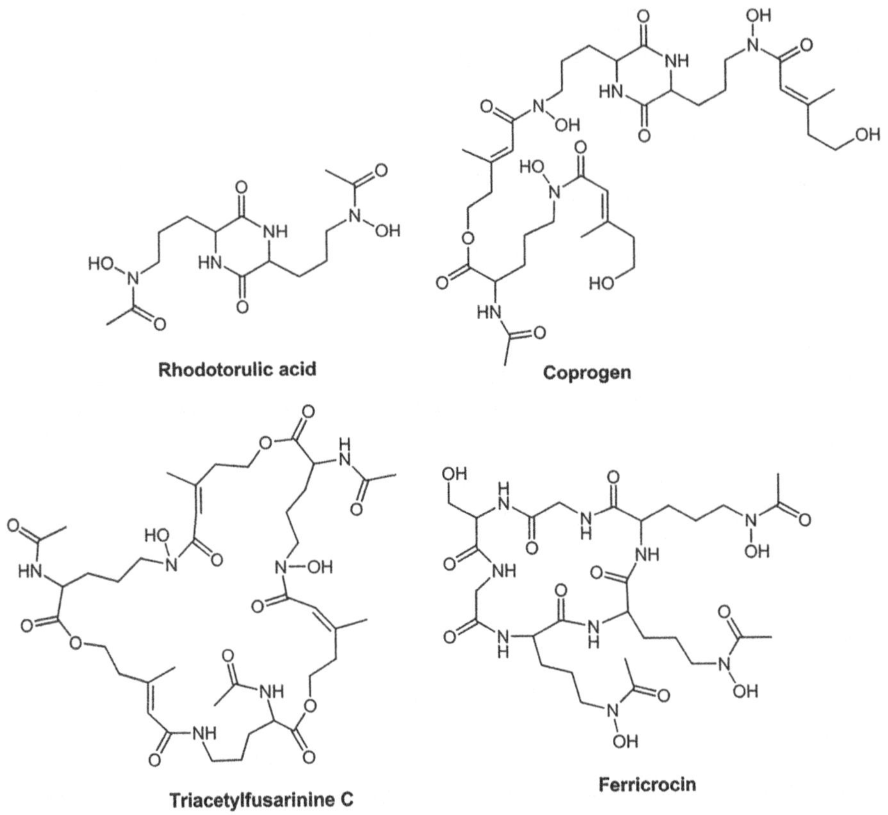

Fig. 11.4 Four representative siderophores of the corresponding hydroxymate families produced by fungi

their medical applications in diagnosing and treating pathogenic infections in humans.

Fusarinine C (FsC), a natural hydroxamate siderophore produced by *Aureobasidium* spp., contains three L-ornithine residues (Fig. 11.5) (Wang et al. 2009b), enabling straightforward chemical modification to design multifunctional derivatives. However, the limited availability of FsC has hindered extensive research on this compound. Although *A. fumigatus* has been employed for FsC production via fermentation, its ubiquitous presence as a saprophytic fungus poses a significant risk, as it is a common airborne pathogen responsible for human infections.

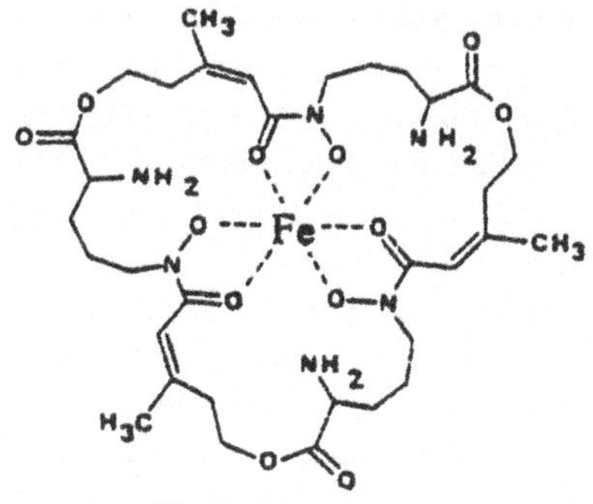

Fig. 11.5 Chemical structure of siderophore fusigen (FsC) produced by HN6.2 strain

11.3.1 Overproduction of Siderophore by Metabolic Engineering

To date, *A. fumigatus* cultivated under the unoptimized conditions yields only about 5 mg of FsC per gram of dry cell weight, a level far from satisfactory. Moreover, the optimized conditions for FsC production by this fungus have yet to be described. Additionally, as previously noted, *A. fumigatus* mycelia negatively impact large-scale fermentation. In conclusion, this fungal strain is unsuitable for industrial FsC production.

Over 300 yeast strains isolated from various marine environments were screened for siderophore production. Among these, 14 strains demonstrate siderophore-producing capability. Notably, strain HN6.2, identified as a member of *A. melanogenum*, exhibits particularly high siderophore yields (Wang et al. 2009a). The optimal medium composition for the maximal siderophore production by HN6.2 strain was determined to be: sucrose (3.0% w/v); NH_4NO_3 (0.3% w/v); K_2HPO_4 (0.3% w/v); citric acid (0.1% w/v); $MgSO_4$ (0.008% w/v); $ZnSO_4$ (0.0002% w/v); L-ornithine (10 mM). The optimal growth conditions include: shaking speed: 200 rpm; initial pH: 6.2; temperature: 28 °C. Under these conditions, the maximum siderophore titer reaches 1.1 mg/mL (0.44 mM) after 120 h of cultivation (Wang et al. 2009a, b).

Addition of L-ornithine to the production medium enhances siderophore production by the HN6.2 strain, while excess Fe^{3+} supplementation reduces its production. Furthermore, Fe^{3+} significantly represses the expression of the *SidA* gene (encoding N^5-L-ornithine hydroxylase), the key gene for hydroxamate siderophore biosynthesis, whereas the added L-ornithine stimulates its expression.

Strain HN6.2 produces a crude siderophore with clear antibacterial activity against two marine animal pathogens, *Vibrio anguillarum* and *Vibrio parahaemolyticus*, as well as the common bacterium *B. subtilis*. However, the crude siderophore shows no antimicrobial activity against *E. coli*, *S. aureus*, the pathogenic yeast *Metschnikowia bicuspidata* WCY in aquatic animals, or the fungi *A. niger* and *Penicillium* sp. Notably, even after heating, the siderophore retains its antibacterial activity against *B. subtilis*, *V. anguillarum*, and *V. parahaemolyticus*. These results suggest that the crude siderophore is highly stable and may be effective in inhibiting marine bacterial pathogens (Wang et al. 2009a).

The crude siderophore was extracted from the culture supernatant using organic solvents, then purified by reverse-phase high-performance liquid chromatography (RP-HPLC). The major peak was collected and analyzed by mass spectrometry. Based on the molecular ion peak and comparison with the known siderophore structures, the siderophore produced by strain HN6.2 was identified as fusigen (FsC), a hydroxamate-type siderophore (Fig. 11.5) (Wang et al. 2009b).

Previous studies have demonstrated that *A. melanogenum* HN6.2 can produce FsC, and its yeast-like morphology facilitates large-scale production of bioproducts. However, FsC production using this strain still faces several challenges: (a) High cost of production**:** High-yield FsC synthesis requires supplementation of expensive L-ornithine; (b) Low specificity: The HN6.2 strain produces multiple siderophores alongside FsC; (c) Insufficient titer: The current FsC yield is too low for practical acquisition. To address these issues, the metabolic pathways of the HN6.2 strain were rewired using the previously established Cre/lox system, guided by the genome sequencing, annotation, and mining (Chap. 5).

After addressing the challenge of L-ornithine self-supply, the pure siderophore FsC was tried to be produced. To achieve this, other components of the extracellular siderophores (FC and HFC) produced by the DO8 strain need to be removed. By deleting the L-ornithine-N5-transacetylase gene (*SidL*) in the DO8 strain, a double-deficient strain, DOL2 (lacking both *OTC* and *SidL* genes), was generalized. The total extracellular siderophore titer of the DOL2 strain increases by nearly 100% compared to that produced by the DO8 strain, reaching a concentration of 1.830 ± 0.033 mM.

However, FC and HFC components still persist in the culture. To completely eliminate these components from the extracellular siderophores, the FC synthase gene (*SidC*) in the DOL2 strain was further deleted, generating the triple-deficient strain DOLC19. HPLC analysis confirms the successful removal of both FC and HFC by the triple-deficient strain DOLC19, and the weight proportion of FsC is estimated to be approximately 97%. Notably, the total extracellular and intracellular siderophore levels produced by the DOLC19 mutant do not differ significantly ($p > 0.05$) from those produced by the DOL2 mutant. To further enhance the biosynthesis of target FsC, the native L-ornithine-N5-oxygenase gene (*SidA*) and anhydromevalonyl-CoA transferase gene (*SidF*) were overexpressed in the DOLC19 mutant. First, the *SidA* gene overexpression in the DOLC19 mutant yields the EA26 strain, which exhibits the highest extracellular FsC titer (1504.1 ± 39.0 mg/L) among all the transformants. Subsequently, the *SidF* gene was overexpressed in the EA26 strain, generating the EAF13 strain (Fig. 11.6). The resulting FsC titer reaches 1704.2 ± 46.8 mg/L (306.6 mg per gram dry cell weight), representing a 61-fold increase compared to that produced by *A. fumigatus* under the unoptimized conditions (Lu et al. 2019b).

In summary, the metabolic rewiring strategy developed here enables self-sufficient L-ornithine production. Subsequently, the component simplification

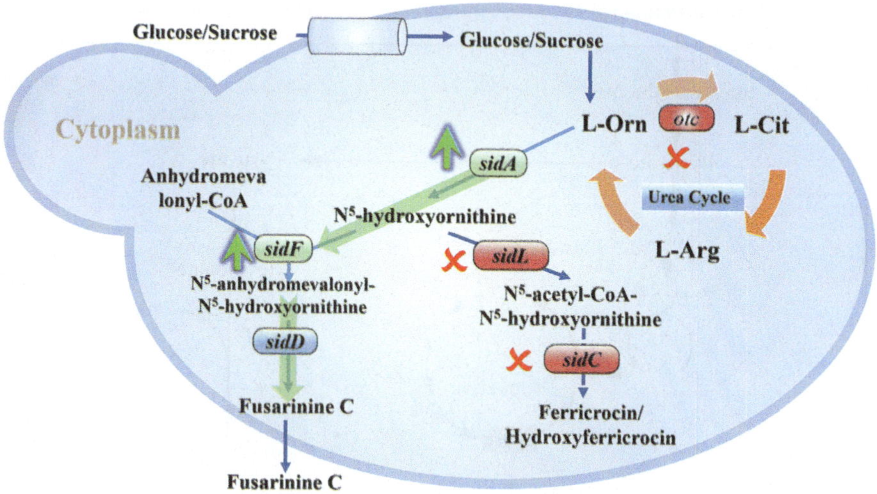

Fig. 11.6 Schematic overview of metabolic rewiring for FsC biosynthesis in the HN6.2 strain. X means deletion

significantly enhances extracellular siderophore FsC titer (1704.2 ± 46.8 mg/L) in the genetically modified strain EAF13 of *A. melanogenum* HN6.2. To date, this represents the highest reported FsC titer produced by fungi and yeasts, with FsC accounting for approximately 97% of the total output (Lu et al. 2019b).

The L-ornithine, mevalonate synthesis, and siderophore synthesis pathways (Fig. 11.7) were further metabolically enhanced in the derepressed *DGSH* strain, where *GrxD*, *SreA*, and *HapX* genes were deleted. This modification generates the AF6 strain, which produces high levels of FsC siderophore even under the iron-rich conditions. This means that Fe^{3+} repression of FsC siderophore biosynthesis by the AF6 strain has been relieved. During the 10-Liter fermentation in a stainless steel fermenter (a system prone to Fe^{3+} release), the strain AF6 yields 1.05 mM extracellular siderophore and 7.4 g/L cell mass within 132 h, achieving a productivity of 5.78 mg/L/h and a yield of 33.2 mg/g of sucrose. Thus, the iron repression-resistant DGSH strain was successfully generated by knockout of the *GrxD*, *SreA*, and *HapX* genes (Kang et al. 2025).

11.3.2 Elucidation of Biosynthesis Pathway of Siderophore by Genome Editing and Synthetic Biology

After whole-genome sequencing, the HN6.2 strain (NCBI accession number: MWII00000000) was found to have a genome size of 26,123,185 bp with a GC content of 50.11%. The genome consists of 43 scaffolds containing 12,587 genes, including 9701 annotated genes. Notably, the *A. melanogenum* HN6.2 genome

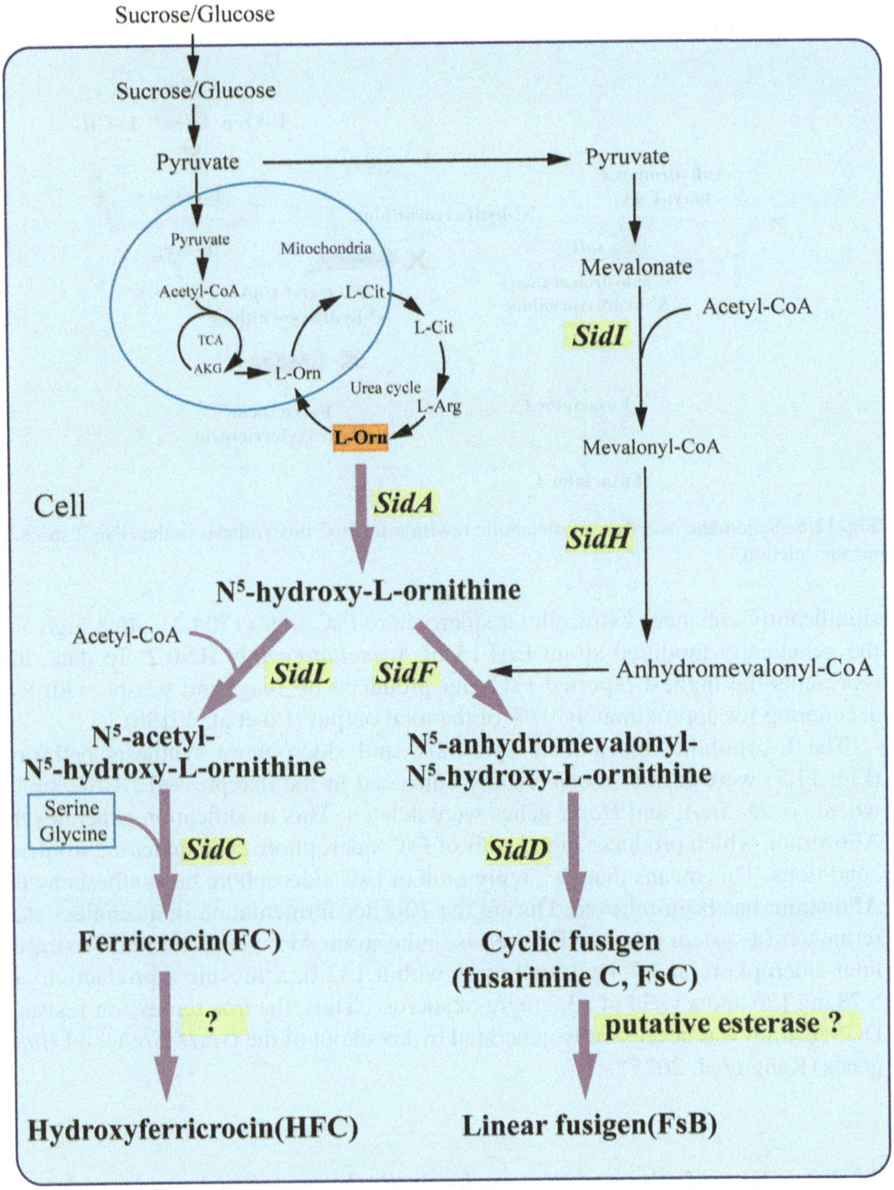

Fig. 11.7 Biosynthesis pathway of FsC

contains the complete siderophore biosynthesis genes, including: (a) *SidA* gene (encoding ornithine oxygenase; accession number: a149A6623); (b) the *SidD* gene (encoding fusarinine C/coprogen NRPS; a149A1333); **c.** the *SidC* gene (encoding ferricrocin/ferrichrome NRPS; a149A6625); (d) the *SidF* gene (encoding transacylase; a149A1334); (e) the *SidH* gene (encoding mevalonyl-CoA hydratase; a149A6744); (f) the *SidI* gene (encoding levalonyl-CoA ligase; a149A1335); (g) the *SidL* gene (encoding transacetylase; a149A0257); (h) the *HmgS* gene (encoding HMG-CoA synthase; a149A2065); (i) the *HmgR* gene (encoding HMG-CoA reductase; a149A7950); (J) the *EstC* gene (encoding siderophore esterase; a149A1338); (k) the *SreA* gene (encoding transcriptional activator; a149A6784); (l) the *HapX* gene (encoding activation of iron uptake; a149A8820); (m) the *MpkC* gene (encoding MAPK signaling pathway; a149A2537). After all these genes are edited, the whole pathway of fusigen biosynthetic pathway is elucadated as shown below.

The biosynthesis of fusigen (Fig. 11.5) begins with the N^5-hydroxylation of L-ornithine, catalyzed by the L-ornithine-N^5-monooxygenase encoded by the *SidA* gene, producing N^5-hydroxy-L-ornithine—the building block for fusigen siderophores. Subsequently, N^5-hydroxy-L-ornithine reacts with anhydromevalonyl-CoA to form N^5-anhydromevalonyl-N^5-hydroxy-L-ornithine, a reaction mediated by the transacylase encoded by the *SidF* gene. Finally, the assembly of three N^5-anhydromevalonyl-N^5-hydroxy-L-ornithine units into fusarinine C (FsC) is catalyzed by non-ribosomal peptide synthetases (NRPSs) encoded by the *SidD* gene. The anhydromevalonyl-CoA is produced from mevalonyl-CoA through the action of mevalonyl-CoA hydratase, encoded by the *SidH* gene. Mevalonyl-CoA, in turn, is synthesized from mevalonate and acetyl-CoA under the catalysis of the mevalonyl-CoA ligase, encoded by the *SidI* gene (Fig. 11.7). L-Ornithine-N^5-monooxygenase participates in the initial step of fungal siderophore biosynthesis, and the expression of the *SidA* gene is readily repressed by Fe^{3+} (Fig. 11.6). The *SidA* gene (accession number: FJ769160), encoding L-ornithine-N^5-monooxygenase, was isolated from both genomic DNA and cDNA of the marine yeast *A. melanogenum* HN6.2 using inverse PCR and RT-PCR. The open reading frame (ORF) spans 1461 bp and encodes a 486-amino-acid protein with a calculated molecular weight of 55.4 kDa and an isoelectric point of 7.79. The intronless promoter region of the *SidA* gene spans nucleotides −28 to −77 and contains one TATA box. Additionally, the upstream region includes two GATA boxes, which serve as binding sites for the transcriptional repressor Sre1, encoded by the *SreA* gene (Chi et al. 2012).

The *SidA* gene in *A. melanogenum* HN6.2 was disrupted by integrating the *HPT* gene (encoding hygromycin B phosphotransferase) into the open reading frame (ORF) of the *SidA* gene via homologous recombination. Among all the obtained disruptants, strain S6 (*ΔsidA*) lost the ability to synthesize both intracellular and extracellular siderophores. Consequently, this mutant cannot inhibit the growth of pathogenic bacteria *V. anguillarum* and *V. parahaemolyticus* as mentioned above. Furthermore, the disruptant S6 fails to grow in iron-depleted medium and seawater medium due to the arrested cell budding because of no produced siderophore. This growth defect stems from insufficient Fe^{3+} in yeast cells when siderophore production is impaired. However, growth is restored in the iron-replete medium supplemented with 10 μM Fe^{3+}/Fe^{2+} (Fig. 11.8). Notably, H_2O_2 in the medium exhibits greater toxicity toward the disruptant S6 than toward its wild-type HN6.2. These results suggest that the siderophores produced by the marine-derived *A. pullulans* HN6.2 can play essential roles in iron chelation, uptake, and concentration to

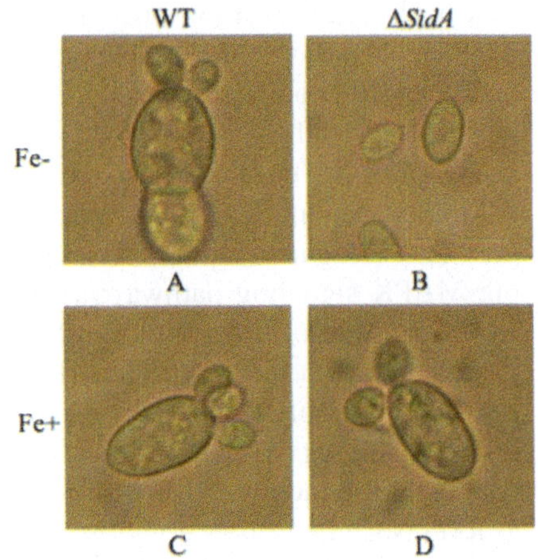

Fig. 11.8 Cell budding of the wild-type HN6.2 and the disruptant S6 grown in the iron-deplete medium (**A** and **B**) and the iron-replete medium (**C** and **D**)

maintain cellular physiological functions (Chi et al. 2012), and the siderophore produced has antioxidant activity.

Furthermore, its genome contains all the genes involved in iron uptake, storage, and transport, including their associated enzymes and proteins. These components closely resemble those found in *A. fumigatus* and *A. melanogenum* CBS110374 (Fig. 11.9) (Lu et al. 2019a). Genome sequencing and analysis reveal that the HN6.2 strain is the first yeast species known to possess all four mechanisms of iron acquisition: (i) siderophore-mediated iron uptake; (ii) reductive iron assimilation; (iii) low-affinity ferrous uptake; (iv) heme utilization (Fig. 11.9). This strain also employs vacuolar iron storage to sequester excess iron and prevent cellular toxicity (Fig. 11.9) (Lu et al. 2019a). Notably, the genome mining indicates that the HN6.2 strain synthesizes ferricrocin siderophores. Further HPLC and Q-Tof-MS analyses confirms that the siderophores produced by this strain include cyclic fusigen (FsC), linear fusigen (FsB), ferricrocin (FC), hydroxyferricrocin (HFC), and an unidentified iron chelator (Fig. 11.7). Additionally, heme utilization in this fungal-like yeast was experimentally validated through knockout of the heme oxygenase gene.

11.3.3 Clarification of Regulation Mechanisms of Siderophore Biosynthesis by Genome Editing and Synthetic Biology

At the same time, the expression of the genes and enzymes involved in siderophore biosynthesis must be tightly and precisely regulated. When the medium lacks Fe^{3+}, the iron regulator, unbound to Fe^{3+}, binds to the *SidA* gene promoter (as described earlier) and activates its expression. Consequently, the yeast cells synthesize and secrete a large quantity of siderophores to chelate and concentrate Fe^{3+} in the medium. The chelated Fe^{3+} is then imported into the yeast cells to support their physiological functions and cell growth. In contrast, when the medium contains

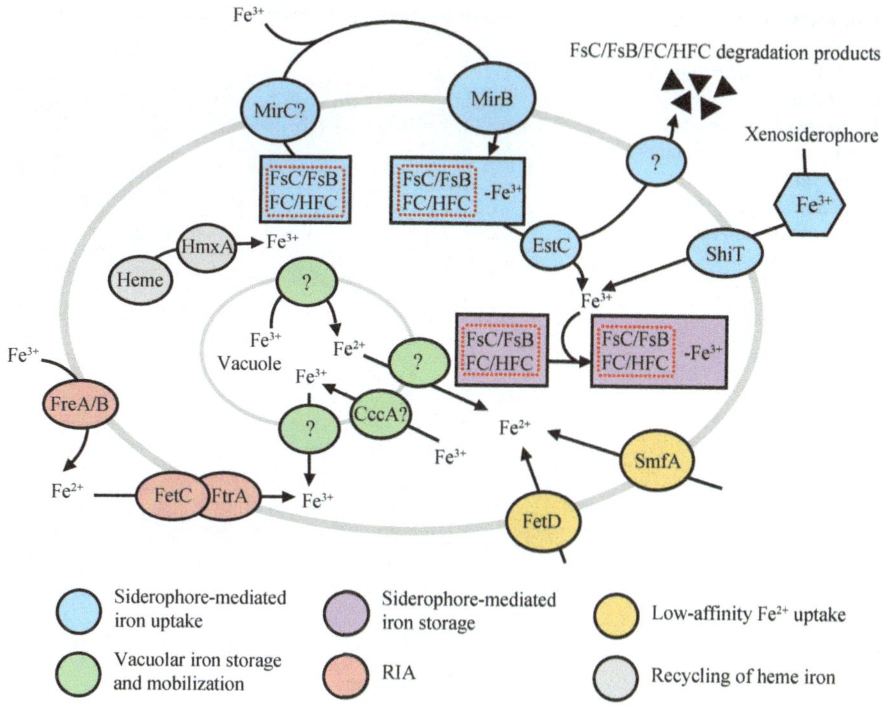

Fig. 11.9 Mechanisms involved in the iron uptake and storage in *A. melanogenum* HN6.2 strain revealed by the genome mining

abundant Fe^{3+}, the iron regulator binds Fe^{3+}, preventing its attachment to the *SidA* promoter. This suppresses the *SidA* gene expression and reduces siderophore biosynthesis.

One major iron regulator, Sre1, is encoded by the *SreA* gene and positively controls siderophore biosynthesis through transcriptional activation. Conversely, another major iron regulator negatively regulates siderophore biosynthesis—a highly conserved GATA-type transcriptional repressor containing zinc fingers that bind to the promoter of the key siderophore biosynthetic gene *SidA*. The *SreA* gene, cloned from both the genomic DNA and cDNA of the marine yeast *A. melanogenum* HN6.2, spans 1002 bp and encodes a 334-amino-acid protein (isoelectric point: 8.6) with a predicted molecular weight of 35.1 kDa. The gene contains a single 51-bp intron and features two putative 5′-HGATAR-3′ sequences in its promoter, indicating autoregulation via its own protein product. The deduced SreA protein includes: (a) Two conserved zinc-finger domains [Cys-(X_2)-Cys-(X_{17})-Cys-(X_2)-Cys)]; (b) Nine Ser (Thr)-Pro-X-X motifs, the characteristic of such regulators; (c) A cysteine-rich central domain localized between the zinc fingers (Chi et al. 2013).

In *A. melanogenum* HN6.2, the *SreA* gene was disrupted by inserting the *HPT* gene into its open reading frame (ORF) via homologous recombination. Among all the disruptants obtained, the disruptant R6 (*Δsre1*) retains the ability to synthesize

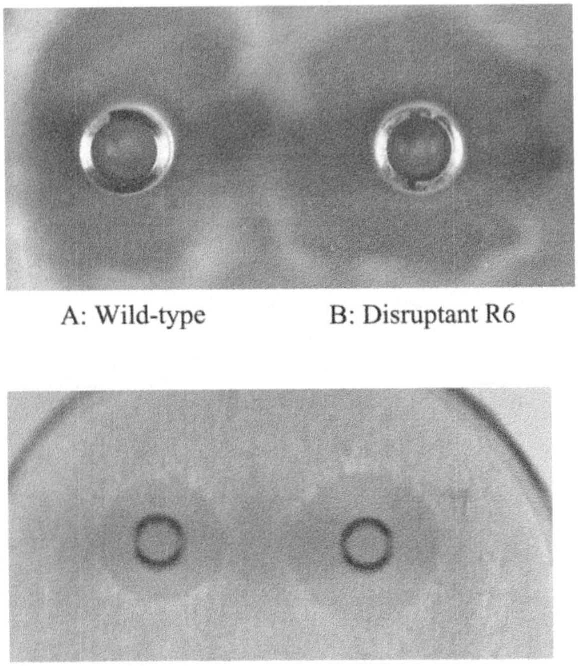

Fig. 11.10 The antibacterial activity against *V. anguillarum* (**A** and **B**) and *V. parahaemolyticus* (**C** and **D**) of the supernatants produced by *A. melanogenum* HN6.2 and its disruptant R6

both intracellular and extracellular siderophores even in the presence of Fe^{3+}. Additionally, the gene *SidA* (encoding L-ornithine N5-oxygenase) shows partial derepression under Fe^{3+}-replete conditions. The disruptant R6 forms brown-colored colonies on iron-replete medium, indicating heavy Fe^{3+} uptake via intracellular siderophores. In contrast, the parent strain (*A. melanogenum* HN6.2) fails to grow in the iron-replete medium containing high Fe^{3+} concentrations (1.5–2.0 mM). Since the disruptant R6 produces high levels of siderophores under the iron-replete conditions, its culture supernatant exhibits antibacterial activity (Fig. 11.10) (Chi et al. 2013). This study presents the first investigation into siderophore biosynthesis and regulation in marine yeasts.

For iron homeostasis, transcriptome analysis reveals that this yeast strain primarily employs two central regulators, SreA and HapX, to modulate iron uptake and storage at the transcriptional level. Additionally, the mitogen-activated protein kinase C gene (*MpkC*) shows transcriptional upregulation under iron-sufficient conditions, suggesting its potential role in repressing siderophore biosynthesis (Lu et al. 2019a; Fig. 11.11).

The GrxD protein is known to sense and transport iron signals by binding to iron-sulfur clusters. It regulates siderophore biosynthesis through two mechanisms: inhibiting siderophore synthesis via iron-sulfur cluster binding and sensing iron-excess signaling. Deletion of the *GrxD* gene partially relieves iron repression, suggesting that GrxD modulates siderophore synthesis by influencing the HapX/SreA

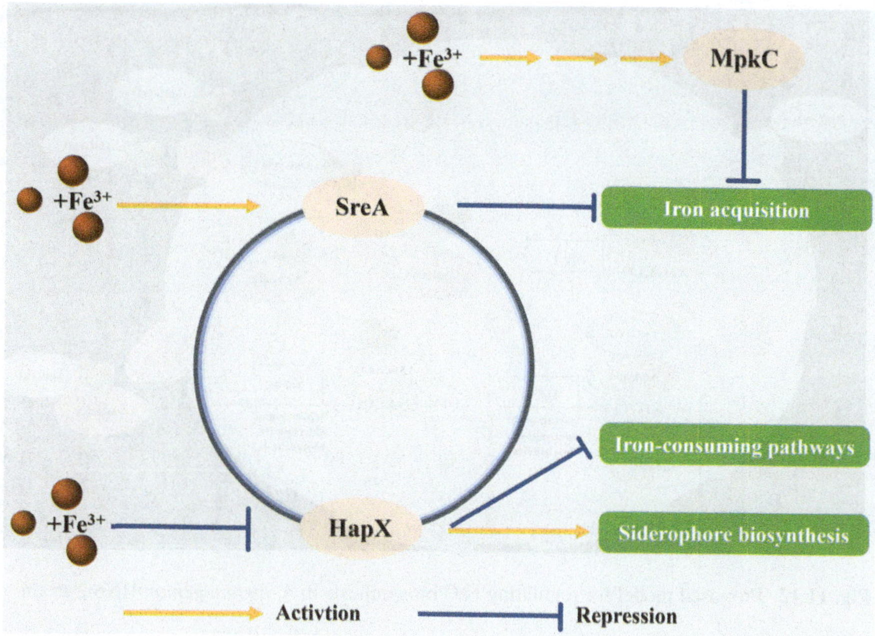

Fig. 11.11 The proposed iron regulation and homeostasis mechanism in the HN6.2 strain based on the genome mining and the transcriptome analysis

regulatory loop (Fig. 11.11). SreA, as previously mentioned, is a transcriptional repressor of the *SidA* expression, while HapX acts as a transcriptional activator (Fig. 11.11). The study reveals no physical interaction between SreA and GrxD proteins (Fig. 11.12). The SreA protein can independently sense iron signals through direct binding to iron-sulfur clusters without requiring the iron-sulfur cluster-GrxD complex. This finding indicates that the iron metabolism regulation mechanism in HN6.2 strain differs fundamentally from the known GrxD-SreA interaction-mediated regulation in *A. fumigatus*. Notably, the strong physical interaction between HapX and GrxD proteins were observed. The subcellular localization of SreA remains in nuclear regardless of iron concentration or GrxD deletion status. Additional results demonstrate SreA's direct binding to the *SidA* gene promoter (Fig. 11.12), suggesting that SreA suppresses siderophore synthesis by inhibiting the *SidA* gene transcription. These findings establish that GrxD and SreA independently regulate siderophore synthesis in response to varying iron availability. Finally, the data show that HapX cannot bind free iron-sulfur clusters but interacts specifically with the [Fe-S]-GrxD complex (Kang et al. 2025) (Fig. 11.12).

The GrxD-HapX complex was found to be localized in the nucleus under iron-deficient conditions, but to be localized in the cytoplasm under iron-rich conditions. However, HapX localizes in the nucleus under iron-rich conditions when the GrxD gene is knocked out. These results suggest that GrxD mediates HapX translocation between the cytoplasm and nucleus in response to iron availability (Kang et al. 2025).

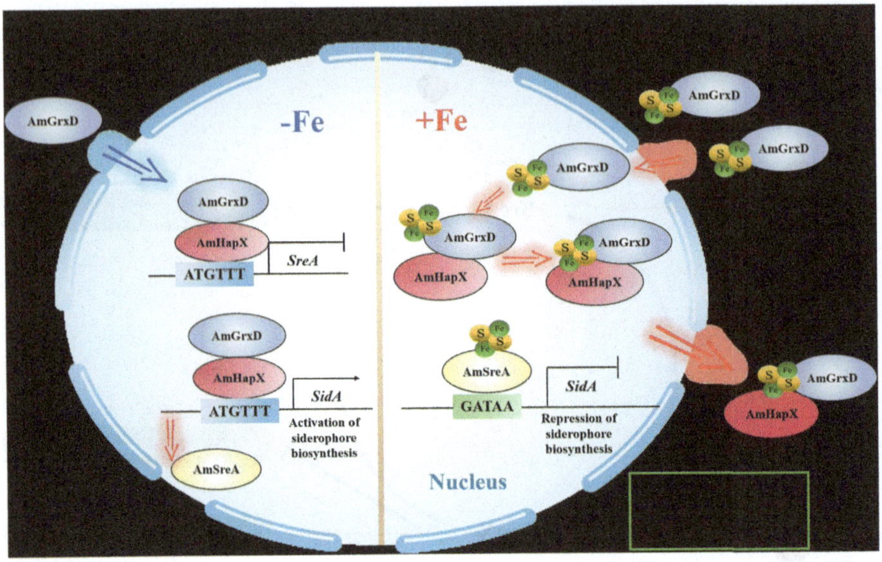

Fig. 11.12 Proposed model for regulating FsC biosynthesis in *A. melanogenum* HN6.2 strain

Additionally, the transcription factor HapX binds to the 5′-ATGTTT-3′ site in the *SidA* gene promoter, demonstrating its direct role in regulating siderophore synthesis by controlling *SidA* transcription (Fig. 11.12). Thus, the HN6.2 strain employs a siderophore regulation mechanism distinct from that of *A. fumigatus* (Kang et al. 2025)

Under the iron-deficient conditions, GrxD senses iron deficiency signals and mediates the nuclear localization of HapX. Subsequently, HapX binds to the *SidA* gene promoter to activate its transcription. In this context, HapX functions as a transcriptional activator, promoting siderophore synthesis during iron scarcity. In contrast, under the iron-rich conditions, GrxD binds to iron-sulfur ([Fe-S]) clusters and detects excess iron signals. The [Fe-S]-GrxD complex facilitates HapX translocation to the cytoplasm, derepressing the *SreA* gene transcription. SreA independently binds [Fe-S] clusters to sense iron levels, directly repressing *SidA* transcription and thereby inhibiting siderophore synthesis (Fig. 11.12) (Kang et al. 2025).

Based on Fig. 11.11, when the *GrxD*, *SreA*, and *HapX* genes are all knocked out, an iron repression-relieving DGSH strain can be obtained as described above. However, deletion of both the *GrxD* and *SreA* genes enables the DGS mutant to produce 0.18 mM extracellular siderophore under the iron-rich conditions. This represents a 50% increase compared to the extracellular siderophore production of the DG mutant (where only the *GrxD* gene was knocked out) (Kang et al. 2025).

11.3.4 Applications of Siderophore

Fungal siderophores serve multiple function. For instance, pathogenic fungi utilize siderophores as virulence factors to acquire iron from host sources during infection. Under aerobic conditions, all microorganisms require iron for essential metabolic processes, including oxygen reduction for ATP synthesis (via the electron transport chain), heme functionalization, and DNA precursor reduction. Despite being a vital nutrient for all life forms, iron's low solubility makes it difficult to obtain. Consequently, microorganisms depend on siderophores—the most effective soluble Fe^{3+} chelators—to assimilate extracellular iron.

Fungal siderophores also play multiple physiological roles, including oxidative stress resistance, asexual/sexual development regulation, iron storage, and protection against iron-induced toxicity during fungal-host interactions. In iron-depleted marine environments, secreted siderophores concentrate iron to support microbial growth. Additionally, the yeast-like fungal strain *A. melanogenum* HN6.2 produces siderophores that inhibit pathogenic bacterial growth in marine animals by outcompeting bacterial siderophores for iron absorption (Wang et al. 2009a, b). Due to their strong metal-chelating ability, siderophores help microbial cells withstand toxic metals (including heavy metals and radioactive elements) in polluted environments. Furthermore, they can be utilized to chelate, concentrate, and recover these hazardous substances for environmental remediation.

When iron is chelated, siderophore-iron complexes are utilized through either reductive iron assimilation or uptake by specific transporters into microbial cells. These complexes enable iron acquisition. Siderophores also serve as molecular imaging agents for detecting fungal and bacterial infections (Lu et al. 2019b). The easy acquisition of FsC enables its exploitation for fungal active targeting. *T. rubrum* is the primary cause of athlete's foot, onychomycosis, and other persistent skin diseases. *C. albicans*, an opportunistic human pathogen, causes candidiasis. In vitro and in vivo assays demonstrate that FsC functions as an active targeting molecule against these two pathogenic fungi—*T. rubrum* and *C. albicans*. The MTT assay confirms that FsC–[Fe^{3+}] chelates exhibit no cytotoxicity toward human L929 cells. These findings support the development of novel FsC-based antifungal agents that actively target *T. rubrum* and *C. albicans* for potential therapeutic applications (Lu et al. 2019b).

Helicobacter pylori is a prevalent human pathogen associated with gastritis, gastric ulcers, gastric malignant lymphoma, and gastric cancer. Therefore, detecting this bacterium with high sensitivity is critical in clinical practice. Due to its high infectivity and pathogenicity, accurately identifying even low levels of *H. pylori* is essential for timely intervention and controlling its spread. Conventional non-invasive diagnostic methods, such as the ^{13}C urea breath test and fecal antigen test, suffer from several limitations, including low sensitivity, poor resolution, susceptibility to interference, and frequent false-positive or false-negative results—especially at low bacterial concentrations. Consequently, improving the detection of low-level *H. pylori* infections remains a major challenge (Wang et al. 2022).

In recent years, optical biosensors have gained substantial research attention due to their rapid, sensitive, and user-friendly pathogen detection capabilities. However, current optical biosensors also still face challenges, such as weak signal intensity, insufficient sensitivity, and interference from impurities when detecting low bacterial levels. Enhancing the sensitivity, accuracy, and interference resistance of the optical biosensors for *H. pylori* detection is, therefore, a crucial area for further research.

To address these challenges, the dual-mode properties of the fungal iron carrier fusarinine C (FsC) obtained above were investigated. Building on these findings, a colorimetric/fluorescence dual-mode biosensor using FsC-conjugated polymers was developed for detecting low concentrations of *H. pylori*. The conjugated polymers, which incorporate numerous FsC signal elements, significantly amplify both colorimetric and fluorescent signals, thereby improving detection sensitivity. The dual-mode system enables cross-validation between signals, reducing background interference and enhancing accuracy. Recognizing the advantages of the colorimetric analysis for real-time monitoring, the biosensor was integrated with smartphones for *H. pylori* detection, primarily using the colorimetric method with fluorescent validation. The major findings are summarized below.

This study demonstrates that fusarinine C (FsC) exhibits (1) photoluminescence, (2) Fe^{3+} complexation for colorimetric detection, and (3) concentration-dependent fluorescence enhancement. Furthermore, the iron ion complex of FsC (FsC [Fe^{3+}]) demonstrates active targeting toward *H. pylori*. Leveraging these properties, the affinity-signal transduction module (FsC-Alg_{25}-FsC [Fe^{3+}]$_{5n}$) was developed. In this module: (a) FsC serves as a dual-mode signal reporter, enabling both fluorescence and colorimetric detection. (b) FsC [Fe^{3+}] functions as an affinity probe, specifically binding to *H. pylori*. The module employs the polycarboxylated polymer Alg as a carrier, which enhances detection sensitivity by loading multiple FsC molecules. This design amplifies both fluorescence and colorimetric signals synchronously by increasing FsC binding to individual *H. pylori* cells (Fig. 11.13) (Wang et al. 2022).

To address the complex matrix interference in fecal samples, the truncated *H. pylori*-specific aptamer H5 was modified with magnetic nanoparticles (MNPs) to form a capture and enrichment module (MNPs@H5). By integrating these modules, a dual-mode biosensor was constructed for *H. pylori* detection. The biosensor employs colorimetric and fluorescent signal outputs, enabling detection within 30 min across a range of 0–10^7 CFU/mL. It achieves a low detection limit (LOD) of 5 CFU/mL in both modes, allowing detection of as few as 10 *H. pylori* cells per sample. When coupled with a smartphone, the biosensor facilitates real-time quantitative detection of *H. pylori* in feces using the colorimetric mode. However, the sensitivity of smartphone cameras restricts recognition below 10 CFU, resulting in an LOD of 10 CFU/mL in the smartphone-assisted mode. The fluorescence mode serves as supplemental confirmation when the colorimetric mode lacks resolution. Overall, this biosensor enables ultrasensitive and precise dual-modal detection of low-level *H. pylori* in the fecal samples (Fig. 11.13) (Wang et al. 2022).

However, the dual-mode detection technology suffers from cumbersome operation. To address this issue, this study investigated the signal sensing mechanism of

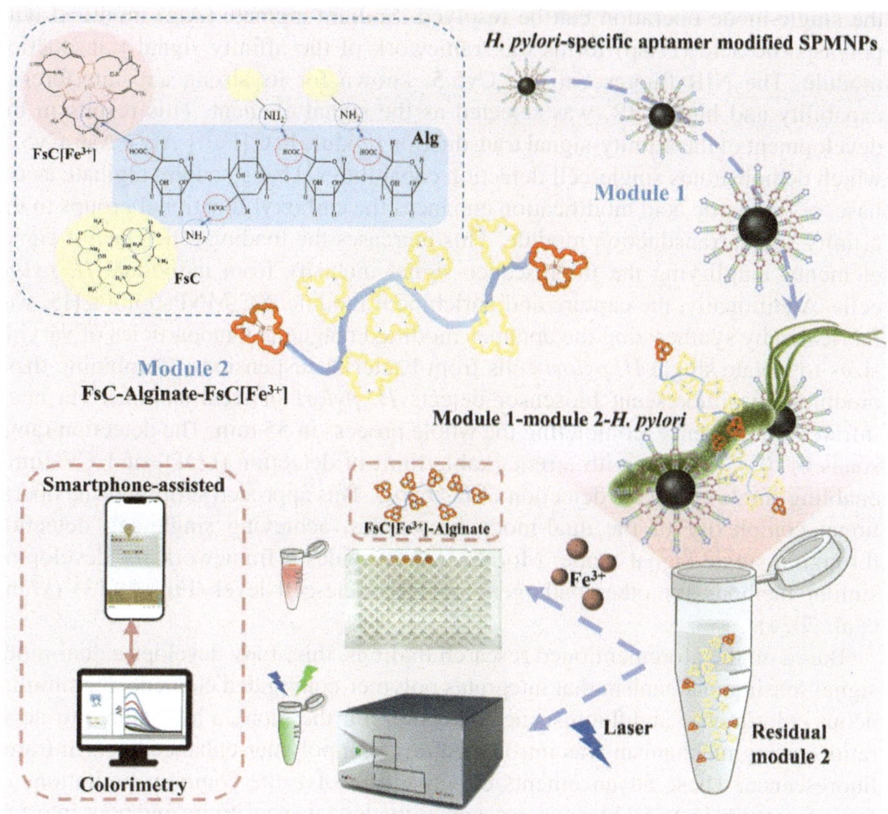

Fig. 11.13 Schematic illustration of the proposed mechanism and workflow of the FsC-powered dual-mode biosensorInspired by natural porphyrins, the authors discovered that fusarinine C (FsC), a water-soluble fungal siderophore, emits intense green fluorescence due to intramolecular and intermolecular conjugation and turns red upon coordination with Fe^{3+}. Based on FsC, the authors of this book developed a dual-mode biosensor. The biosensor consists of two modules: an enrichment module and a signal transduction module. The enrichment module comprises super-paramagnetic nanoparticles (SPMNPs) modified with *H. pylori*-specific aptamers, which are used to capture *H. pylori* from the sample. The signal transduction module consists of alginate (Alg) conjugated with FsC[Fe^{3+}] and a certain amount of Fe^{3+}-free FsC, which converts the captured *H. pylori* into a fluorescent or color signal. The dual-modal colorimetric detection mode, utilizing smartphone-assisted testing, enables highly sensitive detection of *H. pylori* within 20 min; the fluorescent detection mode, with its higher sensitivity, can validate the colorimetric detection results for trace amounts of *H. pylori*. This study presents an FsC-based dual-modal biosensor capable of providing highly sensitive, accurate, and reliable point-of-care rapid detection of *H. pylori* for the first time (Wang et al. 2022)

polymer-based near-infrared (NIR) fluorescence signal elements with high signal-to-noise ratios (SNR). By synthesizing a signal transduction module with a larger load and selecting an NIR fluorescent signal element with a high SNR, the sensitivity limitations and weak anti-interference capabilities of the optical biosensors in the single-mode operation can be resolved. Sodium alginate (Alg) modified with polyaspartic acid (PAsp) forms the framework of the affinity signal transduction module. The NIR fluorescent dye Cy5.5, known for its strong anti-interference capability and high SNR, was selected as the signal element. This results in the development of the affinity-signal transduction module FsC [Fe^{3+}]-Alg-PAsp-Cy5.5, which demonstrates single-cell detection capabilities. Using sodium alginate as the base, polyaspartic acid modification enhances the carboxyl functional groups in the affinity signal transduction module. This increases the loading efficiency of signal elements, amplifying the fluorescence signal intensity from individual *H. pylori* cells. Additionally, the capture and enrichment module, SC-MNPs6 nm@H5, was fabricated by synthesizing the aptamer-modified magnetic nanoparticles of varying sizes to isolate single *H. pylori* cells from bacterial suspensions. Combining these modules, the fluorescent biosensor detects *H. pylori* in fecal samples via near-infrared fluorescence, completing the whole process in 55 min. The detection range spans 0–10^4 CFU/mL, with a remarkable limit of detection (LOD) of 1 CFU/mL, enabling single-cell-level detection of *H. pylori.* This approach simplifies the operational complexity of the dual-mode biosensors, achieving single-cell detection through a single signal mode. Moreover, it provides a framework for developing similar methods for other pathogens at the single-cell level (Fig. 11.13) (Wang et al. 2024).

Based on the aforementioned research findings, this study develops a dual-mode signal sensing mechanism that integrates polymer-conjugated elements for simultaneous colorimetric and fluorescence detection. Furthermore, a high signal-to-noise ratio sensing mechanism was introduced utilizing polymer-enhanced near-infrared fluorescence. These advancements effectively resolve the common limitations of the conventional optical biosensors, such as inadequate sensitivity and poor interference resistance, particularly in detecting low concentrations of pathogenic bacteria. Consequently, this work provides critical solutions and technical support to overcome the challenges associated with achieving ultrasensitive and accurate detection of *H. pylori* in fecal samples (Wang et al. 2024).

11.4 Production, Biosynthesis, Regulation, and Application of L-Piperazic Acid

Piperazic acid (L-Piz), a non-proteinogenic amino acid containing an N–N bond, is a derivative of hexahydropyridazine-3-carboxylic acid. It occurs naturally in various non-ribosomal peptides (NRPs) and hybrid NRP-polyketide (PK) structures (Fig. 11.14). As a key synthetic building block, L-Piz is notably used in producing the antihypertensive drug cilazapril and interleukin-1 converting enzyme (ICE) inhibitors, demonstrating its pharmaceutical significance. Additionally, it is a constituent of numerous bioactive compounds produced by *Streptomyces* species.

Fig. 11.14 Chemical structure of L-piperazic acid (L-Piz)

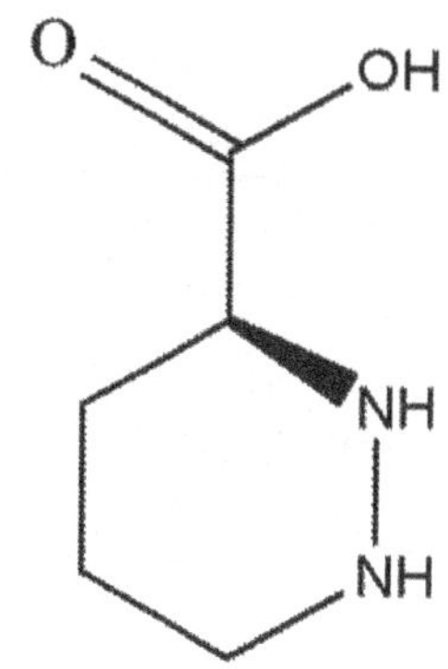

11.4.1 Elucidation of Biosynthesis Pathway of L-Piperazic Acid by Genome Editing and Synthetic Biology

As shown in Fig. 11.14, L-Piz contains an N-N bond in its chemical structure. This bond is also present in several other compounds, including: biosynthesized himastatin (the key gene: *htmC*, accession number: FR823394); Padanamides (the key gene: *padO*, accession number: KC915040); Sanglifehrins (the key gene: *sfaC*, accession number: KC915040); Kutznerides (its key gene: *ktzT*, accession number: EU074211); Matlystatins (the key gene: *matF*, accession number: NZ_KB907224); Gerumycin C (the key gene: *XF36_RS26795*). Under the catalysis of piperazate synthase (encoded by genes in Fig. 11.15), N^5-hydroxyl-ornithine—derived from L-ornithine in *A. melanogenum*—can be cyclized into L-Piz.

11.4.2 Clarification of Regulation Mechanisms by Synthetic Biology and Overproduction of L-Piperazic Acid by Metabolic Engineering

To synthesize L-Piz in the engineered *A. melanogenum* chassis, the codon-optimized piperazate synthase gene *ktzT*—derived from *Kutzneria* sp. 744—was integrated into the genome of the L-ornithine-overproducing DF1 strain, generating the recombinant strain DFK1, which can produce L-Piz. The results confirm successful L-Piz production by the recombinant strain DFK1. Following this success, the influence of different *ktzT* homologs on L-Piz biosynthesis efficiency was examined. The other five piperazate synthase genes were also codon-optimized based on the *A. melanogenum* genome. To determine the best-performing gene, expression cassettes for these five genes were constructed and introduced into the DF1 strain, yielding DFK1–DFK6 (Fig. 11.16). Among these strains, the DFK1 strain produced the highest extracellular L-Piz titer (179 mg/L), indicating that the *ktzT* gene from

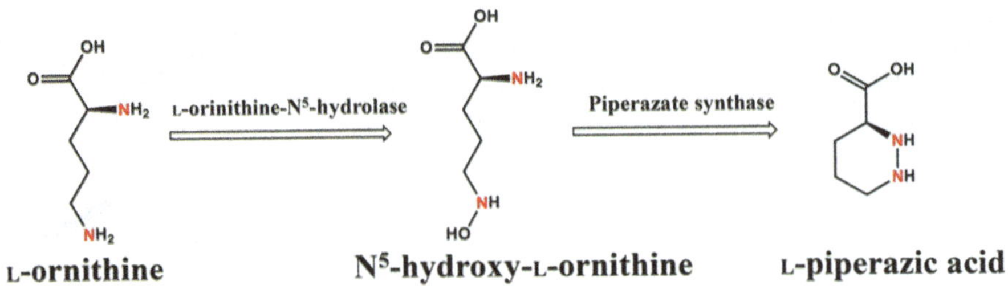

Fig. 11.15 Biosynthesis of L-Piz under catalysis of piperazate synthase

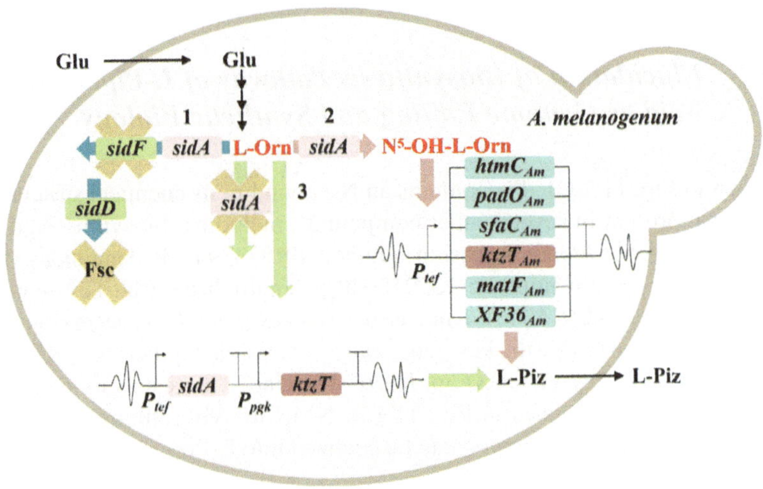

Fig. 11.16 Schematic overview of metabolic rewiring for L-Piz overproduction

Kutzneria sp. 744 is the most suitable for L-Piz production in *A. melanogenum* (Kong et al. 2021).

Following the development of the DFK1 strain, this study aimed to construct an industrial strain for enhanced L-Piz production in large-scale fermentation. To achieve this, the native *SidA* gene in the DF1 strain was deleted, yielding the DFA1 mutant strain. Subsequently, a genetic cascade comprising *SidA* and *ktzT*, each under the control of a constitutive promoter, was introduced into the DFA1 strain. The resulting engineered strain, designated DFAK1 strain (Fig. 11.16), exhibits increased L-Piz production (695 mg/L) and relieves iron-mediated inhibition. In a fed-batch fermentation, the DFAK1 strain produces 1.1 g/L L-Piz, with a yield of 21.0 mg/g glucose and a productivity of 9.2 mg/L/h (Fig. 11.16) (Kong et al. 2021).

To further enhance L-ornithine production by relieving feedback inhibition from L-arginine and L-Piz biosynthesis, the native *ArgB* gene from the AGC pathway was introduced into *A. melanogenum* DFAK1. The gene was integrated into the genome and overexpressed (Fig. 11.16). The resulting DFAK-ArgB6 strain exhibits improved performance, producing 1.3 g/L L-Piz, 4.4 mg intracellular L-ornithine/

mg protein, and 11.6 g/L cell mass in the flask cultures. In contrast, the parent strain DFAK1 yields only 0.69 g/L L-Piz, 3.7 mg intracellular L-ornithine/mg protein, and 9.7 g/L cell mass under the identical conditions (Chen et al. 2023). Additionally, the *ArgB* gene overexpression upregulates the expression of other genes in the AGC pathway. Medium optimization reveals that corn steep liquor supplementation significantly enhances L-Piz production in the engineered DFAK-ArgB6 strain (Chen et al. 2023).

During a fed-batch fermentation with an additional 120 g of glucose added to the optimized medium, the transformant DFAK1 produces 3.0 g/L of L-Piz within 120 h, with a cell mass of 25.5 g/L. This results in a yield of 55.6 mg L-Piz per gram of glucose and a productivity of 24.8 mg/L/h. To our knowledge, this represents the highest reported titer of L-Piz achieved using a microbial cell factory (Chen et al. 2023).

11.4.3 Applications of L-Piperazic Acid

L-Piz is currently produced primarily through chemical synthesis. However, the chemical method causes environmental pollution and is unsustainable. Consequently, employing synthetic biology techniques to overproduce L-Piz using microbial systems holds significant scientific value and practical potential. *A. melanogenum* HN6.2 exhibits strong L-ornithine biosynthesis capacity, exceptional tolerance to extreme environments, genetic tractability, and diverse cellular biochemical composition. Leveraging these attributes, the metabolic engineering and synthetic biology approaches were employed to develop this strain as a platform cell factory, achieving L-Piz production of 3.0 g/L with 86.4% purity.

11.5 Production, Biosynthesis, Regulation, and Applications of Putrescine

Putrescine, a diamine synthesized from L-ornithine via L-ornithine decarboxylase-catalyzed reactions, is critical for protein synthesis, cell division, and proliferation. Structurally, it consists of a saturated four-carbon backbone with two amine groups (Fig. 11.17). While it is widely used in the chemical industry, conventional putrescine synthesis faces challenges such as environmental pollution and unsustainability. As outlined in Sect. 11.1, *A. melanogenum* exhibits robust L-ornithine biosynthesis, extreme environmental tolerance, and high genomic plasticity,

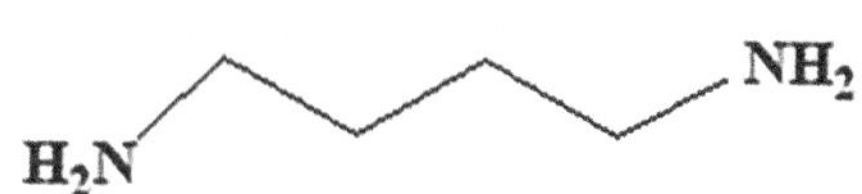

Fig. 11.17 Chemical structure of putrescine

making it ideal for synthetic biology applications. Leveraging these traits, the metabolic engineering and synthetic biology strategies were employed to develop an *A. melanogenum* HN6.2 cell factory for efficient putrescine production by modifying key metabolic pathways.

11.5.1 Elucidation of Biosynthesis Pathway, Regulation Mechanisms of Putrescine by Synthetic Biology, and Overproduction of Putrescine by Metabolic Engineering and Synthetic Biology

As previously stated, L-ornithine undergoes decarboxylation to produce putrescine, catalyzed by L-ornithine decarboxylase (Fig. 11.18).

Through the analysis of L-ornithine-related metabolic genes in the HN6.2 strain, a chassis strain, DF1, was constructed for synthesizing L-ornithine and its derivatives. Genomic analysis identifies the key pathways involved in L-ornithine metabolism, and rate-limiting enzymes are assessed to optimize L-ornithine accumulation. Results indicate that siderophore biosynthesis is the primary pathway consuming L-ornithine, necessitating gene knockout to block this pathway. Consequently, the DF1 strain was successfully engineered to produce L-ornithine and derivatives, including putrescine. As demonstrated above, *A. melanogenum* HN6.2 is a yeast-like fungal strain with high siderophore production. After disrupting siderophore biosynthesis and attenuating the expression of the ornithine carbamoyltransferase gene (*OTC*) (via the weak xylanase gene promoter *PxylP* from *Penicillium chrysogenum* fused to the native ORF of the *OTC* gene; Fig. 11.19), the engineered D-LCFAO-cre strain (with its antibiotic resistance gene excised) produced 2.1 mg of intracellular L-ornithine per mg of protein (Kong et al. 2022).

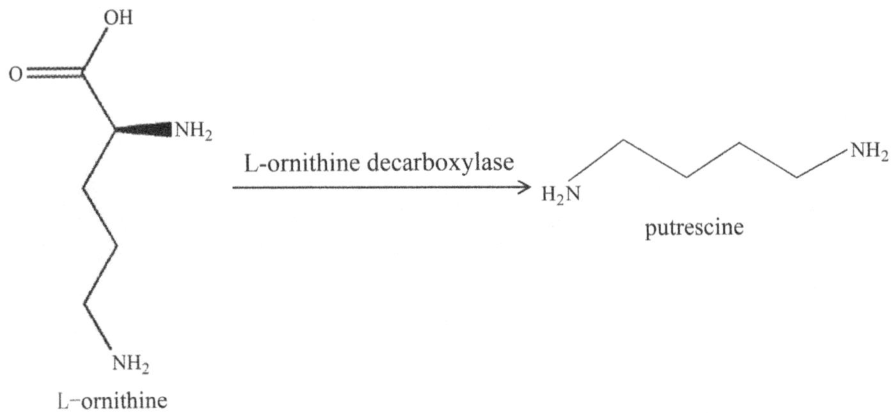

Fig. 11.18 Putrescine biosynthesis by action of L-ornithine decarboxylase

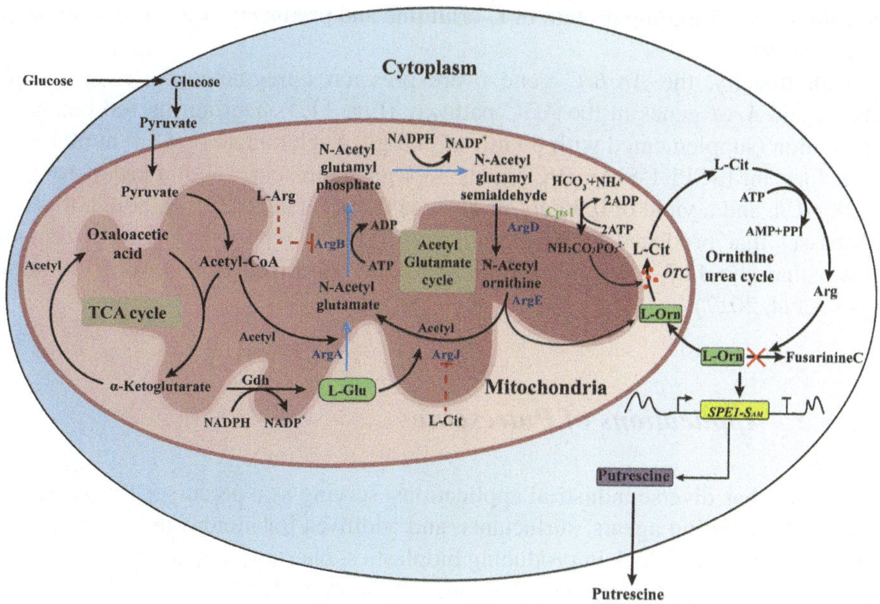

Fig. 11.19 Engineered pathways for high putrescine biosynthesis and production. ∵means knock-down, T means the feed-back inhibition; x means deletion

Based on the *A. melanogenum* genome (available at http://www.kazusa.or.jp/codon/), the following L-ornithine carbamoyltransferase genes were codon-optimized: (a) The *SPE1-S* gene from *S. cerevisiae* S288C (GenBank: 853651); (b) The *speC* gene from *E. coli* strain. K-12 (GenBank: 947457); (c) The *speF* gene from *Enterobacter cloacae* strain GGT036 (GenBank: 60929387); (d) The *SPE1-C* gene from *C. albicans* SC5314 (GenBank: 3639327). After each codon-optimized gene was individually expressed in the mutant strain D-LCFAO-cre, the resulting putrescine production by different recombinant strains was measured.

The study reveals that overexpression of the codon-optimized L-ornithine decarboxylase gene (*SPE1-S* gene, GenBank accession no. OM294021) from *S. cerevisiae* in the mutant strain D-LCFAO-cre enables the transformant E-SPE1-S to produce the highest intracellular levels of L-ornithine (3.6 mg/mg protein) and putrescine (10.5 g/L). These results indicate that the codon-optimized *SPE1-S* gene, encoding L-ornithine decarboxylase, is the most effective for putrescine biosynthesis from L-ornithine.

The N-acetyl glutamate kinase (ArgB/C) in the arginine biosynthesis cluster (AGC) as mentioned above is a rate-limiting enzyme whose activity is inhibited by feedback from L-arginine (L-Arg) in the ornithine-urea cycle (OUC) (Fig. 11.3). To enhance L-ornithine production, the *ArgB/C* gene (encoding bifunctional acetylglutamate kinase/N-acetyl-gamma-glutamyl-phosphate reductase) was overexpressed in the E-SPE1-S strain, generating the *E-SPE1-S-ArgB/C* transformant. This strain accumulates 4.2 mg/mg protein of L-ornithine and produced 21.3 g/L of putrescine (Fig. 11.19).

Additionally, the *ArgB/C* gene overexpression upregulates the *ArgA*, *ArgD*, *ArgE*, and *ArgJ* genes in the AGC pathway (Fig. 11.3). During the fed-batch fermentation (supplemented with 60 mL of 300 g/L L-glutamate (L-Glu) at 60 h), the transformant E-SPE1-S-ArgB/C yields 33.4 g/L putrescine, with a productivity of 0.28 g/L/h and a yield of 0.96 g/g of glucose (Kong et al. 2022). Although this titer surpasses that produced by most engineered strains reported to date, it remains lower than the 42.3 ± 1.0 g/L produced by the optimized *E. coli* WL3110 strain (Noh et al. 2017).

11.5.2 Applications of Putrescine

Putrescine has diverse industrial applications, serving as a precursor for pesticides, metal ion chelating agents, surfactants, and additives for motor oils and fuels. It is also a key building block in producing bioplastics, pharmaceuticals, agrochemicals, and surfactants. For instance, putrescine can be condensed with bio-based adipic acid to synthesize nylon 46, a material with an annual global demand of approximately 10,000 tons and widespread use in the chemical industry. Currently, putrescine and nylon production predominantly rely on chemical synthesis, which has been shown to cause severe environmental pollution, exacerbate global warming, and is unsustainable. Consequently, microbial synthetic biology offers a scientifically significant and highly promising alternative for putrescine production. The bio-based putrescine can then be used to produce nylon 46, a versatile polymer with numerous pharmaceutical, medical, and industrial applications.

11.6 Conclusions and Perspectives

In this chapter, metabolic engineering, genome editing, and synthetic biology are employed to genetically modify L-ornithine metabolism, enabling the construction of strains optimized for synthesizing L-ornithine and its derivatives. These modifications are crucial for the de novo synthesis of siderophores, L-Piz, and putrescine. Subsequently, efficient biosynthetic pathways for L-ornithine, L-Piz, siderophores, and putrescine are designed, reconstructed, and their regulatory mechanisms in *A. melanogenum* HN6.2 are elucidated. Notably, this study achieves the first successful biosynthesis of L-Piz and putrescine in eukaryotic systems. These findings strongly demonstrate that the HN6.2 strain can be widely applied as a cell factory platform for producing L-ornithine and its derivatives—including siderophores, L-Piz, and putrescine—through metabolic engineering and synthetic biology.

References

Chen H, Kong CC, Wei X, Chi Z, Liu GL, Chi ZM (2023) Overproduction of L-piperazic acid by overexpression of *ArgB* gene in *Aureobasidium melanogenum* DFAK1. Proc Biochem 126:1657–1162

Chi Z, Wang XX, Ma ZC, Buzdar MA, Chi ZM (2012) The unique role of siderophore in marine-derived *Aureobasidium pullulans* HN6.2. Biomet 25:219–230

Chi Z, Wang XX, Geng Q, Chi ZM (2013) Role of the GATA-type transcriptional repressor Sre1 in regulation of siderophore biosynthesis in the marine-derived *Aureobasidium pullulans* HN6.2. Biomet 26:955–967

Dong J, Kan B, Liu H, Zhan M, Wang S, Xu G, Han R, Ni Y (2020) CRISPR-Cpf1-assisted engineering of *Corynebacterium glutamicum* SNK118 for enhanced L-ornithine production by NADP-dependent glyceraldehyde-3-phosphate1 dehydrogenase and NADH-dependent glutamate dehydrogenase. Appl Biochem Biotechnol 191:955–967

Kang X, Liu W, Wang X, Aslam M, Liu GL, Chi Z (2025) Iron-derepressed robust production of Fasarinine C siderophore by *Aureobasidium melanogenum*. ACS Synth Biol 14:1525–1637

Kong CC, Wang ZZ, Liu GL, Chi ZM, Ledesma-Amaro R, Chi Z (2021) Bioproduction of L-piperazic acid in gram scale using *Aureobasidium melanogenum*. Microb Biotechnol 14:1722–1729

Kong CC, Wei X, Liu GL, Chi ZM, Chi Z (2022) Metabolic engineering of *Aureobasidium melanogenum* for the overproduction of putrescine by improved L-ornithine biosynthesis. Microbiol Res 260:127041

Lu Y, Liu GL, Jiang H, Chi ZM, Chi Z (2019a) An insight into the iron acquisition and homeostasis in *Aureobasidium melanogenum* HN6.2 strain through genome mining and transcriptome analysis. Funct Integrat Genom 19(1):137–150

Lu Y, Wang H, Wang Z, Cong Y, Zhang P, Liu GL, Liu CG, Chi ZM, Chi Z (2019b) Metabolic rewiring improves the production of the fungal active targeting molecule fusarinine C. ACS Synth Biol 8:1755–1765

Myo Myo Tint K, Wei X, Wang P, Liu GL, Zhang M, Chi ZM, Chi Z (2024) Biotechnological application of *Aureobasidium* spp. as a promising chassis for biosynthesis of ornithine-urea cycle-derived bioproducts. Crit Rev Biotechnol 23:82954

Noh M, Yoo SM, Kim WJ, Lee SY (2017) Gene expression knockdown by modulating synthetic small RNA expression in *Escherichia coli*. Cell Syst 5:418–426

Wang WL, Chi ZM, Chi Z, Li J, Wang XH (2009a) Siderophore production by the marine-derived *Aureobasidium pullulans* and its antimicrobial activity. Biores Technol 100:2639–2641

Wang WL, Chi Z, Liu GL, Buzdar MA, Chi ZM, Gu Q (2009b) Chemical and biological characterization of siderophore produced by the marine-derived *Aureobasidium pullulans* HN6.2 and its antibacterial activity. Biomet 22:965–972

Wang ZZ, Wang HY, Chen XH, Cheng X, Geng J, Wang L, Dong Q, Liu C, Chi ZM, Chi Z (2022) Aptamer-superparamagnetic nanoparticles capture coupling siderophore-Fe^{3+} cavenging actuated with carbon dots to confer an "off-on" mechanism for the ultrasensitive detection of *Helicobacter pylori*. Biosens Bioelectron 193:113551

Wang ZZ, Zhao MX, Wang Y, Liu CG, Chi ZM, Chi Z (2024) Nature-inspired fungal fusarinine C-powered dual-mode fluorometric/colorimetric biosensing of fecal *Helicobacter pylori* with high-reliability. Senso Actuat B-Chem 412:135838

Chapter 12
Aureobasidium spp. as an Effective Antagonist in Biocontrol of Pathogenic Fungi in Crop Plants and Fruits

Abstract Different strains of *A. pullulans* and *A. melanogenum* have high activity against fungal pathogens, which infect various plants and their fruits and vegetables and cause big economic loss and food insecurity. *A. pullulans* and *A. melanogenum* have such potent anti-fungal activity due to their high nutrient and space competition and site exclusion, production of exopolysaccharides, extracellular lytic enzymes, antimicrobial compounds, volatile organic compounds and the extracellular alkaline protease, induction of host resistance, promotion of host plant growth, enhancement of host plant immunity and promotion of colonization of beneficial microbes, and degradation of fungal toxin or choline. So, such *Aureobasidium* agents can be used to replace chemical pesticides to effectively control fungal pathogens on plants, vegetables, and fruits in order to reduce or stop uses of chemical pesticides, to protect environments and ecosystems and to keep food security. Such *A. pullulans* agents have multiple actions on fungal pathogens and can be easily produced and applied in a large scale.

Keywords *A. pullulans* · Anti-fungal activity · Biocontrol · Anti-fungal mechanisms

12.1 Introduction

Pathogenic fungi affecting plants and fruits cause substantial annual losses in crop yield and quality worldwide, along with mycotoxin contamination in food and feed. These impacts lead to significant economic damage and pose risks to human and animal health. As a sustainable alternative to synthetic chemical pesticides, biocontrol agents employ living microorganisms to suppress pathogenic populations, inhibit their growth, or maintain pathogen levels below infection thresholds (Da Costa Arruda et al. 2024). This is crucial for reducing chemical pollution in our ecosystem.

Z. Chi et al., *Biology and Synthetic Biotechnology of Aureobasidium spp.*,
https://doi.org/10.1007/978-981-92-2607-8_12

12.2 Background

It is well documented that numerous crops—including wheat, rice, corn, cotton, peanuts, apples, grapes, pears, and others—are highly susceptible to fungal pathogens. Common fungal pathogens include *Aspergillus* spp., *Fusarium* spp., *Penicillium* spp., *Magnaporthe grisea*, *Ustilaginoidea oryzae*, *Verticillium dahliae*, *Alternaria alternata*, *Phytophthora nicotianae*, *Sclerotinia sclerotiorum*, and *Rhizoctonia solani*. Notably, *A. flavus* produces aflatoxin, causing aflaroot or yellow mold in peanuts (Fig. 12.1), while *A. niger* generates hazardous ochratoxin and induces root rot in the same crop. Additionally, *Fusarium graminearum*, *Fusarium proliferatum*, and *Fusarium oxysporum* not only cause crop diseases but also produce mycotoxins such as fumonisins, trichothecenes, and zearalenone. Post-harvest infections in fruits and vegetables are also primarily attributed to fungal pathogens. Key genera responsible for these diseases include *Penicillium* spp., *Botrytis* spp., *Monilinia* spp., *Rhizopus* spp., *Alternaria* spp., *Aspergillus* spp., *Fusarium* spp., *Geotrichum* spp., *Gloeosporium* spp., and *Mucor* spp.

Botrytis cinerea, the causal agent of gray mold disease, is a major pathogen affecting peach, strawberry, blackberry, and raspberry crops (Fig. 12.2). This fungal pathogen causes significant economic losses both pre- and post-harvest.

Colletotrichum acutatum causes gray mold and anthracnose, respectively. This pathogen is highly destructive, leading to significant economic losses during strawberry cultivation and post-harvest storage.

B. cinerea and *Alternaria alternata* are two major plant pathogens that cause significant economic losses in various horticultural crops worldwide, including plum, apple, quince, sweet cherry, and pear.

P. expansum, known as the cause of postharvest blue mold in various fruits (such as apples, pears, peaches, and sweet cherries), produces several toxic secondary metabolites, including patulin, citrinin, and chaetoglobosins—all of which are potential carcinogens (Fig. 12.3). These fungal toxins pose risks to animal and human health through the feed and food chains. Specifically, patulin produced by *P. expansum*, a lactone, can cause nausea, gastrointestinal damage, kidney dysfunction, tumor growth, genetic mutations, and embryonic developmental defects, making it a significant concern in agriculture and food industries. Additionally,

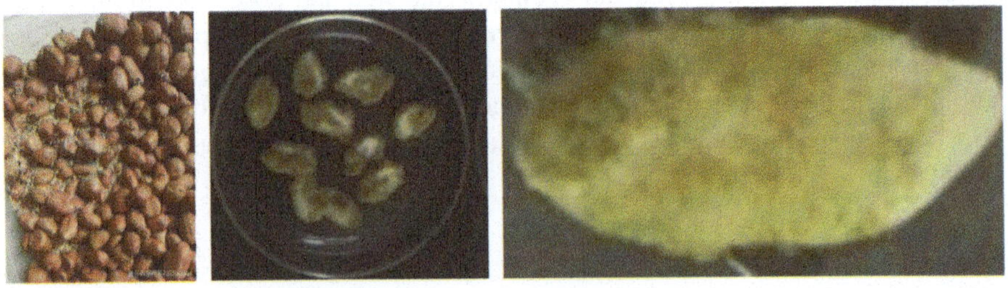

Fig. 12.1 Yellow mold in peanuts by *A. flavus* NRRL3357

A B C D

Fig. 12.2 Gray mold on strawberry (**A**), peach (**B**), tomato (**C**), and grapes (**D**)

A B C

Fig. 12.3 Green mold on orange (**A**), apples (**B**), and grapes (**C**)

P. digitatum causes green mold, the most common postharvest disease in citrus, further raising public health concerns.

Rhizopus stolonifer causes Rhizopus rot in strawberries, peaches, plums, and raspberries (Fig. 12.4). *Geotrichum citriaurantii* (Ferraris) is responsible for post-harvest sour rot in citrus, affecting all species and cultivars worldwide. Recently, *Botryosphaeria* spp., including *Neofusicoccum parvum*, have been reported to cause significant losses in apple production in Trentino-Alto Adige, Italy.

Plant diseases and fungal toxins can significantly reduce crop quality, leading to substantial economic losses in global crop production and grain storage (Fig. 12.5). These toxins remain stable during common food processing, and consumption of toxin-contaminated food poses serious public health risks. Consequently, developing effective strategies to control fungal crop and fruit pathogens and toxin-producing fungi has become a worldwide priority.

Currently, chemically synthesized fungicides (e.g., isoprothiolane, carbendazim, hexaconazole, and difenoconazole) are commonly used to control fungal crop pathogens. Additionally, fungicides like imazalil, thiabendazole, pyrimethanil, and fludioxonil are applied to manage pre- and post-harvest infections caused by *Penicillium* spp. Despite their high efficacy and widespread use, these synthetic fungicides raise significant environmental and public health concerns. The key issues include: (a) Pathogen resistance: Fungal pathogens rapidly develop resistance to the chemical treatments; (b) Persistence: These compounds degrade slowly in the environments; (c) Toxicity: They pose risks to human and animal health; (d) Pollution: Their non-biodegradable nature harms ecosystems, animal and human health.

Fig. 12.4 Rhizopus rot of plant

A B C

Fig. 12.5 Fungal diseases on wheat (**A**), rice (**B**), and corn (**C**)

Growing fungicide-resistant pathogen strains, coupled with consumer demand for reduced chemical residues in food, has driven a global shift toward safer alternatives. Biofungicides—efficient, low-toxicity, and eco-friendly—are now prioritized to address these challenges.

One of the novel environmentally friendly solutions for sustainable and effective fungal plant disease control is the utilization of biopesticides. Extensive research has demonstrated that biopesticides offer several advantages over synthetic chemicals. For example, they are biodegradable, environmentally safe for humans, animals, and ecosystems, and exhibit lower toxicity due to their targeted action against fungal pathogens. Additionally, the diverse modes of action of biopesticides reduce the likelihood of fungal pathogens developing resistance. Thus, identifying a bio-fungicide with efficacy and consistency comparable to the chemical fungicides is critically important. So far, many bio-fungicides produced by various species of bacteria and fungi have been discovered. Although they offer advantages over chemical pesticides, available bio-fungicides also have many drawbacks, such as low yield, poor bioavailability, instability, difficult purification, and high costs. These limitations hinder their further development and application in agriculture. Despite the ecological and environmental hazards of the chemical pesticides, they remain widely used in agriculture because current bio-fungicides are less effective and inconsistent compared to their chemical counterparts.

In recent years, *Aureobasidium* spp. have been recognized as effective antagonists in the biocontrol of pathogenic fungi across multiple plant and fruit pathosystems. Compared to other antagonistic yeasts—such as *Wickerhamomyces anomalus*, *Pichia membranifaciens*, *Meyerozyma guilliermondii*, *Metschnikowia pulcherrima*, and *Hanseniaspora uvarum*—as well as *Trichoderma* spp. fungi (Zhang et al. 2011), *Aureobasidium* spp. offer several advantages over these and other bio-fungicides. For instance, they can be cultivated and mass-produced inexpensively and efficiently. Additionally, due to their wide distribution, high environmental tolerance, and adaptability to harsh conditions—such as extreme temperatures, dry surfaces, and oxidative stress—their morphological and phenotypic plasticity and polyextremotolerance make them suitable for diverse ecosystems as mentioned in Chaps. 2 and 4. Furthermore, *Aureobasidium* spp. effectively colonize plant and fruit surfaces through biofilm formation, enzyme secretion and utilize multiple mechanisms to suppress pathogen growth as noticed below. Notably, they constitute a key component of the fruit and plant core microbiome, suggesting their significant role in fruit and plant health and disease resistance. *Aureobasidium* spp. contribute to stabilizing the microbial ecosystem on fruit surfaces and enhance fruit resilience to abiotic stresses (e.g., mechanical injury and environmental fluctuations) that may predispose fruits to microbial infections. This improves fruit quality and extends storage life (≥6 months). These yeasts exhibit environmental friendliness, leave no toxic residues, require simple application methods, and demonstrate effective biocontrol against fungal and bacterial pathogens. Genetically stable with simple nutritional requirements, *Aureobasidium* spp. rapidly colonize fruit and vegetable surfaces, including injured areas. They show broad-spectrum efficacy against pathogens, resistance to pesticides, and generally do not produce metabolites harmful to animal and human health. Additionally, metabolites such as liamocin, massoia lactone, killer toxins, and siderophores (Zhang et al. 2021; Wang et al. 2009; Kang et al. 2022) exhibit antifungal and antibacterial properties. *Aureobasidium* spp. are effective against postharvest fungal pathogens in various fruits, including citrus, peach, apple, table grape, kiwifruit, pear, and strawberry. For instance, *A. pullulans* effectively controls *Monilinia* spp. on stone fruits, *B. cinerea* on grapes, strawberries, and kiwifruit, and *P. expansum* on apples.

It has been well documented that *Alternaria* spp., *Cladosporium* spp., *Didymella* spp., and *Fusarium* spp. affect leaves and fruits of olive, causing serious diseases of the plant. The endophytic fungus *A. pullulans* Aljabal-Oman-A isolated from olive leaves in Oman can produce antifungal metabolites and volatile organic compounds, which can cause hyphal deformation, shrinkage, and electrolyte leakage from the pathogenic fungi *Fusarium equiseti* and *Alternaria tenuissima* and 100% suppression of leaf spot symptoms of olive (Al-Fahdi et al. 2026).

Currently, multiple biocontrol products utilizing *A. pullulans* are commercially available for agricultural use (e.g., BoniProtect®, BlossomProtect®, and Botector®). These products serve as highly promising alternatives to the synthetic chemicals in viticulture and horticulture during both pre- and post-harvest stages (Bziuk et al. 2025).

12.3 Anti-fungal Activity of *Aureobasidium* spp.

As a biocontrol agent, *A. pullulans* effectively controls post-harvest molds, including *B. cinerea*, *Colletotrichum acutatum*, and *Penicillium* spp. As discussed in Chap. 3, *Aureobasidium* spp. are generally poor fermenters, rendering them unsuitable for alcohol and wine production. However, many *Aureobasidium* spp. exhibit rapid growth and adaptability to harsh environments, allowing them to colonize fruits, vegetables, and plant surfaces. This characteristic makes them promising candidates for controlling fungal pathogens on fruits, vegetables, and plants. For instance, one *A. pullulans* strain effectively suppresses the grapevine trunk disease pathogen *Diplodia seriata*, while another strain is used to manage gray mold (*B. cinerea*) in grapes (Fig. 12.2).

Currently, three strains of *A. pullulans* have been commercially applied to control plant pathogens. Two strains, DSM 14940 and DSM 14941, serve as active ingredients in three products:

- Blossom Protect® for fire blight control on apple and pear.
- Botector® for *Botrytis* spp. management across various crops (including grapes).
- Boni Protect® for suppressing storage rots caused by *Neofabraea* spp., *Penicillium expansum*, *Monilinia fructigena*, and *B. cinerea* in pipfruit.

The third strain, CG163, is the active component of Aureo® Gold, used to combat bacterial canker in kiwifruit caused by *Pseudomonas syringae* pv. *actinidiae* (Psa) (Cornish et al. 2025). Thus, *Aureobasidium* spp. have emerged as key biological control agents for pre- and post-harvest fruit crop disease management in this world.

Many native *Aureobasidium* spp. can serve as both biocontrol agents and modulators of the epiphytic microbiome when applied at different fruit development stages. For example, *A. pullulans* strain A69, isolated from apple surfaces, exhibits strong antagonistic activity against *P. expansum* by producing volatile and diffusible metabolites, effectively suppressing lesion development in fruit. When applied in pre-harvest apple orchards, strain A69 not only reduces post-harvest decay incidence but also alters the microbial community by:

- Decreasing fungal diversity.

- Enriching beneficial bacteria and yeast genera with antagonistic activity against key postharvest pathogens.
- Promoting deterministic selection in fungal communities.
- Fostering competitive microbial interactions.

These changes result in a more stable and resilient microbiome (Yeka Zhimo et al. 2025).

Epiphytic populations of *A. pullulans* on apple and pear flowers commonly exceed 1×10^4 CFU/flower. Notably, this yeast-like fungus colonizes the hypanthial surface within the floral cup—the primary infection site of the fire blight pathogen *Erwinia amylovora*. Studies suggest that *A. pullulans* may competitively inhibit *E. amylovora* by acidifying the floral cup microenvironment and upregulating host defense genes in hypanthial tissue (Kunz et al. 2024).

A. pullulans strain NBB 7.2.1 exhibits strong antagonistic activity against several plant pathogens and high competitiveness against any other yeasts on apple fruits. Although *A. pullulans* NBB 7.2.1 displays similar growth and antagonistic activity to *A. pullulans* EXF-150, it inhibits *F. oxysporum* NRRL 26381/CL57 more effectively than *A. melanogenum* CBS 110374, *A. namibiae* CBS 147.97, and *A. subglaciale* EXF-2481 (Paula Rueda-Mejia et al. 2021). The genome of *A. pullulans* NBB 7.2.1 encodes potential biocontrol genes, including predicted secreted hydrolases (e.g., glycosylases, esterases, and proteases) and enzymes involved in secondary metabolite synthesis. These secondary metabolites and hydrolases are hypothesized to contribute to plant protection and antagonistic effects against pathogens by degrading fungal cell walls and plant-derived compounds. Additionally, proteases and aminopeptidases may play a role in the biocontrol activity of *A. pullulans* NBB 7.2.1 at the molecular level. However, further verification—through genomic editing and gene deletion or overexpression studies—is required to confirm their involvement in the biocontrol phenotype (Paula Rueda-Mejia et al. 2021).

A. pullulans strains L1 and L8 were demonstrated in vitro and in vivo to compete with *Monilinia laxa*, the causative agent of stone fruit brown rot, by secreting siderophores (1.2 and 1.4 mg/mL, respectively), thereby preventing post-harvest peach decay. As discussed previously in the last chapter, siderophore biosynthesis is suppressed by Fe^{3+}. Therefore, the two strains reduce both mycelial growth and conidial germination of *M. laxa* only under low iron concentrations (5.0 and 10.0 μg/L $FeCl_3$), demonstrating greater efficacy under iron-limited conditions. In in vivo assays, *A. pullulans* strains L1 and L8 significantly inhibit pathogen virulence, reducing lesion diameter in peaches by an average of 83.5% and 84.4%, respectively, across tested iron concentrations (5.0, 10.0, and 20.0 μg/L $FeCl_3$). Higher iron levels (20.0 μg/L) diminish the antagonists' effectiveness while enhancing pathogen aggressiveness. Thus, *A. pullulans* L1 and L8 compete with *M. laxa* for iron via siderophore production, revealing new biocontrol mechanisms. Both strains effectively limit iron availability, reducing fungal pathogen virulence. These findings offer promising applications for biocontrol strategies in agriculture (Di Francesco and Baraldi 2021a).

At the same time, *A. pullulans* strains L1 and L8 serve as an alternative strategy against *Neofusicoccum parvum*, a pathogen affecting apple wood and fruit. These strains can reduce *N. parvum* conidia germination by nearly 50% and completely inhibit sporulation in the wood of 'Gala' and 'Fuji' apple cultivars. Additionally, they preserve cellulose content, slow wood degradation, and colonize the wood surface (Rusin et al. 2019).

It also has been reported that cold-tolerant and metabolically active strains of *A. pullulans* (L1 and L8) can effectively control stem-end rot in kiwifruit during long-term storage under the controlled atmospheres (2.0% O_2, 4.5% CO_2). Their biocontrol mechanism involves nutrient competition with pathogens and inhibition of *B. cinerea* conidial germination (Di Francesco et al. 2018).

A. pullulans L1 strain also can promote tomato growth, increasing root length, stem diameter, and leaf area by 19.1% and 27.3%, respectively. It also acts as a biocontrol agent against *R. solani* (80% inhibition), a major soil-borne tomato pathogen. Additionally, it reduces the soil concentration of Fe^{2+}, Mn, Cu, and Zn by nearly 50% within 30 days of inoculation (Di Francesco and Baraldi 2021b). The *A. pullulans* can further indirectly enhance the growth of soybean and bean plants through hydrolytic enzyme production, nutrient competition, siderophore and antifungal compound synthesis, and heavy metal reduction in soil.

A. melanogenum also exhibits the most significant growth inhibition in the radial inhibition assay, achieving 60%, 100%, and 70% inhibition against *P. expansum*, *P. digitatum* (green mold), and *P. italicum* (blue mold). These pathogens can produce toxic patulin and citrinin and are the major postharvest pathogens in apples and oranges, respectively. The mode of action of the biocontrol agents includes: (A) Competition for nutrients and space; (B) Superior colonization of plant surfaces; (C) Multiple mechanisms that inhibit pathogen growth (such as mycotoxin absorption and degradation, production of antifungal compounds like antibiotics, siderophores, and limocin); (D) Secretion of cell wall-hydrolyzing enzymes (e.g., chitinase, β-1,3-glucanase, and protease); (E) Induction of host defense mechanisms by producing defense-related enzymes, thereby reducing pathogen populations and enhancing plant health (Gomomo et al. 2025).

However, among *A. pullulans*, *A. melanogenum*, and *A. subglaciale*, *A. pullulans* shows the strongest antagonistic activity against *B. cinerea* (the causal agent of grey mold in tomatoes), while *A. subglaciale* and *A. melanogenum* are ineffective and unsuitable for biocontrol applications (Di Francesco et al. 2021; Di Francesco et al. 2020). But *A. subglaciale* can be used as a biocontrol agent for long period of storage of the post-harvest fruits because it can tolerate cold, being much greater than that of *A. pullulans* by producing soluble metabolites and volatile organic compounds.

The yeast-like fungus *A. pullulans*, isolated from the surfaces of fruit, leaves, and branches of "Rocha" pear trees, significantly reduces postharvest decay caused by *P. expansum* when applied as a wound treatment. This effect is observed under storage conditions and during simulated shelf-life (Cebola Lidon et al. 2012).

Additionally, the endophytic fungus *Aureobasidium* sp. JRF1 colonizes *Arabidopsis* and tomato seedlings, enhancing their growth through the combined effects of the strain and its metabolites, including volatile organic compounds (Jiang et al. 2025).

A. pullulans strain AP-SLU6 effectively controls gray mold (caused by *B. cinerea*) in strawberry fields. Field treatments involving foliar sprays of *A. pullulans* (10^7 CFU/mL and 10^6 CFU/mL) with sugar beet extract reduce postharvest disease severity, extend fruit shelf life, and increase total fruit yield by 53% compared to the control (Iqbal et al. 2023).

As stated above, *P. expansum* can produce the mycotoxin patulin, which causes blue mold disease in apples. *Aureobasidium* sp. UC14, isolated from the wild olive plants in the Friuli Venezia Giulia region, can suppress *P. expansum* growth and patulin production. Both volatile and non-volatile metabolites from *Aureobasidium* sp. UC14 reduce *P. expansum* growth by an average of 50%. At an optimal cell concentration of 1×10^8 cells/mL, *Aureobasidium* sp. UC14 completely inhibits blue mold symptoms in apple fruits. Additionally, *Aureobasidium* sp. UC14 persistently colonizes apples, reducing disease severity by 82.5–89.8% and patulin production by 96.2–98.1% (Cignola et al. 2024).

The strains *A. pullulans* GE17 (isolated from apples) and *Metschnikowia guilliermondii* KL3 (isolated from lemons) demonstrate strong antagonistic activity against two major postharvest pathogens: *P. expansum* DSM62841 (blue mold) and *P. digitatum* DSM2750 (green mold). Among these, *A. pullulans* GE17 alone shows the highest inhibition of mycelial growth, achieving 83.4% inhibition against *P. expansum* and 74.7% against *P. digitatum*. Additionally, when applied as mixed cultures at a concentration of 10^8 cells/mL, these two yeasts inhibit spore germination of both pathogens by 86–95% (Agirman and Erten 2020).

The yeast-like fungal strain *A. pullulans* (ACBL-77), isolated from Tangor Murcott leaves in an orchard in Colina, São Paulo, produces killer toxins (β-1,3-glucanase and chitinase) that effectively control green mold (*P. digitatum*) and sour rot (*G. citriaurantii*) in oranges, with a success rate of 91% (Moura et al. 2020). Volatile compounds—such as alcohols, ketones, alkanes, and aldehydes, including phenylethyl alcohol and 2-methyl-1-propanol—secreted by the biocontrol yeasts *A. pullulans* M13 and C10 and encapsulated in hydrogel spheres, exhibit strong inhibitory effects on mycelial growth, suppressing *B. cinerea* (70.3%) and *Colletotrichum acutatum* (52.4%) in strawberries (Tulukoglu-Kunt et al. 2024). Additionally, the volatile compounds 3-methyl-1-butanol, 2-methyl-1-propanol, ethanol, and 2-phenylethanol, produced by *A. pullulans* strains KR605651.1 and KR605653.1 (isolated from wine grapes), show high antifungal activity against *B. cinerea* and *A. alternata* (Yalage Don et al. 2021a, b).

The non-toxic and safe strain *A. pullulans* S2 effectively reduces gray mold incidence and lesion diameter in tomatoes caused by *B. cinerea*, as well as the natural decay rate of post-harvest tomatoes. Its mechanisms include environmental adaptability, host resistance induction (via enhanced PPO, POD, and SOD enzyme activities as stated below), biofilm formation, and production of antimicrobial volatile organic compounds (VOCs) such as 2-phenylethanol, 2-heptanol, and acetic acid octyl ester (Shi et al. 2023). Additionally, exposure to a mixture of *A. pullulans* and its VOCs (ethanol, 2-methyl-1-propanol, 3-methyl-1-butanol, and 2-phenylethanol) induces electrolyte leakage, inhibits spore germination and mycelial growth in *B. cinerea* and *A. alternata*, triggers ROS accumulation, generates superoxide radicals, and alters fungal cell wall structures (Yalage Don et al. 2021a, b).

The obtained healthy mushrooms also must be cultivated using low-environmental-impact systems that exclude synthetic fungicides, thereby preventing toxicological risks and pathogen resistance. *Trichoderma pleuroti* and *Trichoderma atroviride* cause green mold disease in *Agaricus bisporus* and are common fungal pathogens in mushroom cultivation. Volatile organic compounds (VOCs) produced by *Aureobasidium* spp. strains Ap1 and Uc14 exhibit strong antifungal activity, inhibiting *Trichoderma* spp. growth by up to 30%. Therefore, *Aureobasidium* spp. strains can potentially serve as biocontrol agents against *Trichoderma* spp. mycelial growth without affecting mushroom development (Di Francesco et al. 2024).

The fungal pathogen *Geotrichum citriaurantii* is a major disease affecting citrus crops. Adding ammonium sulfate (1%) to the yeast-like fungal culture can enhances biofilm production by *A. pullulans* strain ACBL-77 (isolated from citrus), which exhibits antagonistic activity against *G. citriaurantii*. This treatment also improves the survival of the yeast-like fungus on wounded citrus fruit surfaces and causes deformation of the pathogen's hyphae (Nadjara Klein and Cristina Kupper 2018).

A. pullulans strain PA-2, isolated from infected poplar leaves in Pingan District, Haidong City, Qinghai Province, China, and its fermentation filtrate exhibit potential as a microbial herbicide. They effectively control target weeds—including *Galium aparine* var. *tenerum*, *Chenopodium album*, *Malva crispa*, *Polygonum lapathifolium*, and *Avena fatua*—in crops. Notably, no significant reduction in plant height is observed in wheat, faba bean, or barley (Guo et al. 2020).

A. pullulans LS-30, isolated from apples, exhibits significant antagonistic activity against *B. cinerea*, *P. expansum*, *R. stolonifer*, and *A. niger* on table grapes, as well as against *B. cinerea* and *P. expansum* on apple fruit. This antagonism occurs through nutrient competition and the production of extracellular exo-chitinase and β-1,3-glucanase (Castoria et al. 2001).

However, only live *A. pullulans* PL5 exhibits antifungal activity against various postharvest fruit pathogens, including *M. laxa* (on plums and peaches) and *B. cinerea* and *P. expansum* (on apples) (Zhang et al. 2010).

It was reported that *A. pullulans* could be transmitted to strawberry flowers by bees, leading to reduced post-harvest *B. cinerea* infections in strawberries. *F. equiseti* is an emerging phytopathogen causing field and postharvest losses in the smallholder farming systems. The endophytic *Aureobasidium* sp. BGP-C02, isolated from the healthy cardamon (*Amomum subulatum* Roxb) leaves in eastern Nepal, can inhibit growth of *F. equiseti*, reaching approximate 75% over 9 days and induce its hyphal abnormalities by producing amylase, cellulase and VOC. In the greenhouse experiments, *Aureobasidium* sp. BGP-C02 can reduce disease incidence in the biocontrol-treatmented tomato compared to the control. Under pathogen-free conditions, biocontrol treatment using *Aureobasidium* sp. BGP-C02 can lead to a numerical increase in shoot dry weight of tomato (Bir Rai, et al. 2026)

However, so far most studies have been only focused on fruit crops, while the potential impact of *A. pullulans* on cereals and legumes remains understudied. Post-harvest research has also predominantly examined fruits, and the effects of *A. pullulans* on insects and their microbiomes have yet to be assessed. In addition, future work should further investigate *A. pullulans* interactions with native plant microbiomes.

12.4 Mechanisms of Anti-fungal Activity

Although the mechanisms underlying the efficacy of *Aureobasidium* spp. as a biocontrol agent are not fully understood, recent research has made significant progress in elucidating the mechanisms of their antifungal activity, as described below. Notably, multiple mechanisms typically contribute to the antifungal effects of *Aureobasidium* spp. Therefore, it is impossible for *Aureobasdium* spp. to be developed to be drug resistance,

12.4.1 Nutrient and Space Competition and Site Exclusion

Due to their rapid growth rate and high cell density, *Aureobasidium* spp. exhibit a strong competitive ability for space and nutrients (e.g., carbohydrates, amino acids, iron, and nitrogen), which are essential for pathogen growth. This allows them to colonize and form biofilms on plant and fruit surfaces more effectively, inhibiting pathogenic bacteria and fungi through niche exclusion. Notably, competition for nitrogen directly impacts pathogenic fungi, as their spore germination depends on this nutrient. Additionally, *Aureobasidium* spp. form multicellular biofilms embedded in a protective matrix, enhancing surface adhesion and stress resistance in their cells. Thus, biofilm formation serves as a key biocontrol mechanism for these antagonistic yeast-like fungi.

12.4.2 Exopolysaccharide Production

As established in Chap. 7, *Aureobasidium* spp. are known to produce both pullulan and β-glucans. Pullulan functions as a protective agent for cell membranes, while β-glucans trigger plant cellular responses to fungal attacks. Together, these compounds facilitate biofilm formation by *Aureobasidium* spp. on plants, vegetables and fruits.

12.4.3 Production of Extracellular Lytic Enzymes

Different strains of *Aureobasidium* spp. can secrete various extracellular lytic enzymes, including glycosylases, esterases, proteases, pectate/pectin lyases, cellulases, xylanases, cutinases, and fructofuranosidases as stated in Chap. 6. These enzymes can actively degrade the pathogenic fungal hyphae and cell walls.

12.4.4 Induced Host Resistance

As an indirect mechanism to prevent infections, *Aureobasidium* spp. can enhance systemic acquired resistance in plants against pathogenic fungi and bacteria. This is achieved by: (A) Increasing reactive oxygen species (ROS) levels; (B) Boosting the activity of defense enzymes involved in oxidative damage repair and disease resistance in plants; (C) Stimulating host defense responses to induce disease resistance; (D) Causing transient increases in β-1,3-glucanase, chitinase, and peroxidase activities of plants against pathogenic fungi.

These mechanisms represent the most intriguing aspect of *A. pullulans*-mediated biocontrol, providing long-lasting protection against diverse pathogenic microorganisms. For example: (a) Polyphenol oxidase (PPO), an oxidative enzyme, is linked to fruit tissue browning and plant systemic resistance against pathogens; (b) Peroxidase (POD), a PR-9 family protein, hydrolyzes H_2O_2 and plays a critical role in lignin biosynthesis; (c) Superoxide dismutase (SOD) converts superoxide anions (O_2^-) into H_2O_2 and O_2, mitigating oxidative stress in plants. Thus, POD, PPO, and SOD activities are all associated with induced pathogen resistance in plants. Excessive ROS disrupts cellular detoxification mechanisms, altering the redox balance and initiating lipid peroxidation in membranes. This free radical chain reaction increases pathogenic fungal cell membrane permeability, ultimately causing membrane disruption. Under aerobic conditions, ROS primarily originates from mitochondrial respiration, with enzyme complex I of the respiratory chain contributing significantly to ROS release.

12.4.5 Promotion of Host Growth

Certain *Aureobasidium* spp. strains can promote tomato plant growth by enhancing root length, stem diameter, and leaf area through multiple mechanisms. These include: (A) Producing phytohormones (e.g., indole-3-acetic acid for auxin biosynthesis and zeatin for cellular division); (B) Controlling plant pathogens; (C) Increasing bioavailablity of Fe (II), Mn, Cu and Zn and reducing pathogen virulence by producing siderophore and acting as a biocontrol agent against soil-borne pathogenic *R. solani*; (D) Increasing phosphate solubility in soils and nutrient availability/uptake;

Additionally, *A. pullulans* indirectly supports soybean and bean plant growth by: (A) Secreting hydrolytic enzymes; (B) Competing for nutrients; (C) Producing siderophores and antifungal compounds; (D) Reducing soil heavy metal content.

12.4.6 Enhancement of Host Immunity

Certain biocontrol agents, such as specific yeasts and *Aureobasidium* spp., can induce the host to produce pathogen response (PR) proteins, thereby enhancing resistance to the pathogens. This immune response is facilitated by upregulated expression of pathogenicity-related (PR) genes encoding chitinase, peroxidase, and glucanase, along with a reduction in pH. Consequently, these changes suppress disease by reducing virulence/effector gene expression and disrupting bacterial chemotaxis (Kunz et al. 2024).

12.4.7 Degradation of Fungal Toxin or Choline

The applied *Aureobasidium* spp. can mitigate fungal pathogen damage through multiple mechanisms: (A) Degrading mycotoxins produced by pathogens; (B) Binding Toxin; (C) Suppressing toxin biosynthesis-related gene expression; (D) Directly disrupting pathogen cell membranes; (E) Interfering with the pathogens' energy supply; (F) Reducing pathogen inoculum. Even, choline, a compound that stimulates the growth of *Fusarium graminearum*, can be degraded by *A. pullulans* strain AS 55 which can reduce the severity of the fungal pathogen on wheat by 84% (Schisler et al. 2006). *A. melanogenum* DAP1 isolated from the nutmeg leaves has the ability to decontaminate 50.4% of afatoxin B1 (AFB1) and inhibit 58.3% of *A. flavus* growth (Harahap et al. 2026).

12.4.8 Enhanced Colonization of Beneficial Microbes

The applied *Aureobasidium* spp. promote and regulate the colonization of beneficial microbes on plant and fruit surfaces, reducing infection levels during fruit storage. As the key modulators of the plant surface microbiome, they stabilize the microbial ecosystem on fruit surfaces and enhance resistance to abiotic stresses, including mechanical injury, UV radiation, rainfall, moisture, temperature fluctuations, pH changes, oxygen level variations, and wind exposure. Consequently, these effects can improve fruit quality and extend its shelf life.

12.4.9 Production of Antimicrobial Compounds

A. melanogenum can produce various antifungal compounds, including antibiotics, siderophores, exophilins, liamocins, massoia lactone, killer toxins (β-1,3-glucanase and chitinase), free fatty acids (Zhang et al. 2022), pullulan, and other polysaccharides. These compounds act by damaging the cell walls of pathogenic fungi. Pullulan and other polysaccharides synthesized by *Aureobasidium* spp. enhance biofilm formation by the yeast-like fungi, thereby improving their adhesion to plants and fruits, as previously described. Additionally, under the iron-limiting conditions (low Fe^{3+}), *A. melanogenum* produces and secretes siderophores. These molecules can strongly chelate environmental iron or compete with the pathogens for iron uptake, inhibiting the growth of pathogenic bacteria and fungi. Liamocin, produced by *Aureobasidium* spp., can specifically kill *Streptococcus* spp. Additionally, massoia lactone, a compound released from liamocin, exhibits strong antifungal activity against *A. flavus* and other pathogenic fungi. It damages the fungal cell membrane, disrupts cellular integrity, and induces pore formation on the cell surface, ultimately causing cellular necrosis in the pathogenic fungi and their spores. Notably, when loaded into nanoemulsions, massoia lactone demonstrates enhanced efficacy. It actively targets both spores and hyphae of *A. flavus*, disrupting their cell walls and membranes, thereby inhibiting fungal growth, reducing conidia production, and suppressing aflatoxin biosynthesis on peanuts. These findings suggest that nanoemulsion-encapsulated massoia lactone holds great potential for controlling *A. flavus* growth and aflatoxin contamination in food and feed (Wang et al. 2024). Furthermore, massoia lactone derived from liamocin produced by *A. melanogenum* M39 not only suppresses hyphal growth and spore germination in fungal crop pathogens but also induces cell membrane pore formation, ergosterol depletion, elevated reactive oxygen species (ROS) levels, and intracellular component leakage. Consequently, it leads to cellular necrosis and cell death in multiple fungal pathogens. Direct contact between massoia lactone and *Fusarium graminearum* spores inhibits the development of *Fusarium* head blight symptoms in the infected wheat. Due to its potent antifungal activity and multifaceted mode of action, massoia lactone shows promise as an eco-friendly biofungicide (Zhang et al. 2021). As discussed in Chap. 5, various genome-editing techniques have significantly improved liamocin and siderophore biosynthesis in *Aureobasidium* spp. Killer toxins, which are low-molecular-weight proteins or glycoproteins, are lethal to susceptible microbial cells, but do not affect the producing organism (Liu et al. 2015). β-1,3-Glucanase hydrolyzes β-1,3 and β-1,6-glucans in the cell walls of the pathogenic fungi, while chitinases cleave β-1,4-N-acetylglucosamine bonds in chitin, a structural component of filamentous fungal cell walls.

12.4.10 *Inhibition of the Spore Germination Rate and the Germ Tube Elongation of the Fungal Pathogens*

The fungal pathogens' spore germination rate and germ tube elongation can be effectively inhibited by *Aureobasidium* spp. Both parameters serve as crucial indicators of pathogen activity.

12.4.11 *Production of the Volatile Organic Compounds*

Volatile organic compounds (VOCs) have high potential to trigger various plant responses, including the production of intracellular reactive oxygen species (ROS), lipid peroxidation, electrolyte loss, and the activation or regulation of plant immunity and growth. *Aureobasidium* spp. produce non-volatile metabolites capable of inducing systemic resistance in plants against pathogen growth. *Aureobasidium*-derived VOCs, which are low-molecular-weight molecules (<300 Da) with low polarity and high vapor pressure at room temperature, exhibit diverse chemical compositions (e.g., alcohols, aldehydes, hexanes, benzene derivatives, heterocyclic compounds, hydrocarbons, ketones, phenols, and esters). These compounds have been shown to inhibit *Penicillium* spp. growth and/or mycotoxin production (Cignola et al. 2024). Additionally, a mixture of volatile metabolites—including phenylethyl alcohol (derived from phenylpyruvate), 2-methyl-1-propanol (from valine), ethanol (from pyruvate), and 3-methyl-1-butanol (from leucine) (Fig. 12.6)—secreted by *A. pullulans* induces ROS accumulation, electrolyte leakage, and cell membrane disruption in the fungal pathogens *B. cinerea* and *A. alternata*, thereby reducing disease incidence. 1-Propanol-2-methyl, a compound with broad-spectrum antifungal activity, has been reported to inhibit the radial mycelial growth of *Aspergillus carbonarius*, impair conidial germination and hyphal membrane integrity, and suppress the growth of *P. expansum* and *Penicillium nordicum*. As a potential quorum-sensing molecule, it may also influence biological control efficacy, filamentous growth, adhesion, and biofilm formation. Additionally, Lei et al. (2022) demonstrated that 2-phenylethanol enhances the biofilm formation and adhesion capabilities of *Debaryomyces nepalensis* on jujube fruit surfaces, thereby improving its biocontrol efficiency.

12.4.12 *Production of the Extracellular Alkaline Protease*

As mentioned in Chap. 6, the extracellular alkaline protease produced by the antagonist *A. pullulans* PL5 can inhibit *M. laxa* infection on stone fruits, the infection of *B. cinerea* and *P. expansum* on pome fruits.

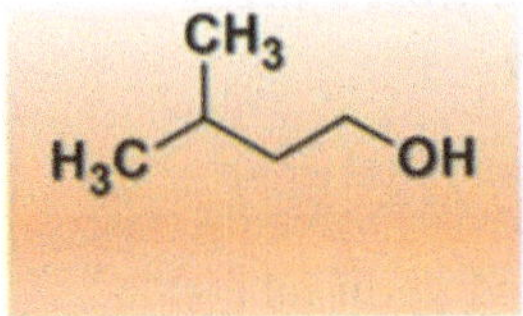

3-Methyl-1-butanol

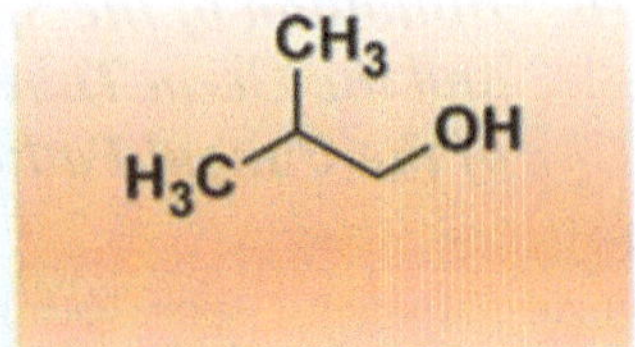

2-Methyl-1-prppanol

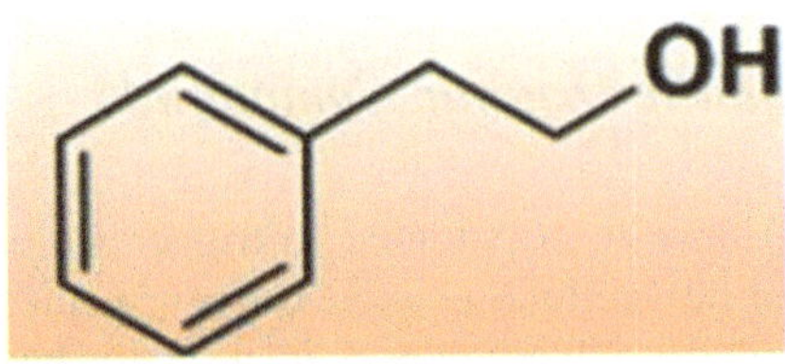

2-phenylethanol

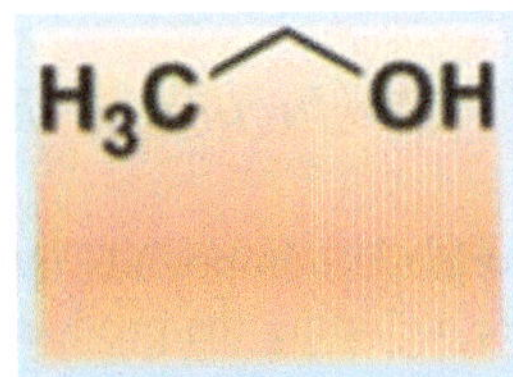

Ethanol

Fig. 12.6 The chemical structures of some anti-fungal volatile metabolites produced by *A. pullulans*

Therefore, *Aureobasidium* spp. serve as effective biocontrol agents against pathogenic fungi in crops and fruits. These microorganisms provide a renewable, soil-safe, sustainable, and eco-friendly alternative to the conventional chemical fungicides, contributing to food and feed security and clean environments (Yeka Zhimo et al. 2025). All the mechanisms of antifungal activity are summarized in Fig 12.7.

12.5 Industrial Production of the Effective Biocontrol Agent

The industrial production of *Aureobasidium* spp. begins with fermenting active fungal cells in an optimized medium of fermenter under controlled growth conditions. Co-formulants may influence the product's efficacy and performance, either by enhancing microbial potential or by introducing side effects against the target pathogen. Once high cell biomass is achieved, downstream processing involves stabilizing the cells through refrigeration, freeze-drying, or dehydration. The stabilized biocontrol agent can then be commercialized in various formulations, such as dried powder, granules, liquid suspensions, or oil- and water-based products.

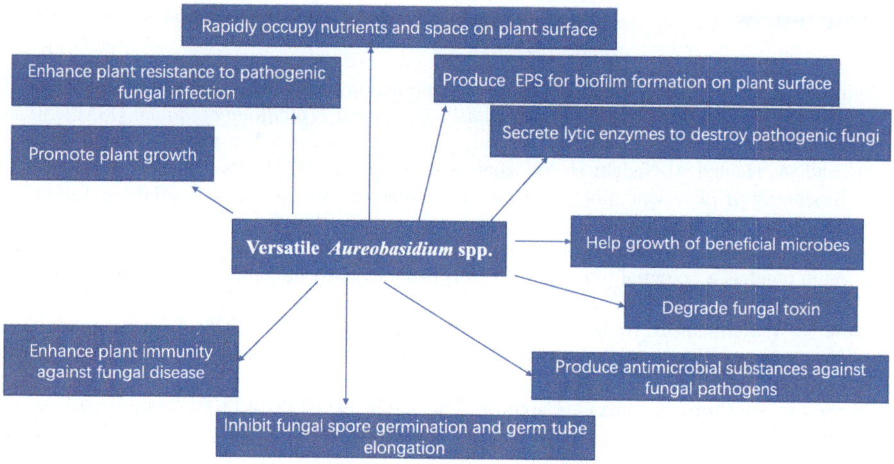

Fig. 12.7 Mechanisms of antifungal activity of *Aureobasidium* spp

12.6 Conclusions and Perspectives

However, numerous unanswered questions remain regarding the use of *Aureobasidium* spp. as a biocontrol agent:

(A) How does its composition correlate with the reduced postharvest decay rates?
(B) How does applying *Aureobasidium* spp. affect the abundance and diversity of other beneficial microbes and microbial pathogens over time?
(C) How do *Aureobasidium* spp. colonize fruit surfaces, and how do they interact with other microbial strains in the epiphytic fruit microbiome at different developmental, ripening, and storage stages?
(D) How do these interactions contribute to disease suppression?
(E) How can the biosynthesis of antifungal and antibacterial compounds (e.g., siderophores, liamocins, and massoia lactone) be stimulated in both native and engineered *Aureobasidium* spp. on fruit, vegetable and fruit surfaces?
(F) How can these bioactive substances be optimized to target pathogens more effectively and selectively?
(G) How can low-cost culture substrates be used to produce high biomass of biocontrol agents to minimize production costs?
(H) How can genomic characterization of active yeast-like fungal strains be linked to their antifungal activities?

When developing a formulated product, key criteria—including antagonistic efficacy, field performance, environmental impact, production costs, and marketability—also must be considered. A deeper understanding of microbial dynamics in the epiphytic microbiome, along with the effective action of antifungal and antibacterial compounds, is essential to optimize *Aureobasidium* spp. applications and enhance biocontrol efficacy.

References

Agirman B, Erten H (2020) Biocontrol ability and action mechanisms of *Aureobasidium pullulans* GE17 and *Meyerozyma guilliermondii* KL3 against *Penicillium digitatum* DSM2750 and *Penicillium expansum* DSM62841 causing postharvest diseases. Yeast 37(9-10):437–448

Al-Fahdi A, Hamed Al-Nadabi H, Velazhahan R, Al-Hattali W, Hussain S, Al-Sadi AM (2026) Biocontrol of olive leaf spot pathogens by endophytic *Aureobasidium pullulans*. Egyptian J Biol Pest Cont 36:11

Bir Rai G, Kayastha S, Dahal S, Subba B, Sailesh Malla S (2026) *Aureobasidium* sp. from cardamom plant as a potential biocontrol for *Fusarium equiseti* infecting tomato plants. Biol Cont 219:106077

Bziuk N, Wassermann B, Bickel S, Omidvar R, Manica A, Berg G (2025) *Aureobasidium pullulans*: a microbiome-based perspective from global biomes to edible plant tissues. Front Plant Sci 16:1652366

Castoria R, De Curtis F, Lima G, Caputo L, Pacifico S, De Cicco V (2001) *Aureobasidium pullulans* (LS-30) an antagonist of postharvest pathogens of fruits: study on its modes of action. Postharv Biol Technol 22:7–17

Cebola Lidon F, Cochicho Ramalho J, Pais IP, Paula Ramos A, José Santos M, Daniel Arrabaça J, da Graça Barreiro M (2012) Fungistatic action of *Aureobasidium pullulans* on *Penicillium expansum* in "Rocha" pear: implications for oxidative stress during fruit storage. Int J Pest Manag 58(1):41–52

Cignola R, Zucchinali S, Firrao G, Francesco AD (2024) Aspects of the biocontrol activity of *Aureobasidium* spp. strain against *Penicillium expansum* of apple. Ann Appl Biol 184(3):307–313

Cornish DA, Schipper MM, Oldham JM, Yu J, Vanneste JL (2025) The biological control agent for bacterial canker of kiwifruit, in Aureo® Gold, is a strain of *Aureobasidium pullulans* identifiable by novel SCAR marker primers. Biol Control 202:105709

Da Costa Arruda GC, Kovács M, Pomázi A (2024) Biological control mechanisms of non-*saccharomyces* yeasts against filamentous fungal pathogens within Rosaceae family. Int J Agron 1:1538220

Di Francesco AD, Baraldi E (2021a) How siderophore production can influence the biocontrol activity of *Aureobasidium pullulans* against *Monilinia laxa* on peaches. Biol Control 152:104456

Di Francesco A, Baraldi E (2021b) How siderophore production can influence the biocontrol activity of *Aureobasidium pullulans* against *Monilinia laxa* on peaches. Biol Cont 152:104456

Di Francesco A, Mari M, Ugolini L, Baraldi E (2018) Effect of *Aureobasidium pullulans* strains against *Botrytis cinerea* on kiwifruit during storage and on fruit nutritional composition. Food Microbiol 72:67–72

Di Francesco A, Di Foggia M, Zajc J, Gunde-Cimerman N, Baraldi E (2020) Study of the efficacy of *Aureobasidium* strains belonging to three different species: *A. pullulans*, *A. subglaciale* and *A. melanogenum* against *Botrytis cinerea* of tomato. Ann Appl Biol 177(2):266–275

Di Francesco A, Sciubba L, Bencivenni M, Marzadori C, Baraldi E (2021) Application of *Aureobasidium pullulans* in iron-poor soil. Can the production of siderophores improve iron bioavailability and yeast antagonistic activity? Ann Appl Biol 180(3):398–406

Di Francesco A, Moret E, Cignola R, Garagozzo L, Torelli E, Di Foggia M (2024) Yeasts volatile organic compounds (VOCs) as potential growth enhancers and molds biocontrol agents of mushrooms mycelia. Fung Biol 128(4):1859–1867

Gomomo Z, Fanadzo M, Mewa-Ngongang M, Justin Hoff J, van der Rijst M, Lucky Mokwena L, Jeremy J, Boonzaier JJ, Rodney S, Hart RS, du Pless HW (2025) Biocontrol potential of non-*saccharomyces* yeasts against post-harvest *Penicillium* species in apples and oranges. Agric Res. https://doi.org/10.1007/s40003-025-00890-3(0123456789

Guo Q, Cheng L, Zhu H, Li W, Wei Y, Chen H, Guo L, Weng H, Wang J (2020) Herbicidal activity of *Aureobasidium pullulans* PA-2 on weeds and optimization of its solid-state fermentation conditions. J Integr Agri 19(1):173–182

Harahap I, Maryam R, Idwan Sudirman L, Indri Astuti R (2026) Whole-genome analysis reveals the genetic potential of yeast Aureobasidium melanogenum DAP1 for aflatoxin B1 decontamination and Aspergillus flavus inhibition in nutmeg seeds. Intern J Food Microbiol 453:111728

Iqbal M, Andreasson E, Stenberg JA (2023) Biological control of strawberry diseases by *Aureobasidium pullulans* and sugar beet extract under field conditions. J Plant Pathol 105:933–941

Jiang Z, Peng F, Qi Y, Li Q (2025) Plant growth-promoting effects and possible mechanisms of a plant endophytic fungus *Aureobasidium* sp. JRF1. Plant physiol Biochemist 222:109724

Kang XX, Jia SL, Wei X, Zhang M, Liu GL, Hu Z, Chi Z, Chi ZM (2022) Liamocins biosynthesis, its regulation in *Aureobasidium* spp., and their bioactivities. Crit Rev Biotechnol 42(1):93–105

Kunz S, Zeng Q, Johnson B (2024) History, efficacy, orchard ecology, and mode of action of *Aureobasidium pullulans*, the microbial agent in blossom protect, for suppression of fire blight of pome fruit. J Plant Pathol 106:871–881

Lei XM, Deng B, Ruan CQ, Deng LL, Zeng KF (2022) Phenylethanol as a quorum sensing molecule to promote biofilm formation of the antagonistic yeast *Debaryomyces nepalensis* for the control of black spot rot on jujube. Postharv Biol Technol 185

Liu GL, Chi Z, Wang GY, Wang ZP, Li Y, Chi ZM (2015) Yeast killer toxins, molecular mechanisms of their action and their applications. Crit Rev Biotechnol 35(2):222–234

Moura VS, Pollettini FL, Ferraz LP, Mazzi MV, Kupper KC (2020) Purification of a killer toxin from *Aureobasidium pullulans* for the biocontrol of phytopathogens. J Basic Microbiol 61(2):77–87

Nadjara Klein M, Cristina Kupper K (2018) Biofilm production by *Aureobasidium pullulans* improves biocontrol against sour rot in citrus. Food Microbiol 69:1–10

Paula Rueda-Mejia M, Nägeli L, Lutz S, Hayes RD, Varadarajan AR, Grigoriev IV, Ahrens CH, Freimoser FM (2021) Genome, transcriptome and secretome analyses of the antagonistic, yeast-like fungus *Aureobasidium pullulans* to identify potential biocontrol genes. Microb Cell 8(8):184–202

Rusin C, Di Francesco A, Di Foggia M, D'Aquino S, Rombolà A, Tugnoli V, Botelho RV, Baraldi E (2019) An emerging problem affecting apple production: *Neofusicoccum parvum*. *Aureobasidium pullulans* L1 and L8 strains as an alternative control strategy. Biol Control 134:157–162

Schisler DA, Khana NI, Boehm MJ, Lipps PE, Slininger PJ, Zhanga S (2006) Selection and evaluation of the potential of choline-metabolizing microbial strains to reduce Fusarium head blight. Biol Control 39:497–506

Shi Y, Zhao Q, Xin Y, Yang Q, Dhanasekaran S, Zhang X, Zhang H (2023) *Aureobasidium pullulans* S2 controls tomato gray mold and produces volatile organic compounds and biofilms. Posthar Biol Technol 204:112450

Tulukoglu-Kunt KS, Moret E, Khomutovsk N, Stenberg J, Alessandra Di Francesco A (2024) Efficacy of volatile organic compounds by *Aureobasidium pullulans* immobilized on hydrogel spheres against strawberry fungal pathogens. J Sci Food Agric 104:8380–8386

Wang WL, Chi ZM, Chi Z, Li J, Wang XH (2009) Siderophore production by the marine-derived *Aureobasidium pullulans* and its antimicrobial activity. Bioresour Technol 100:2639–2641

Wang P, Zhang M, Zhao SF, Zhang ZR, Liu GL, Chi Z, Chi ZM (2024) Liamocins overproduction via the two-pH stage fermentation and anti-*aspergillus flavus* activity of Massoia lactone. Biotechnol J 19:2300675

Yalage Don SM, Gambetta JM, Steel CC, Schmidtke LM (2021a) Elucidating the interaction of carbon, nitrogen, and temperature on the biosynthesis of *Aureobasidium pullulans* antifungal volatiles. Environ Microbiol Rep 13(4):482–494

Yalage Don SM, Schmidtke LM, Gambetta JM, Steel CC (2021b) Volatile organic compounds produced by *Aureobasidium pullulans* induce electrolyte loss and oxidative stress in *Botrytis cinerea* and *Alternaria alternata*. Res Microbiol 172:103788

Yeka Zhimo V, Kumar Sharma V, Feygenberg O, Salim S, Susan Whitehead R, Spadaro D, Freilich S, Wisniewski M, Droby S (2025) Changes in microbial assembly and ecological processes following application of *Aureobasidium pullulans* on apple fruit surface. Intern J Food Microbiol 441:111299

Zhang D, Spadaro D, Garibaldi A, Lodovica Gullino M (2010) Efficacy of the antagonist *Aureobasidium pullulans* PL5 against postharvest pathogens of peach, apple and plum and its modes of action. Biol Control 54:172–180

Zhang D, Spadaro D, Garibaldi A, Maria Lodovica Gullino M (2011) Potential biocontrol activity of a strain of *Pichia guilliermondii* against grey mold of apples and its possible modes of action. Biol Control 57(3):193–201

Zhang M, Gao ZC, Chi Z, Wang Z, Liu GL, Li XF, Hu Z, Chi ZM (2021) Massoia lactone displays strong antifungal property against many crop pathogens and its potential application. Microb Ecol 84:376–390

Zhang M, Wang Z, Chi Z, Liu GL, Chi ZM (2022) Metabolic engineering of *Aureobasidium melanogenum* 9–1 for overproduction of liamocins by enhancing supply of acetyl-CoA and ATP. Microbiol Res 265:127172